# BIOLOGY
## A CRITICAL THINKING APPROACH

# ROBERT D. ALLEN

## VICTOR VALLEY COLLEGE

BIOLOGY
A CRITICAL
THINKING
APPROACH

**WCB** **Wm. C. Brown Publishers**

Dubuque, IA   Bogota   Boston   Buenos Aires   Caracas   Chicago
Guilford, CT   London   Madrid   Mexico City   Sydney   Toronto

## Book Team

Editor  *Carol J. Mills*
Developmental Editor  *Diane E. Beausoleil*
Production Editor  *Carla D. Kipper*
Designer  *Christopher E. Reese*
Art Editor  *Rachel Imsland*
Photo Editor  *Janice Hancock*
Permissions Coordinator  *LouAnn K. Wilson*

## Wm. C. Brown Publishers
A Division of Wm. C. Brown Communications, Inc.

Vice President and General Manager  *Beverly Kolz*
Vice President, Publisher  *Kevin Kane*
Vice President, Director of Sales and Marketing  *Virginia S. Moffat*
Vice President, Director of Production  *Colleen A. Yonda*
National Sales Manager  *Douglas J. DiNardo*
Marketing Manager  *Craig Johnson*
Advertising Manager  *Janelle Keeffer*
Production Editorial Manager  *Renée Menne*
Publishing Services Manager  *Karen J. Slaght*
Royalty/Permissions Manager  *Connie Allendorf*

## Wm. C. Brown Communications, Inc.

President and Chief Executive Officer  *G. Franklin Lewis*
Senior Vice President, Operations  *James H. Higby*
Corporate Senior Vice President, President of WCB Manufacturing  *Roger Meyer*
Corporate Senior Vice President and Chief Financial Officer  *Robert Chesterman*

Copyedited by Julie Bach

Cover credit © Bill Ivy/Tony Stone Images
Photo research by Toni Michaels

The credits section for this book begins on page 293 and
is considered an extension of the copyright page.

Library of Congress Catalog Card Number: 94–71609

ISBN 0–697–14749–5

Printed in the United States of America by Wm. C. Brown Communications, Inc.,
2460 Kerper Boulevard, Dubuque, IA 52001

10  9  8  7  6  5  4  3  2  1

THIS BOOK IS DEDICATED TO MY MANY STUDENTS OVER THE YEARS WHOSE STRUGGLES, FRUSTRATIONS, MISTAKES, AND ULTIMATE SUCCESSES REVEALED HOW CRITICAL THINKING SKILLS ARE LEARNED AND HOW THEY SHOULD BE TAUGHT.

# CONTENTS

*Preface xv*

## SCIENTIFIC INVESTIGATION    1

## THE CELLULAR AND CHEMICAL BASIS OF LIFE    11

## CELLS    13

## PART IV

### EVOLUTION AND ECOLOGY  209

# DIVERSITY OF LIFE   279

# LIST OF TABLES

# PREFACE

 *iology: A Critical Thinking Approach* presents the fundamental concepts of biology and develops students' critical thinking skills to apply these concepts.

When I first started emphasizing critical thinking in my biology teaching more than 25 years ago, students exhibited great difficulty with interpreting data, analyzing experimental results, making predictions, identifying assumptions, and applying concepts to solve new problems. My colleagues and I were concerned by these difficulties and were determined to find ways to help students improve their skills. Thus we began to search for learning activities and teaching materials that would improve students' thinking skills. Quickly it became apparent that there was no suitable textbook that combined the presentation of basic concepts and the development of thinking skills.

To fill the void, we began to develop effective teaching strategies and materials, fully expecting that a useful textbook would become available because teacher interest nationwide continued to increase. However, while many new biology titles have entered the general biology market over the years, none has truly addressed this issue. *Biology: A Critical Thinking Approach* is my response to this need.

Most instructors will view this as a nonmajor's text. Nonmajors need the basic concepts without excessive detail or terminology, and they need to apply those concepts to understand themselves and the world around them. *Biology: A Critical Thinking Approach* provides the means for concept and skill development.

Additionally, biology majors may find the book an ideal complement to their main text. Science majors need a solid conceptual foundation of the discipline and should begin developing the thinking and analytical skills necessary for further study in biology. *Biology: A Critical Thinking Approach* can serve as an excellent beginning.

Although traditional in its comparative approach, *Biology: A Critical Thinking Approach* is nontraditional in depth of coverage. You will not encounter in this text the detail and terminology found in more traditional textbooks. Instead *Biology: A Critical Thinking Approach* emphasizes the basic and important concepts without immersing students in excessive detail, thus maintaining its focus on thinking skill development.

Even though the text is relatively brief, its difficulty should not be underestimated. Throughout the text students are expected to analyze results, make interpretations, and criticize experimental procedures using the experimental results and data that are given in each chapter.

## ORGANIZATION OF THE TEXT

*Biology: A Critical Thinking Approach* comprises 23 chapters beginning with cell biology and physiology, and moving to ecology and evolution.

Although critical thinking skills can be developed regardless of a textbook's sequencing of topics, beginning with cells has advantages. Doing so allows student to focus on making concrete observations, which enables them to make clear interpretations. Once students develop sound observation and interpretation skills, they are prepared to tackle the more difficult abstractions posed in evolution and ecology. Another advantage of this organization is that the study of evolution and ecology requires a thorough understanding of genetics, physiology, protein synthesis, and reproduction.

## INTRODUCTION

This first chapter is important because it establishes the way in which analysis and interpretation are used throughout the book. Scientific investigation is presented as the means biologists use to answer questions about the living world. The procedures of hypothesis formation, prediction, experimental design, and interpretation are introduced as the essential parts of scientific investigation.

# PART I—THE CELLULAR AND CHEMICAL BASIS OF LIFE

Cell theory is one of the basic principles of biology and is a logical starting point for a biology textbook. Chapters 2 through 5 focus primarily on energy, as well as the catalytic action of enzymes, and diffusion across cell membranes. Later chapters build on these concepts.

# PART II—PHYSIOLOGY OF PLANTS AND ANIMALS

The major physiological systems in organisms are covered in chapters 6 through 11. Primary emphasis is placed on the application of basic concepts such as diffusion, osmosis, energy capture and release, and the action of enzymes. With the exception of the chapter on photosynthesis, the physiology of plants and animals is integrated and compared within individual chapters. This integration emphasizes for students the conceptual similarities between plants and animals and provides comparisons that illustrate variety among organisms.

# PART III—REPRODUCTION AND INHERITANCE

Chapters 12 through 16 include molecular biology and population genetics, as well as cell division and Mendelian inheritance. A primary goal of this section is to interrelate protein synthesis, chromosome behavior in cell division, and in populations.

# PART IV—EVOLUTION AND ECOLOGY

Chapters 17 through 19 logically follow population genetics and cover the mechanisms of selection and speciation as well as the long range implications of evolution. Numerous examples and investigations integrate molecular genetics and physiology as part of the evidence for evolution. Chapters 20 through 22 emphasize ecosystem structure, energy flow, and matter cycling. Chapter 23, on diversity of life, presents the principles of five-kingdom classification.

# LEARNING AIDS FOR THE STUDENT

Questions are presented throughout each chapter at points where they relate directly to the information being discussed. These, in addition to other questions, are presented at the end of each chapter as critical thinking exercises. Students then are expected to analyze, interpret, and predict. To develop critical thinking skills requires practice. These exercises are designed to provide the necessary, extended practice that is essential for developing critical thinking skills.

At the end of each Part are exercises entitled Writing Across the Disciplines. These questions integrate biological concepts with other disciplines such as history, law, literature, and political science.

# ANCILLARY MATERIALS

The **Instructor's Manual** presents my rationale for writing *Biology: A Critical Thinking Approach*. Each chapter begins with a short description of the content as well as a listing the most common and most important student difficulties. Suggestions are offered for dealing with and overcoming these difficulties. Further, analysis of each exercise in the chapter illustrates the reasoning and the concept application that should be used to solve the exercises. The **Test Item File** includes all the questions from Microtest III, the computerized testing software.

The **Critical Thinking Student Workbook** serves as a valuable aid in developing critical thinking skills. Basic procedures are described for distinguishing observations from interpretations, analyzing data, and establishing a line of reasoning to solve the exercises. From each chapter one exercise is presented, along with two answers—one well-done and one poorly done. These comparisons provide students with a model while completing the exercises.

The **Critical Thinking Case Study Workbook** provides additional practice in the use of analytical skills and with forming and testing hypotheses, making predictions, and critically evaluating results. The workbook also presents three additional biological concepts.

**Fifty transparency acetates** and **50 masters** featuring key illustrations from the text are available free to all adopters.

# ACKNOWLEDGMENTS

It is amazing how many people are involved in the production of a textbook, and this book owes a vote of thanks to many individuals. Vice-president and publisher Kevin Kane, and acquisitions editor Carol Mills provided the encouragement and continuing support to pursue a different kind of textbook focusing on thinking skill development. The efforts of many others at Wm. C. Brown were essential. Diane Beausoleil assisted with the final developmental work; Carla Kipper managed the production; Julie Bach's copyediting is an outstanding contribution to the text; and Toni Michaels, photo editor, located the excellent photographs.

Many reviewers devoted a great deal of time to the manuscript. They corrected errors where necessary and made numerous suggestions that were a significant influence on the final form of the textbook.

Most of all, thanks to my family whose tolerance and assistance made this book possible. Thank you to my wife, Gloria, for keying countless manuscript pages, and to my two sons, Ryan for his many contributions of drawings and sketches for the artwork and Kevin for his good cheer and pleasantness throughout the frustrating times.

# REVIEWERS

**Harold S. Adams**  Dabney S. Lancaster Community College
**Olukemi Adewusi**  Ferris State University
**Anne Adkins**  University of Winnipeg
**Tom Amlung**  Belleville Area College
**Kenneth W. Andersen**  Gannon University
**Sarah F. Barlow**  Middle Tennessee State University
**Charles H. Bennett**  Kentucky State University
**Char A. Bezanson**  St. Olaf College
**George C. Boone**  Susquehanna University
**John F. Britt**  University of Dayton
**Anna Brownell**  Chapman University
**Virginia G. Carson**  Chapman University
**Carolyn Chambers**  Xavier University
**Robert Creek**  Eastern Kentucky University
**Roger M. Davis**  University of Maryland
**Dana Demmans**  Finger Lakes Community College
**Ernest F. DuBrul**  University of Toledo
**Larry E. Eason**  Pasco-Hernando Community College
**Betty Eidemiller**  Lamar University at Orange
**DuWayne Englert**  Southern Illinois University at Carbondale
**Wayland L. Ezell**  St. Cloud State University
**John W. Ferner**  Thomas More College
**Don W. Finnie**  Colorado Northwestern Community College
**Chet S. Fornari**  DePauw University
**John C. Frandsen**  Tuskegee University
**Peggy Goldman Fuller**  Bossier Parish Community College
**Francis E. Gardner, Jr.**  Columbus College
**Grant A. Gardner**  Memorial University of Newfoundland
**Michael H. Gipson**  Oklahoma Christian University
**Ben Golden**  Kennesaw State College
**Nancy Goodyear**  Bainbridge College
**Eileen Gregory**  Rollins College
**Ruth Gyure**  Edgewood College
**Rebecca Halyard**  Clayton State College
**John P. Harley**  Eastern Kentucky University
**Patricia Hauslein**  St. Cloud State University
**David J. Hicks**  Manchester College
**Nelda W. Hinckley**  John A. Logan College
**Fred D. Hinson**  Western Carolina University
**Rita A. Hoots**  California State University at Sacramento
**Herbert House**  Elon College
**Dallas D. Johnson**  North Hennepin Community College
**Vincent Johnson**  Saint Cloud State University
**J. Michael Jones**  Culver-Stockton College
**Stephen R. Karr**  Carson-Newman College
**Dennis Kingery**  Metropolitan Community College

**Debra A. Kirchhof-Glazier**  Juniata College
**Jessie W. Klein**  Middlesex Community College
**Tim Knight**  Ouachita Baptist University
**Helen G. Koritz**  College of Mt. St. Vincent/Manhattan College
**Lynn O. Lewis**  Mary Washington College
**Lurana K. McCarron**  Hudson Valley Community College
**Julie K. Meents**  New Mexico State University—Alamogordo
**Debra Meuler**  Cardinal Stritch College
**Richard W. Miller**  Butler University
**Lewis M. Milner**  North Central Technical College
**Don Mumm**  Southeast Community College
**William Nicholson**  University of Arkansas—Monticello
**Thomas E. Oldfield**  Ferris State University
**John Olsen**  Rhodes College
**Lynne J. Osborn**  Middlesex Community College
**David Parker**  Northern Virginia Community College—Alexandria
**Gail R. Patt**  Boston University
**John Phythyon**  St. Norbert College
**Helen K. Pigage**  United States Air Force Academy
**Mary Ann Polasek**  Cardinal Stritch College
**Kathryn Stanley Podwall**  Nassau Community College
**Laura E. Powers**  Colorado State University
**Lorraine C. Priess**  Wayne County Community College
**John H. Pritchett, III**  Florence-Barlington Technical College
**Pattle Pun**  Wheaton College
**Michael L. Rodgers**  Southeast Missouri State University
**Lee E. Roecker**  Berea College
**Rob Ross**  Linn-Benton Community College
**Douglas P. Schelhaas**  University of Mary
**Gail B. Schiffer**  Kennesaw State College
**Linda M. Simpson**  University of North Carolina—Charlotte
**Walter B. Sinnamon**  Central Wesleyan College
**Gilbert D. Starks**  Central Michigan University
**John E. Stencel**  Olney Central College
**Barbara L. Stewart**  J. Sargeant Reynolds Community College
**David J. Stroup**  Francis Marion University
**Gerald Summers**  University of Missouri—Columbia
**Charles Lee Swendsen**  Warren Wilson College
**Annehara K. Tatschl**  Johnson County Community College
**David E. Taylor**  University of South Carolina—Spartanburg
**William R. Teska**  Furman University
**Harry Womack**  Salisbury State University
**Richard Wright**  Valencia Community College
**Don Yeltman**  Davis & Elkins College

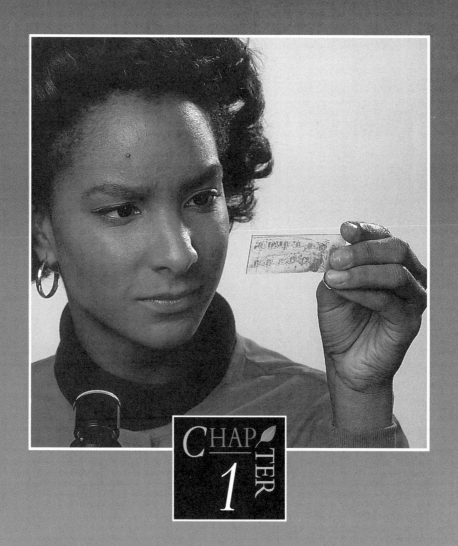

CHAPTER 1

# SCIENTIFIC INVESTIGATION

ow do we know what we know about biology? Biologists study living plants and animals and do experiments that yield facts about biology. How are these studies and experiments done? What rules are followed? In other words, what is the process of scientific investigation?

To begin understanding the answers to these questions, consider that **scientific investigation** is a search for explanations. In the field of biology, these explanations are answers to questions about living things. The process starts with questions we might have about events or observations. For example, one question might be:

How do plants use sunlight to make food?

This is a complex question to answer and would require a great many experiments to do so. A question must be sufficiently simple so that it can be investigated with a single experiment. Some specific questions would be:

What colors of sunlight will produce the most food material by green plants?
How much sunlight is required?

These specific questions answer a small part of the more general question. If we investigate a great many of these simple questions, gradually we will find an answer to the general question.

Science progresses by combining the results of small experiments to answer large, general questions. This is like putting together an enormous jigsaw puzzle, with each piece representing a single experiment. With enough hard work, careful thought, trial and error, and some luck, we can fit these pieces together to reveal a bigger picture (fig. 1.1).

Biology is more difficult than this simplified model suggests, however. Each piece of the puzzle must be put in place by carefully investigating a single specific question. How is this process of investigation carried out? What steps must we follow? Why are these steps necessary? To begin answering these questions, let us look at some specific experiments.

## A GENERAL QUESTION

Suppose we start with the general question, "What controls the growth and body form of insects?" Answering this question can do much more than just satisfy our curiosity. It can have valuable applied results. If we understand how insect growth is controlled, we can control the development of harmful insects and prevent the crop damage these insects cause.

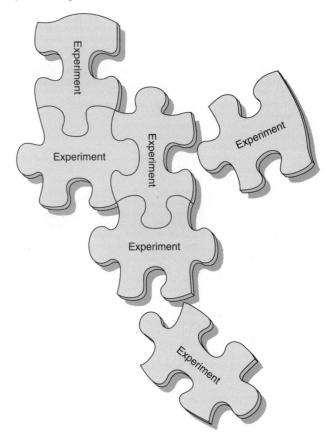

**FIGURE 1.1**   Individual experiments fit together like pieces of a puzzle to give a bigger picture—a more complete answer to the general question being asked.

## A SPECIFIC QUESTION

A more specific question is based on observations of a moth found in Arizona that lays eggs only on Arizona oak trees. The caterpillars hatching from the eggs eat only the leaves and flowers of the oak tree. Although all young caterpillars appear identical just after they hatch, those hatching early in the spring grow to closely resemble the flowers of the tree, called catkins, while those hatching later in the summer resemble the twigs of the tree (fig. 1.2). From these observations, a simple question is immediately obvious: What causes the caterpillars to develop into two different forms?

We can propose several answers to this question. A possible answer is called a **hypothesis.**

## STATING A HYPOTHESIS

Identifying and carefully stating a hypothesis will guide our investigation of the question and help organize the experiment. In this example, three possible hypotheses are:

*Hypothesis 1*   The temperature of the environment affects the caterpillar form.
*Hypothesis 2*   The food eaten by the caterpillars affects their form.

FIGURE 1.2 Different developmental forms of the caterpillar. (a) Early spring form on a catkin. The caterpillar closely resembles the catkins. (b) The twig-resembling form of the caterpillar appearing later in the summer.

a.

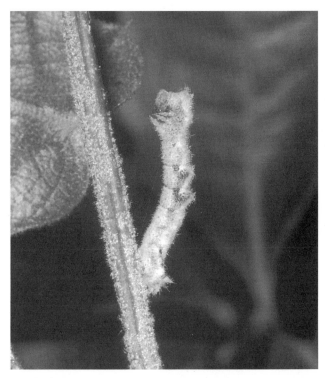

b.

*Hypothesis 3* The amount of daylight to which caterpillars or moth eggs are exposed affects caterpillar form.

All of these hypotheses seem reasonable explanations and we could pick any of these to test. Suppose we first test hypothesis 1. In early spring, the average temperature in Arizona is 15° C, while in summer it is 25° C. Young caterpillars growing at these different temperatures might grow into different forms.

How can the effects of this variable be examined?

## MAKING PREDICTIONS

Using hypothesis 1, we can make a **prediction.** Predictions are an essential part of scientific investigation because they indicate more precisely what should be done in an experiment and what results to expect. For hypothesis 1, the following prediction can be made:

*Prediction 1:* *If* temperature affects caterpillar form, *then* raising young caterpillars at different temperatures should result in catkin forms developing at 15° C and twig forms at 25° C.

We can now do experiments to test this prediction. How should we conduct these experiments? The precise steps to be taken in the experiment and the tasks to be performed are described in the **experimental design.**

## THE EXPERIMENTAL DESIGN

In any experiment, it is important to hold all variables but one constant. In experiment 1, we used the eggs from a single female moth and separated them into two groups. The young caterpillars hatching from these eggs were all fed catkins, but one group was raised at 15° C (characteristic of early spring) while the other group was raised at 25° C (characteristic of summer). The prediction and experimental design are shown in figure 1.3.

In this experiment, diet, source of eggs, and light were held constant; they are the **controlled variables.** Temperature was varied, so it is the **experimental variable.** It is very important to control (if possible) all the variables except the one being investigated. If two or more variables are allowed to change in an experiment, it will not be possible to determine which variable caused the effect.

In this experimental design, why did we use the eggs from only a single female moth?

FIGURE 1.3 Experimental design. One group of caterpillars was raised at 25° C, the other group at 15° C. Both groups received the same food (catkins) and the same amount of light. If temperature affects body form, then the catkin form should develop at 15° C and the twig form at 25° C.

FIGURE 1.4 Experimental results. Contrary to the predicted results, observations of the experimental results showed that both groups developed into catkin forms. Since predictions and results do not agree, the hypothesis is rejected.

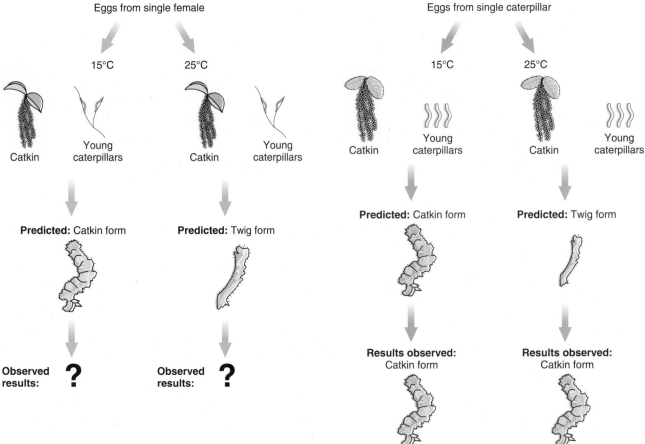

## RESULTS

When this experiment is carried out, what **results** or experimental outcomes do we observe? In this experiment, the results are the effects of temperature on moth form, shown in figure 1.4. These results are the **data** (observations and information) collected from the experiment. What do the data mean? How can we interpret these results?

## INTERPRETATION

An **interpretation** tells what the data or results mean. It explains the data. In this experiment, the data show that the results are *not* in agreement with the hypothesis because the same form developed at both 25° C and 15° C. Hypothesis 1 is not a valid explanation and must be discarded.

An essential part of this process is establishing a **line of reasoning** that leads from the experimental results to the interpretation and discusses any considerations that might influence our interpretation. For example, what other reasonable hypotheses have *not* been tested? In these experiments on caterpillar body form, should hypotheses 2 and 3 on the effects of food and light be tested? Do the results in experiment 1 clearly indicate that temperature is

not a factor affecting form, or is there possibly some problem with the experimental design?

The interpretation is much more than a simple restatement of the results; it must consider as many factors as possible that affect the answer to the original question and the explanation of the results. The exercises at the end of this chapter and the other chapters in the book are designed to give you extensive practice in establishing lines of reasoning. Pay careful attention to these exercises and you will learn how to give clear and complete explanations.

We can now test another hypothesis as a possible explanation of caterpillar body form. Suppose we test hypothesis 2. We can make the following prediction:

*Prediction 2:* If the food eaten by the caterpillars affects body form, *then* caterpillars feeding on different food material will grow into different forms.

To test prediction 2, we can conduct an experiment very similar to experiment 1. In experiment 2, temperature and other variables are held constant, but caterpillars are fed either catkins or leaves as shown in figure 1.5.

## Is It Possible to Prove a Hypothesis?

The short answer is no. It is not possible to prove with absolute certainty and beyond doubt that a hypothesis is "true." Why not? And if we cannot prove a hypothesis, what is the purpose of doing an experiment to test it?

Consider the procedure described in this chapter. A hypothesis is stated, a prediction is made based on the hypothesis, and an experiment is designed and performed. If the experimental results and interpretation are consistent with the prediction, then we can say the hypothesis is supported.

It is possible, however, that the experimental results we obtain are consistent with other hypotheses that we have not tested or even thought of. If we continue to make different predictions based on the original hypothesis and test these, we can test the hypothesis in different ways. If all the tests support the hypothesis, then our confidence in the "truth" of the hypothesis increases, as illustrated in figure 1. At the same time, the likelihood of another, better hypothesis that answers the question decreases. If even one of the tests does not support the hypothesis, however, then the hypothesis must be rejected and is disproven (fig. 1b). If this happens, it is a clear indication that there must be some other explanation (hypothesis) for the experimental results that we have not considered.

Scientists are constantly searching for tests to disprove hypotheses. If their hypotheses continue to be supported under rigorous testing, their confidence increases even more.

To make this process work, remember to do the following:

1. Search for different predictions and experiments to test the hypothesis.
2. Search for alternative interpretations of experimental results. Other interpretations will suggest new ways to test the hypothesis.

FIGURE 1    (a) If a hypothesis is supported by different predictions based on the hypothesis then we gain increasing confidence in the "truth" of the hypothesis. (b) If even one prediction is not supported by the experimental results, then the hypothesis must be rejected.

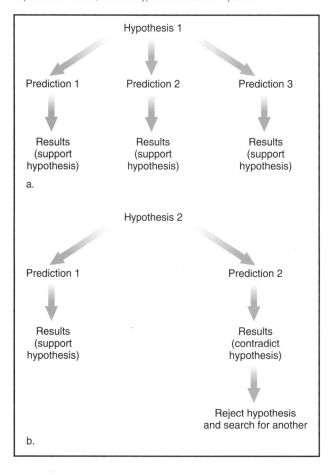

These results show that caterpillars fed only catkins develop into catkin forms, while those fed only leaves develop into twig forms. Since the results are consistent with prediction 2, hypothesis 2 is supported.*

Now we must make more predictions and perform additional experiments as further tests. If the experiments continue to support the hypothesis, we can increase our confidence that the hypothesis is a valid answer to the question. However, we can never "prove" beyond doubt that the hypothesis is "true" because new information may surface in the future (Thinking in Depth 1.1).

---

*The experiments are reported in more detail by Erick Green, 1989, "A Diet-Induced Developmental Polymorphism in a Caterpillar," *Science* 243: 643–48.

In experiment 2, what was the experimental variable? What would happen if caterpillars hatching in the early spring (when young caterpillars normally develop into catkin forms) were fed only oak leaves?

Let us take a moment to summarize the steps in scientific investigation. They are presented in table 1.1. This table will be a valuable reference for you. As you analyze experiments later in the textbook, refer to it often. When experiments are described, all the steps listed in the table may not be specifically named. Try to identify all seven steps so that you more completely understand each experiment.

## TABLE 1.1　Scientific Investigation

| Step | Purpose | Example |
|---|---|---|
| General question | Serves as a starting point for a broad area of investigation | What controls the development of insects? |
| Specific question | Provides a more detailed question that can be tested | What causes caterpillars to develop into different forms in spring and summer? |
| Hypothesis | A possible answer to the specific question; indicates what variables to look at in the experiment | The food eaten by the caterpillars affects their form. |
| Prediction | Indicates what should be done in an experiment and what results to expect if the hypothesis is "true" | If the food eaten by the caterpillars affects body form, then caterpillars feeding on different foods will grow into different forms. |
| Experimental design | Tells what to do in the experiment, i.e., how to conduct the experiment | Collect eggs from a single female moth and divide these into equal groups. Feed catkins to one group and leaves to the other. |
| Results | The data and observations from the experiment | See figure 1.5. |
| Interpretation | Explains what the data and results mean | Something in the diet controls the development and form of the caterpillars. |

**FIGURE 1.5**　Two groups of caterpillars were fed different food (catkins or leaves). If food type affects body form, then the two groups should develop into different forms. In this case, the observed results are in agreement with the prediction and the hypothesis is supported.

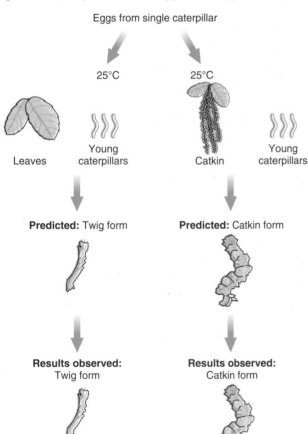

Eggs from single caterpillar

25°C — Leaves — Young caterpillars — **Predicted:** Twig form — **Results observed:** Twig form

25°C — Catkin — Young caterpillars — **Predicted:** Catkin form — **Results observed:** Catkin form

If young caterpillars were fed catkins and leaves and the temperature were varied, what form would you predict?

## FURTHER TESTING THE HYPOTHESIS

If it is important to test the hypothesis in as many ways as possible, what further tests could be done on our caterpillars? Experiment 2 suggests that some difference between catkins and leaves controls caterpillar form. What could this difference be? If further tests support the hypothesis, then some specific food material should be found in catkins or leaves that controls caterpillar form.

Chemical analysis of catkins and leaves shows that catkins are high in protein but low in tannin, a chemical that is poisonous to many animals. Leaves, on the other hand, are relatively low in protein but high in tannin. Do proteins or tannins affect caterpillar form?

In a third experiment, four groups of young caterpillars were fed one of the following diets:

| Diet | Food |
|---|---|
| 1 | Catkins only |
| 2 | Catkins and leaves |
| 3 | Catkins and added tannins |
| 4 | Leaves only |

Based on experiment 2, we would expect caterpillars fed only catkins to develop into catkin forms and those fed only leaves to develop into twig forms. What would you predict for caterpillars fed diets 2 or 3?

## ASSUMPTIONS

**Assumptions** are statements or assertions accepted as true without any evidence. They represent the weakest part of an experiment. In the interpretation of experiment 3 on caterpillar development, the experimenters apparently assumed that tannins do not interfere with protein digestion. But they had no indication or evidence that this is so.

Why was this assumption made if it can cause serious doubts about the interpretation? Most often such assumptions are made because checking them with further experiments would be very difficult or impossible. Rather than attempting to perform these difficult experiments, scientists acknowledge the assumptions and discuss how they affect the interpretation.

Sometimes assumptions are unrecognized, and these are more serious. Unrecognized assumptions can cause investigators to neglect interpretations that could be good explanations or to overlook serious weaknesses in their experiments.

When analyzing experimental designs and results, it is always essential to ask the following:

What assumptions are being made?
How do these assumptions affect the interpretation?

---

*Prediction 3:* *If* tannins cause caterpillars to develop into twig forms, *then* diets 2 and 3 should cause development into twig forms.

The results of this experiment are shown in table 1.2.

The most likely interpretation of these data is that tannin causes development into twig forms, because in every diet with high tannin (diets 2, 3, and 4), most of the caterpillars were twig forms. This is the interpretation chosen by the scientists who first conducted this experiment.

Even though the interpretation makes sense to us, it is very important to consider what other interpretations we could make from these results. A different interpretation may prove to be the best one after further analysis and investigation. Some biologists, in fact, have argued for a different interpretation.★ They point out that adding tannins or leaves in diets 2 and 3 may dilute the protein in catkins, or the tannins may interfere with protein digestion. They argue that the first investigators incorrectly *assumed* that adding tannins or leaves did not change the amount of protein consumed by caterpillars (Thinking in Depth 1.2). The cause of development into twig forms may actually be decreased protein, not increased tannin. If the dilution effect or interference with protein digestion is important, as the second investigators argue, then their interpretation may be the better of the two. Determining which interpretation is best would require additional experiments.

---

★Stanley H. Faith and Kyle E. Hammond, 1989, "Caterpillars and Polymorphisms: Technical Comments," *Science* 246: 1639.

| Diet | Food | Percent Twig Forms |
|------|------|--------------------|
| 1 | Catkins only | 6 |
| 2 | Catkins and leaves | 88 |
| 3 | Catkins and added tannins | 94 |
| 4 | Leaves only | 94 |

**TABLE 1.2** **Effects of Diet on Caterpillar Form**

If young caterpillars were fed leaves and tannins, what form would you predict?

How could you design an experiment to show whether the dilution of protein by added tannin is a better interpretation of experiment 3?

The results of experiments 1, 2, and 3 lead to further questions, such as

What do the tannins do inside the caterpillar to cause development into a twig form?
Is there a survival advantage to caterpillars resembling twigs in the summer and catkins in the early spring?

Investigating these questions requires additional experiments. As described earlier in the chapter, further experiments provide more pieces to the puzzle (see fig. 1.1). We have already seen how a few pieces—experiments 1, 2, and 3—fit together. As we ask more questions and perform more experiments, the "picture" of how insect development is controlled becomes more and more complete.

## SUMMARY

*This introductory chapter on **scientific investigation** presents a number of key concepts and ideas that will be used throughout the book. In every chapter you will be required to interpret experimental **results**, make **predictions**, identify **assumptions**, and evaluate **experimental designs**. In particular, you should remember and be able to apply the following information.*

1. An investigation starts with a specific question.
2. A **hypothesis** is a "best guess" answer to the question.
3. A **prediction**, based on the hypothesis, can be made that indicates what *should* happen under certain conditions if the hypothesis is valid.
4. The prediction is a guide to test the hypothesis; an **experimental design** clearly specifies *how* to conduct the experiment.
5. The experimental design should control all variables except one, the **experimental variable.**
6. The **results** of the experiment are the **data** from the experiment.
7. The **interpretation** explains the results; it tells what the results mean.
8. **Assumptions** made when making the interpretation should be identified and, if possible, tested.
9. Investigators should look for alternative interpretations and consider as many as possible before deciding which one might be the best.
10. Several predictions leading to different experiments should be made from the same hypothesis. If several different experimental results support the hypothesis, the greater is our confidence in the "truth" of the hypothesis.

## REFERENCES

Green, Erick. 1989. "A Diet-Induced Developmental Polymorphism in a Caterpillar." *Science* 243: 643–48.

Faith, Stanley H., and Kyle E. Hammond. 1989. "Caterpillars and Polymorphisms: Technical Comments." *Science* 246: 1639.

References 1 and 2 are the original research on caterpillars feeding on the Arizona oak. In addition to the experiments described in this chapter, experiments are also described on the effects of light period and the reproduction of catkin-fed and leaf-fed caterpillars.

Power, Mary E. 1990. "Effects of Fish in River Food Webs." *Science* 250: 811–14.

Reference 3 is the original research described in exercises 7–11 at the end of the chapter. In this reference, additional data can be found on algae consumption, herbivore populations, and digestive system contents of predators.

Ayala, Francisco J., and Bert Black. 1993. "Science and the Courts." *American Scientist* 81, no. 3: 230–39.

Reference 4 describes misconceptions regarding the interpretation of scientific experiments and the scientific process.

## EXERCISES AND QUESTIONS

1. In the experiment designed to investigate the effects of temperature on caterpillar development, why were the eggs from only a single female used (page 3)?
   a. To select eggs that would produce caterpillars that would develop only into twig forms.
   b. To control any variability due to eggs from different females.
   c. Because a single female moth will lay more than enough eggs to perform the experiment.
   d. Because a single female moth lays its eggs in only one location.
   e. Because the experiment was designed to use one female moth.

2. In the experiment investigating the effects of food on caterpillar development (experiment 2), what was the experimental variable (page 5)?
   a. Type of food consumed.
   b. Temperature.
   c. Total number of caterpillars observed.
   d. Size of the caterpillars.
   e. The source of eggs.

3. Which of the following experimental results would you predict if caterpillars hatching in the early spring were fed only oak leaves (page 5)?
   a. All caterpillars would develop into catkin forms.
   b. Development would depend on the temperature.
   c. Half would develop into twig forms and half into catkin forms.
   d. All caterpillars would develop into twig forms.

4. If young caterpillars were fed leaves and tannin, what form would you predict (page 7)?
   a. All twig forms.
   b. Form cannot be predicted since caterpillars were not fed this diet.
   c. All catkin forms.
   d. Caterpillars would not develop into either twig or catkin forms.

5. If young caterpillars were fed catkins and leaves and the temperature were varied, what form would you predict (page 6)?

    a. All twig forms.
    b. Form cannot be predicted since caterpillars were not fed this diet.
    c. All catkin forms.
    d. Caterpillars would not develop into either twig or catkin forms.

6. How could you design an experiment to show whether the dilution of protein by added tannin is a better interpretation of experiment 3 (page 7)?

Use the following information to answer questions 7 through 11.

## Food Relationships in Rivers*

The question of which animals are consumed by other animals in a natural environment has been investigated in a California river. A scientist examined the relations between midge fly larvae (immature forms of an insect living in the river), predatory insects, and fish. The scientist hypothesized that midge fly larvae are eaten by predatory insects and very young fish, while large fish eat predatory insects and very young fish, as shown in the diagram.

```
large              predatory insects              midge
fish —— eat ——→         and          —— eat ——→ fly
                   very young fish               larvae
```

Closed wire cages were constructed and placed in the river so that plants (algae), insects, and very young fish (fish fry) could move in and out of the cages through the wire mesh of the cage walls. The cages were of two types: Large fish were placed in *enclosure* cages and could not move in or out; they remained in the cages throughout the experiment. Large fish were not placed in *exclosure* cages, and no large fish were present in these cages during the experiment. After five weeks, the number of midge larvae, predatory insects, and fish fry in the cages were counted. The results are shown in table 1.3.

In a second experiment, cages of smaller mesh size were used so that insects and fish fry could not move in or out of the cages but midge fly larvae could. Three types of cages were prepared:

**Type 1** stocked with four predatory insects per cage
**Type 2** stocked with four fish fry per cage
**Type 3** not stocked with predatory insects or fish fry

After twenty days, the number of midge fly larvae in each cage was counted. These results are shown in table 1.4.

---

* These experiments and others in more detail are reported by Mary E. Power, 1990, "Effects of Fish in River Food Webs," *Science* 250: 811–14.

### TABLE 1.3

| Number of Animals per Cage | Enclosure Cages | Exclosure Cages |
|---|---|---|
| Midge fly larvae | 34.0 | 4.0 |
| Predatory insects | 0.1 | 1.0 |
| Fish fry | 0.0 | 30.0 |

### TABLE 1.4

| In Cage With | Number of Midge Fly Larvae |
|---|---|
| Predatory insects | 12 |
| Fish fry | 11 |
| No predators | 48 |

7. Which of the following is the best statement of the results in the first experiment?

    a. Predatory insects and fish fry probably consume midge fly larvae.
    b. The experimental design used two kinds of cages: enclosures with large fish and exclosures without large fish.
    c. Large fish in the enclosure cages can eat midge fly larvae as well as fish fry.
    d. Greater numbers of midge fly larvae are present when large fish are present in the cage.
    e. Larger fish consume predatory insects and fish fry.

8. Which of the following is the best interpretation of the first experiment?

    a. Midge fly larvae eat predatory insects and fish fry while large fish eat midge fly larvae.
    b. Large fish eat predatory insects and fish fry while predatory insects and fish fry eat midge fly larvae.
    c. Predatory insects and fish fry eat each other, as well as large fish.
    d. Cages inhibit the growth of midge fly larvae, fish fry, and predatory insects.
    e. Large fish confined in cages will eat any other animal present in the cage with them.

9. Based on the second experiment, which of the following would be the best prediction if further investigations are carried out?

    a. Midge fly larvae will eat either fish fry or predatory insects when confined with these animals.
    b. If large fish are confined with midge fly larvae, the stomach contents of the fish will contain midge fly larvae.

c. If predatory insects or fish fry are confined in wire cages, the growth of these animals will be inhibited.

d. Predators can be found that eat midge fly larvae, fish fry, and predatory insects.

e. The stomach contents of predatory insects and fish fry will contain midge fly larvae.

10. Which of the following would be an acceptable alternative interpretation of the results from the first experiment?

a. Waste products from predatory insects and fish fry are poisonous to large fish.

b. The presence of fish fry stimulates the growth of midge fly larvae.

c. Waste products from large fish kill predatory insects and fish fry.

d. The presence of midge fly larvae stimulates the growth of fish fry.

e. Waste products from midge fly larvae can kill large fish.

11. In designing the second experiment, why was it important to prevent predatory insects and fish fry from moving in or out of the cages?

a. Only one predator (predatory insects or fish fry) should be in any cage.

b. A better experimental design would have placed both fish fry and predatory insects in each cage.

c. Moving in or out of the cages may have allowed predatory insects or fish fry to be eaten by large fish.

d. A larger mesh was used on the first experiment, allowing other animals to move in and out of the cages.

e. A smaller mesh was used for the cages in the second experiment to prevent movement in and out of the cages.

# PART
# I

# THE CELLULAR AND CHEMICAL BASIS OF LIFE

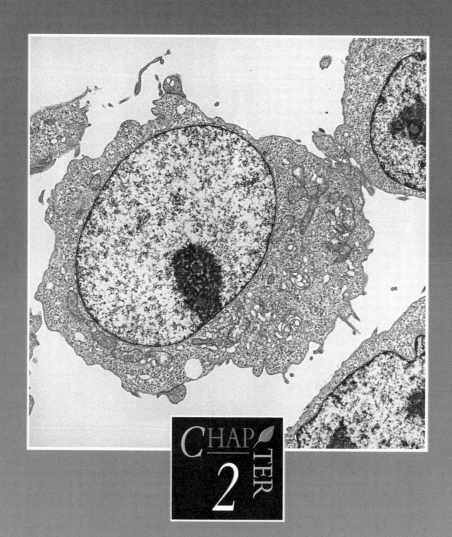

# CELLS

a.

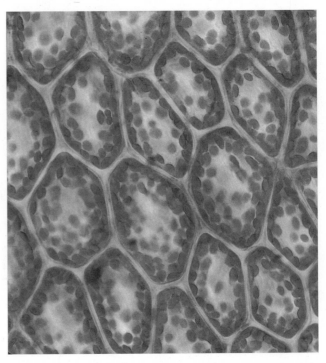

b.

I n chapter 1, we emphasized that scientific investigations start with observations of events or objects that we can see, hear, touch, or smell. These observations lead to questions, which lead to hypotheses and experiments. Much of the early work in the field of biology involves making observations and describing the appearance and structure of living things. If we can increase and improve the observations we make, we can generate more questions, more experiments, and more explanations about living things. How can we do this?

One way to improve observations is to magnify what we see with our eyes. If we could see in more detail how living things are constructed, we could ask why they are constructed in the ways we observe. We could then ask what functions certain structures perform.

## CELL OBSERVATION

Magnifying what we see requires specialized instruments. The light microscope is one instrument that has been exceptionally useful. To see the difference a microscope can make, compare the two photos of a leaf in figure 2.1. In the first photo, we see only the relatively smooth surface of leaves. In the second photo, we see details that we cannot see with the naked eye.

In a microscope, light passes through the specimen to be observed. This light then passes through a series of lenses (objective lenses and eyepiece) that magnify and focus the image as shown in figure 2.2.

FIGURE 2.2 In the microscope, light passes through the specimen being observed, through a series of lenses that magnify the image, and into the eye at the eyepiece.

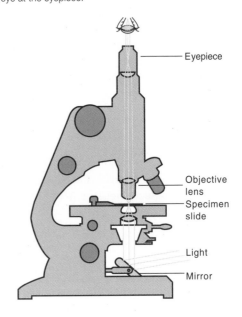

— Eyepiece

— Objective lens
— Specimen slide

— Light

— Mirror

Look again at figure 2.1. What can you see in figure 2.1b that you cannot see in figure 2.1a? When magnified, the leaf appears to be composed of blocklike structures that are not visible to the unaided eye. Other kinds of living matter have a similar appearance when seen under a microscope. For example, onion skin and animal tissue also have blocklike structures (fig. 2.3).

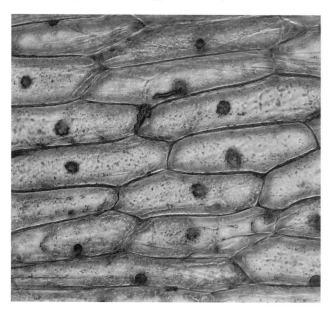

a.

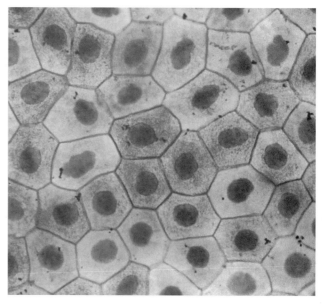

b.

Early scientists using primitive microscopes invented around 1600 A.D. first observed these blocklike structures. Robert Hooke, one of these scientists, called them **cells.** The word comes from the Latin *cella,* which means "small room." Are all living organisms composed of such cells? How would we find out?

The best way to answer this question is to examine all living matter under a microscope. We can do that easily with tissues such as leaves and onion skin, but how do we examine living matter so thick that light cannot pass through it? How do we examine, for instance, a piece of wood, a slice of apple, or part of a muscle?

To make these observations, we must slice these living materials so thin that light can pass through them and we can view them under a microscope. The process is more difficult than it sounds. First the material to be observed is **embedded** in a substance such as paraffin. (The paraffin maintains the cells' shape when the material is sliced.) After it is embedded, the material is **sectioned,** or sliced, using a **microtome** (fig. 2.4).

The microtome firmly holds the embedded material and moves it across the edge of the extremely sharp microtome blade. After each pass across the blade edge, the specimen is advanced a short distance (the thickness of a section) and again moved across the blade. Even fairly hard material, such as wood and bone, can be sectioned in this way.

These sections are placed on a microscope slide and are usually **stained** with one or more stains that, in effect, dye the structures of the cell and make them more obvious when viewed under the microscope.

Scientists are concerned that when they prepare living tissues for slides, the embedding, sectioning, and staining

FIGURE 2.4    The microtome is an instrument used for slicing thin sections of embedded material. The sections stick together as they are sliced to form a ribbon of sequential slices.

Microtome knife                    Embedded material

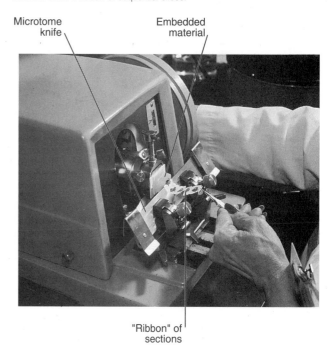

"Ribbon" of sections

may create observed cellular structures that are not actually part of the living cell. These structures are called **artifacts.** How could scientists determine whether a structure they observe is an artifact or real? One technique is to treat many samples of a tissue with a variety of embedding and staining techniques. If an observed structure appears in all the samples, it is more likely to be considered real.

FIGURE 2.5 In the electron microscope, electrons are produced by the filament. The lenses are electromagnets that focus the electrons passing through the specimen to the observed.

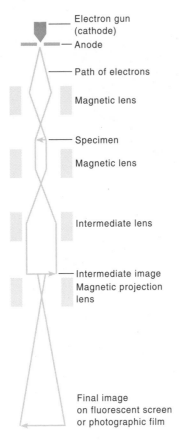

- Electron gun (cathode)
- Anode
- Path of electrons
- Magnetic lens
- Specimen
- Magnetic lens
- Intermediate lens
- Intermediate image Magnetic projection lens
- Final image on fluorescent screen or photographic film

A student observed a section of plant tissue that had been embedded, sectioned, and stained. She noticed that not all the cells had the same internal structures. Why was this an expected rather than unexpected observation?

FIGURE 2.6    Two views of a typical cell. (*a*) An electron micrograph showing complex organelle structure, indicated by dark lines and darkened areas. (*b*) A light microscope image of a stained cell.

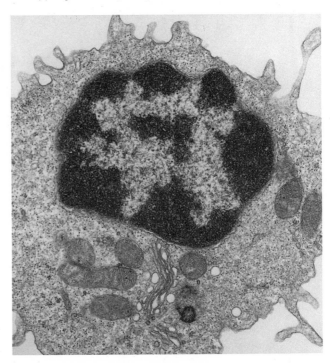

a.

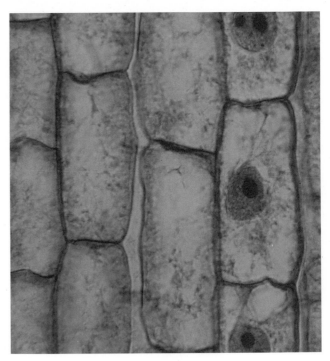

b.

Even greater magnification is possible with the electron microscope. Rather than focusing light rays as in the light microscope, the electron microscope focuses beams of electrons (that can behave as light rays) as shown in figure 2.5.

Preparing material for observation under the electron microscope requires techniques very similar to those used for the light microscope. Material must be stained, but in this case the stains absorb electrons rather than light. After being embedded and sectioned, specimens are ready for observation. Compare the views of cells using the light microscope and the electron microscope in figure 2.6. The image from the electron microscope is characterized by the appearance of dark lines and bodies showing cell structure.

As biologists observed more and more living material, it became apparent that the cellular structure is a fundamental characteristic of all living organisms. Some organisms may be composed of only a single cell, while others, such as the

human body, are composed of trillions of cells. These repeated observations gradually led to the generalization termed the **cell theory:** All organisms are composed of cells. In chapter 12 we will investigate a major concept in the cell theory—that all cells come from preexisting cells.

More than 150 years passed from the invention of the microscope to the time when biologists accepted that all organisms are composed of cells. Why should this have taken so long?

Cells occur in a wide variety of shapes and sizes (fig. 2.7) but have many features in common. A "typical" animal cell constructed from images of many different cells is shown in figure 2.8. This typical cell contains many structures, or **organelles,** that have a characteristic appearance and function. Although different kinds of cells have different organelles, this cell illustrates the most common ones. Because of its complex organelle structure, it is called a **eukaryotic cell.**

In later chapters we will see that different types of cells vary a great deal. For example, muscle cells and nerve cells have very different internal structures. The organelles described in the following section are, however, found in most cells.

## ORGANELLES AND CELL STRUCTURES

- The cell is surrounded by a thin **plasma membrane.** This membrane regulates and controls movement of material in and out of the cell and is considered in detail in chapter 3.
- The **cytosol** is the fluid portion of the cell that surrounds and bathes the structures in the cell interior. It is not considered an organelle, but we mention it here because it is such a major component of the cell contents.
- The **nucleus** is the "control center" of the cell and usually appears as a large spherical body in the cytosol. Genetic material is contained in the nucleus and is responsible for controlling traits of eukaryotic organisms. The nucleus will be considered in more detail in the chapter on genetics.
- The **endoplasmic reticulum** appears as a series of layered membranes, usually covered with small dark bodies called ribosomes. The endoplasmic reticulum (and ribosomes) are important in the synthesis of certain molecules by the cell. It will be discussed in more detail in the chapter on protein synthesis.
- **Mitochondria** appear as oblong bodies with inner foldings of membranes. These organelles function in energy production inside the cell and are discussed in detail in chapter 6.
- The **Golgi complex** appears as a stack of membrane-bound sacs. These sacs are associated with the processing and packaging of some substances produced by the cell.

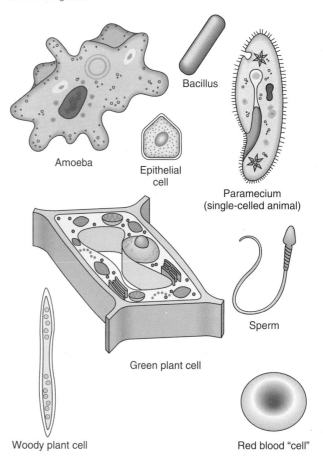

**FIGURE 2.7**  Cells are observed in a variety of cell shapes. In some cases, such as the Amoeba and the Paramecium, the single cell comprises the whole organism.

Amoeba

Bacillus

Epithelial cell

Paramecium (single-celled animal)

Green plant cell

Sperm

Woody plant cell

Red blood "cell"

- **Lysosomes** are saclike bodies that appear to digest material inside the cell. These are discussed in the chapter on digestion.

## Plant Cells

In a typical plant cell, additional specialized structures and organelles are observed. Several of these are shown in figure 2.9.

- **Cell walls** are the structure that appears as a thick layer surrounding the cell. The wall provides protection and support.
- **Chloroplasts** contain pigments (predominantly green in green plants) and appear as bodies enclosed by a membrane with extensive folding and stacking of interior membranes. These organelles function in photosynthesis and are discussed in chapter 7.
- **Amyloplasts** can appear in a variety of shapes and sizes as storage sites for starch.
- The **central vacuole** is a large, fluid-filled vacuole observed in mature plant cells. This vacuole is surrounded by a membrane and may serve as a storage area for a variety of materials, especially water.

FIGURE 2.8 Diagram of a "typical" animal cell showing electron micrographs of organelles and the corresponding diagrams of structures in the cell.

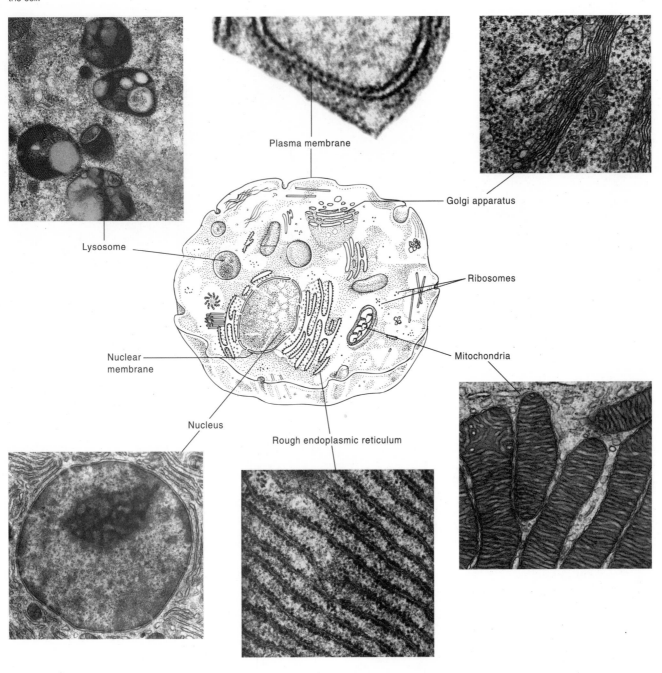

Plasma membrane

Golgi apparatus

Lysosome

Ribosomes

Nuclear membrane

Mitochondria

Nucleus

Rough endoplasmic reticulum

A student argued that after cells have been embedded, sectioned, and stained, they are dead and that observations of dead material should not be accepted as a good representation of living cells. How would you counter the student's argument?

## Prokaryotic Cells

Not all cells contain the complex organelles found in plant and animal cells. The bacteria and cyanobacteria comprise the **prokaryotic cells** (meaning they do not have a nucleus) and have little apparent internal structure (fig. 2.10). A wall surrounds the cell, but the interior contains only dispersed

FIGURE 2.9 Diagram of a "typical" plant cell showing electron micrographs of organelles and structures with corresponding diagrams of structures in the cell.

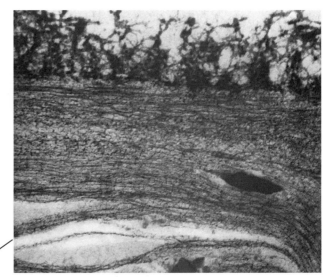

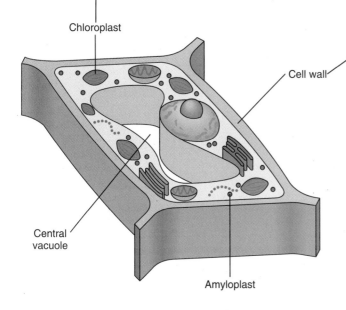

Chloroplast

Cell wall

Central vacuole

Amyloplast

genetic material and ribosomes in the cytosol. A plasma membrane is present inside the cell wall, but none of the complex organelles found in eukaryotic cells are found in prokaryotic cells.

## CELL FRACTIONATION

Earlier in this chapter we asked the question, how do we know whether an observed structure in a cell is real or an artifact? The method we proposed was to treat many samples of a particular tissue with a variety of embedding and staining techniques. A less time-consuming method is **cell**

**fractionation,** which means breaking apart the cell and separating the internal structures. We will use cell fractionation in the following experiment.

Our question about cell structures can be stated as a hypothesis.

> *Hypothesis*  Organelles observed in cells are real structures, not artifacts.

Using this hypothesis and our knowledge of cell fractionation, we can make a prediction.

> *Prediction*  If cell organelles are real, then they can be separated. These separated organelles should retain the appearance they have within the cell.

Now we can outline an experimental design using cell fractionation.

### Experimental design
- Break open a cell and release its contents.
- Spin the broken cell preparation in a centrifuge to separate the cell components (Thinking in Depth 2.1).
- Examine the separated components under a light microscope or electron microscope and look for organelles.

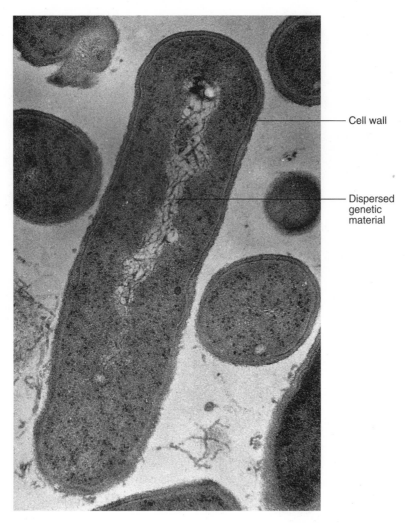

Cell wall

Dispersed genetic material

After we have completed the experiment, we examine the results.

> *Results*    The broken cell preparation has divided into layers, and within the layers we can see the organelles and structures that we observed in the intact cell. For example, we find cell nuclei in one layer, mitochondria in another layer, membrane fragments with attached ribosomes in a third layer, and so on.

Finally, we interpret the results.

> *Interpretation*    Since the cell organelles have maintained their structure and appearance during cell separation, they most likely are real structures in the living cell and not artifacts.

Observations using the microscope and cell fractionation are two **independent techniques** that can help reveal how cells look and what structures are present. Using inde-

A biologist, examining a cell under a microscope, found a structure that appeared to be an organelle, but she could not detect the structure when she examined the material after centrifugation. How could it be argued that this organelle might be real and exist inside living cells?

pendent approaches greatly strengthens the argument that the observed organelles and structures are real since we now have two lines of evidence supporting interpretations of cell structure.

It is essential to remember that the diagrams of cell structure, such as figures 2.8 and 2.9, are *interpretations* based on the *observations* obtained from microscopic examination and cell fractionation. Although the interpretations presented in this chapter have been tested many times, they are still subject to assumptions and the possibility of alternative interpretation.

## CELL FRACTIONATION

Cells can be broken using a homogenizer, a sophisticated blender that mechanically breaks open the cells and releases the cell contents. This material is placed in a centrifuge tube and spun at different speeds. As figure 1a shows, at lower speeds and shorter durations the heavier organelles, such as nuclei, settle to the bottom of the tube. These are removed for examination and the tube is spun again at a higher speed to separate other organelles. As this process is repeated, different organelles can be removed for study.

An alternative separation procedure is shown in figure 1b. The broken cell preparation can be carefully placed on top of a sugar solution that has been prepared to form a density gradient, with the greatest density at the bottom of the tube. When spun at high speeds for a long period of time, the organelles will settle to the area of density equal to their own density. These layers can then be separated for study.

FIGURE 1    Cell fractionation and separation of organelles. (a) After gentle grinding in a homogenizer, broken cells are spun at increasing speeds to separate organelles. (b) Broken cells may be placed at the top of a sucrose gradient and spun for a long period to force organelles to layers of the same density.

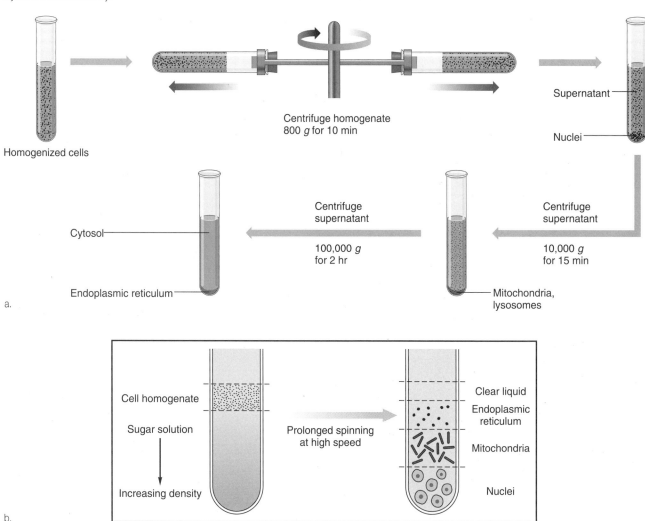

A scientist using a microscope detected a new structure in the cells of a rare species of fish. He interpreted this structure as a new, undiscovered kind of organelle. He then proposed to include this organelle in textbook diagrams of cell structure. What should be done before this supposed organelle is written up in textbooks?

What else can we learn about cell structure and organelles? Cell fractionation offers the possibility of further investigating the composition and function of cell organelles. Since we can separate and purify organelles, we can study their chemical composition in more detail. Two questions present themselves immediately: What is the chemical composition of the organelles and structures, and will the composition provide clues to the function of the organelles? These and other questions about the chemical nature of the cell will be considered in the next chapter.

## SUMMARY

*The information on cell structure and the techniques for studying cells discussed in this chapter will be used in many parts of this book. As different topics of biology are presented, the cells, organelles, and cell functions that are pertinent to the topic will be covered in greater detail, building upon and elaborating on information given here. As you continue reading, pay special attention to the key concepts presented in this chapter.*

1. More detailed and informative observations of biological material can be made by magnifying the material using a microscope.

2. Material too thick to allow light to pass through it must be **embedded, sectioned,** and **stained** in order to observe its cellular structures with the microscope.

3. Repeated microscopic observation of biological material led to the generalization that all living organisms are composed of cells, a conclusion called the **cell theory.**

4. Observations of cell interiors reveal the following structures or organelles in a typical cell:

   - nucleus
   - cytosol
   - endoplasmic reticulum
   - plasma membrane
   - ribosomes
   - mitochondria
   - Golgi complex
   - lysosomes

5. Plant cells usually contain additional organelles and structures:

   - cell wall
   - chloroplasts
   - central vacuole
   - amyloplasts

6. **Prokaryotic cells** do not possess a nucleus or most other cellular organelles; **eukaryotic cells** are the "typical" cells described in 4 and 5 above.

7. **Cell fractionation** is a technique that allows cell organelles to be studied in more detail. Cells are ruptured and the resulting preparation is spun in a centrifuge to separate the cell contents.

8. **Artifacts** are structures in a cell that are not real but are created when material is treated and prepared for observation. Before scientists conclude that observed structures are "real," they must present strong arguments and tests to indicate that these structures are not artifacts.

## EXERCISES AND QUESTIONS

1. A student observed a section of plant tissue that had been embedded, sectioned, and stained. She noticed that only one-fourth of the cells in the section appeared to contain a nucleus. Which of the following would most likely explain this observation (page 16)?

   a. Every cell contains a nucleus, but one-fourth of the cells are not sectioned.
   b. Many other organelles besides a nucleus are present in a cell.
   c. Each section is only one-fourth as thick as a cell.
   d. A nucleus is probably the largest organelle in the cell.

2. Which of the following preparations should show nuclei in all cells?

   a. Embedded cells that were sliced into thinner sections.

   b. Bacteria cells.
   c. Embedded and sectioned cells that are not stained.
   d. Broken cells that have been prepared for cell fractionation.
   e. Material that shows whole cells.

3. A cell is observed to contain a nucleus, mitochondria, and chloroplasts. From this information you can conclude that the cell is

   a. a prokaryotic cell.
   b. a plant cell.
   c. a bacteria.
   d. an animal cell.
   e. a cell from the human body.

4. More than 150 years passed from the invention of the microscope to the time when biologists accepted that all organisms are composed of cells. Which of the following best explains why this conclusion took so long (page 17)?

    a. Many observations on a great many different kinds of organisms had to be made to develop a generalization.
    b. Microscopes had to be invented that could detect organelles in a variety of cells.
    c. A generalization such as the cell theory requires techniques in addition to microscopic observation.
    d. Microscopes powerful enough to detect bacteria had to be invented to generalize the cell theory.

5. If an embedded cell is sliced on a microtome, which of the following structures would always be cut by the microtome knife?

    a. amyloplast
    b. nucleus
    c. mitochondria
    d. plasma membrane
    e. Golgi complex

6. A student argued that after cells have been embedded, sectioned, and stained, they are dead and that observations of dead material should not be accepted as a good representation of living cells. Which of the following is the best counter to the student's argument (page 18)?

    a. "Dead" material can be considerably different from living material.
    b. Preserved "dead" material can closely resemble living structures.
    c. Cells can be considered "dead" only after they are sliced with a microtome.
    d. As soon as a sample of material is removed from the whole organism, it can be considered "dead."

7. A biologist discovered that when a broken cell preparation was centrifuged, all the organelles separated to the bottom of the centrifuge tube. Which of the following would most likely explain this result?

    a. The material was observed before it had been centrifuged.
    b. The material was centrifuged at a low speed for a short period of time.
    c. The cell material was placed in the centrifuge tube before the cells were broken.
    d. The material was centrifuged for a long period of time at a high speed.
    e. The material was placed in the centrifuge tube after the cells were broken.

8. A biologist predicted that when a broken plant cell preparation was centrifuged, parts of the cell wall would be the first structures found at the bottom of the centrifuge tube. Which of the following is the biologist most likely assuming?

    a. Cell walls were not broken when the cell was broken.
    b. Cell wall material is heavier than other structures or organelles.
    c. Walls completely surround the cell before the cell is broken.
    d. Cell walls are the only structures that can be separated by centrifugation.
    e. No other cell structures or organelles were present in these plant cells.

9. A biologist, examining a cell under a microscope, found a structure that appeared to be an organelle, but she could not detect the structure when she examined the material after centrifugation. How could it be argued that this organelle might be real and exist inside living cells (page 20)?

    a. This particular organelle sticks to the sides of the centrifuge tube.
    b. The biologist is mistaken and is observing an artifact.
    c. If an organelle observed with the microscope is real, it should be found after centrifugation.
    d. Artifacts can be present in centrifuged material.

10. A scientist using a microscope detected a new structure in the cells of a rare species of fish. He interpreted this structure as a new, undiscovered kind of organelle. He then proposed to include this organelle in textbook diagrams of cell structure. Which of the following should be done before this supposed organelle is written up in textbooks (page 22)?

    a. Observe more cells from this fish.
    b. Study the biology of this fish in more detail.
    c. Look for the structure in cells of other organisms.
    d. The biologist must be mistaken (it is not a new organelle) since the organelle has likely been previously discovered.

11. Which of the following best indicates why a diagram of a cell is an interpretation of cell structure?

    a. Diagrams show how structures appear when magnified.
    b. Both diagrams and photographs using the microscope are views of the cell.
    c. Cells are treated and prepared in various ways before observation with the microscope.
    d. A diagram will always leave out or modify some of the cell's observed structures.

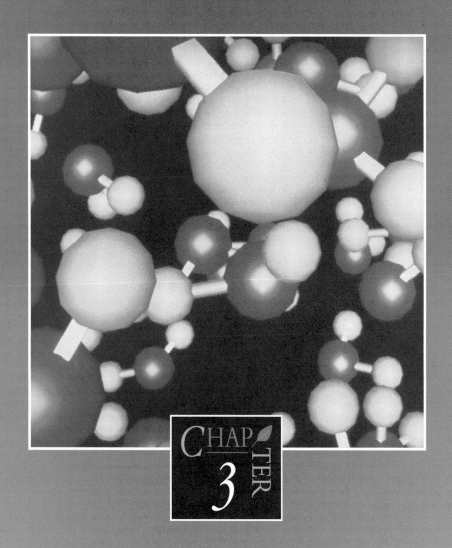

# ATOMS, MOLECULES, AND ENERGY

C ells are made of biological components called organelles, as we learned in the previous chapter. These components are so small they can be examined only with microscopes. But if our knowledge of cells was restricted to the physical structure of organelles, we would know very little about biological organisms. Fortunately, we are able to investigate cells at an even more detailed level.

Researchers have discovered that all cells and their components are made up of chemicals, and the chemical composition of cells can be analyzed. Can chemical composition provide clues to how cells and organelles function?

Before we answer that question, we must understand some basic principles of chemistry. Once we understand these principles, we should be able to determine how organelles carry out their respective tasks. No matter what questions we ask about cells, our interpretations and conclusions must ultimately take into account the chemical organization of cells.

## ATOMIC STRUCTURE

**Matter** is any material that occupies space and has mass. Matter can be liquid, solid, or gas; it can be living or nonliving. All matter on earth is composed of one or more **elements,** materials that cannot be further divided into any simpler element. There are eighty-eight naturally occurring elements on earth, all of which are represented by symbols. For example, carbon, oxygen, and lead are elements represented by the symbols C, O, and Pb.

Though both living and nonliving matter is made of elements, they differ greatly in the types and amounts of elements they contain. Table 3.1, for example, shows the elements that make up the earth's crust and those that make up a human body. What differences are most striking? Hydrogen is much more common in living material than in nonliving material, as are phosphorus, sulfur, and chlorine. The most common elements in the human body are hydrogen, carbon, nitrogen, and oxygen. Why do you think this is so?

To answer this question we must first discuss atomic structure. **Atoms** are the basic units that make up elements. The word atom comes from the Greek *atomos* meaning "indivisible." They are minute particles that cannot be seen, and they are the smallest unit of an element. For example, carbon atoms make up the element carbon, and oxygen atoms make up the element oxygen.

Atoms are constructed of three basic particles: protons, electrons, and neutrons (fig. 3.1). **Protons** are positively charged particles that are grouped in the atom's **nucleus,** or center. **Neutrons** are also grouped in the nucleus but they have no electrical charge. **Electrons** are negatively charged particles that are dispersed around the nucleus in layers or **orbitals.** The number of electrons in an atom always equals the number of protons, so atoms always have no **net** electrical charge; they are neutral. The number of protons and

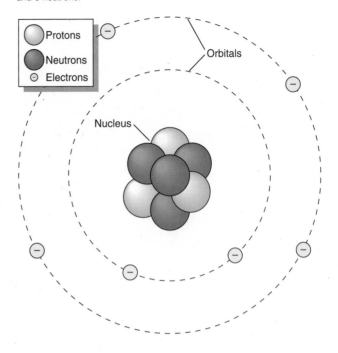

FIGURE 3.1    Structure of the atom: protons and neutrons are grouped in the nucleus. Electrons occupy orbitals around the nucleus. In the example of carbon shown here, there are 6 protons, 6 electrons, and 6 neutrons.

**TABLE 3.1    Abundance of Elements in the Earth's Crust and the Human Body**

| Elements | Percent Composition of the Earth's Crust (not including water) | Percent Composition of the Human Body |
|---|---|---|
| Hydrogen | — | 60.3 |
| Carbon | 11 | 10.5 |
| Nitrogen | 11 | 2.4 |
| Oxygen | 62.6 | 25.5 |
| Sodium | 2.6 | .7 |
| Phosphorus | — | .13 |
| Sulfur | — | .13 |
| Chlorine | — | .13 |

electrons in an atom determines the identity of an element. For example, hydrogen has one proton and one electron while carbon has six of each. Each element has its own unique number of protons designated as the atomic number. Table 3.2 shows the **atomic number** and **chemical symbol** of the most common elements in living organisms.

## MOLECULAR COMPOSITION

An important factor in understanding the chemical makeup of cells is molecular composition. **Molecules** are formed by two or more atoms linked together, either atoms of the same elements or atoms of different elements. For example,

*The Cellular and Chemical Basis of Life*

**FIGURE 3.2** Covalent bonds. The water molecule is composed of one molecule of oxygen and two molecules of hydrogen. Each chemical bond shares one electron from oxygen (X) and one electron from hydrogen (.).

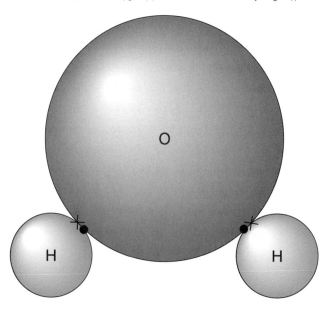

**FIGURE 3.3** Structural formulas for water ($H_2O$). (*a*) In simplest form, the structure can be shown as lines (covalent bonds) connecting the chemical symbol. (*b*) A slightly more realistic structure shows the angle between the bonds of hydrogen and oxygen.

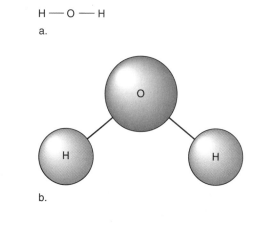

**FIGURE 3.4** Ionic bond. When an electron is transferred from sodium to chlorine, the sodium becomes a positively charged sodium ion and chlorine becomes a negatively charged chloride ion. These ions are attracted to each other by the electrical charge.

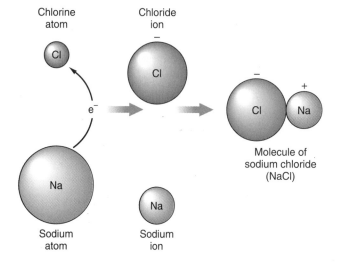

| TABLE 3.2 | Common Elements in Living Organisms | |
| --- | --- | --- |
| **Element** | **Chemical Symbol** | **Atomic Number (number of protons)** |
| Hydrogen | H | 1 |
| Carbon | C | 6 |
| Nitrogen | N | 7 |
| Oxygen | O | 8 |
| Sulfur | S | 16 |
| Phosphorus | P | 15 |

a carbon dioxide molecule is made up of carbon atoms and oxygen atoms. The links between atoms are called **chemical bonds;** the electrons in the outer orbitals of atoms determine the nature of these chemical bonds.

In living organisms, there are three types of chemical bonds: covalent, ionic, and hydrogen. **Covalent bonds** are formed when two atoms share an electron in their outer orbitals. The atoms are bonded by the electron they share, as shown in figure 3.2. When chemists want to designate in writing a molecule formed by a covalent bond, they use the symbols for the elements joined by a line. The line indicates that electrons are shared by the atoms. Scientists can also write the **structural formula** for elements, such as the formula for water, written in figure 3.3. Water is formed from two atoms of hydrogen and one atom of oxygen joined by covalent bonds. In text, the formula for water can be written $H_2O$.

The number of bonds that can be formed by an element is important. For example, hydrogen can form only one bond, oxygen two bonds, and carbon four bonds. The number of bonds determines the number of other elements that can bind to the atom, and thus the kinds of molecules and molecular structures that can be formed.

**Ionic bonds** are formed when atoms gain or lose electrons. Consider, for example, common table salt, or sodium chloride (fig. 3.4). A sodium atom gives an electron to a chlorine atom. Since electrons carry a negative charge, the transfer of the electron results in a net charge for the atoms: the sodium atom becomes positively charged and the chlorine atom becomes negatively charged. Now the two atoms are called **ions.** They become linked or bonded by the electrical attraction of their opposite charges, and an ionic bond is formed.

## FIGURE 3.5

**FIGURE 3.5** Hydrogen bonds. (a) Hydrogen bonds shown by dots, can be formed between molecules such as water. The hydrogen, with a slight positive charge, is attracted to the oxygen, with a slight negative charge. (b) In larger molecules, parts of the molecule that are oppositely charged can be attracted and become hydrogen bonded.

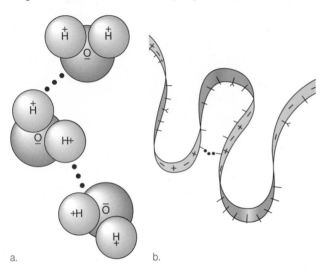

a.                    b.

**Hydrogen bonds** are usually formed between hydrogen atoms in one molecule and oxygen or nitrogen atoms in another molecule. As shown in figure 3.5, in covalent bonds, a slight positive charge can remain on a hydrogen atom and a slight negative charge on an oxygen or nitrogen atom. When these molecules come in contact, the positively charged hydrogen can form a bond with the negatively charged oxygen or nitrogen, similar to ionic bonds. The differences are that hydrogen bonds are much weaker and they occur between molecules or different parts of the same molecule, not between atoms.

From the relatively few elements found in living organisms, thousands of different molecules can be formed. Fortunately, understanding the nature and functions of these molecules is simplified because most of them can be grouped into four major classes of biological molecules with common characteristics. The four classes of biological molecules described in the following section are produced only by living organisms; they are almost always found in living organisms or their remains. Such molecules are the building blocks of all living things.

## BIOLOGICAL MOLECULES

One of the most common types of biological molecules is **carbohydrates,** molecules formed from carbon, hydrogen, and oxygen. Carbohydrates include substances familiar to most people such as sugar, starch, and cellulose. The formula for glucose, a type of sugar and thus a

**FIGURE 3.6** Glucose molecule. Glucose is composed only of carbon, hydrogen, and oxygen. This composition indicates the molecule is a carbohydrate.

Glucose

carbohydrate, is shown in figure 3.6. Note that glucose is made up entirely of carbon, hydrogen, and oxygen atoms.

Glucose molecules and other sugar molecules can be bonded together to form other substances. For example, a glucose molecule can be bonded with fructose, another kind of sugar, to form sucrose (fig. 3.7a). If many glucose molecules are bonded together, they form cellulose or starch (fig. 3.7b and c).

When smaller molecular subunits are combined into more complex molecules, the results are called **polymers.** Cellulose and starch are examples of polymers. The smaller subunits such as glucose are called **monomers.**

In cellulose, a polymer, what are the monomers? Are the monomers in starch and cellulose the same?

**Proteins** are another class of biological molecule; they are made up of carbon, hydrogen, nitrogen, oxygen, and a few other elements. Like carbohydrates, they are polymers. They consist of monomers called **amino acids** (fig. 3.8). There are twenty different amino acids in protein polymers, and they all have the same basic structure except for the group attached at the R position on the molecule. When these amino acids are joined in different sequences, they form different proteins. Because there are twenty amino acids, a great number of proteins is possible. Proteins form structural parts of the body such as hair and fingernails, or they form enzymes that play a primary role in chemical reactions that are discussed in the next chapter.

*The Cellular and Chemical Basis of Life*

FIGURE 3.7    Linkage of glucose and other sugars to form larger molecules. (*a*) One molecule of glucose combines with one molecule of fructose to form sucrose. (*b*) Cellulose is a chain of many glucose molecules. (*c*) A starch (glycogen) is a branched chain of glucose molecules.

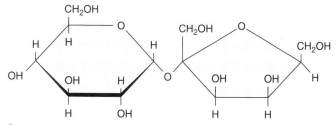

a. Sucrose

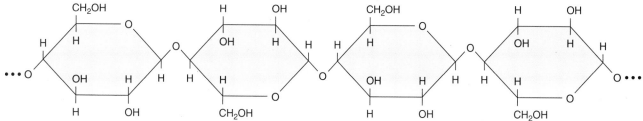

b. Cellulose

c. Starch

 Many different kinds of proteins can be produced, but only one kind of cellulose. Why?

 Would you expect living organisms to have more proteins or more nucleic acids? Why?

**Nucleic acids** are a third class of polymers found in living cells (fig. 3.9). They combine carbon, hydrogen, oxygen, nitrogen, and phosphorus to make DNA and RNA, which we will discuss in chapters 13 and 14. Nucleic acids are formed from monomers called **nucleotides.** Four different nucleotides make up the nucleic acids.

A fourth class of biological molecule is **lipids,** composed almost entirely of carbon, hydrogen, and oxygen, and in some cases phosphorus. Lipids are not polymers. They occur in a variety of forms, but their primary characteristic is that they are insoluble in water. The fats in animal tissues are an example of lipids.

FIGURE 3.8 Amino acid structure. The basic amino acid structure is shown at the top of the diagram. The twenty common amino acids vary in terms of the group at "R". Four of the twenty are shown as examples.

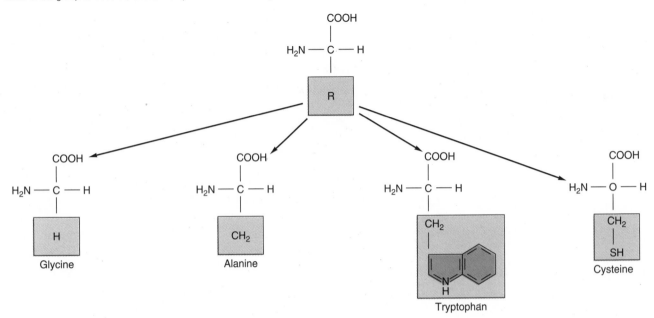

Although **water** is not a biological molecule, it has a major influence on biological functions, including cell functions, so we will discuss it here. Recall the structure of the water molecule shown in figure 3.3. One "end" of the molecule, the oxygen end, is slightly negatively charged, while the other end, the hydrogen end, is positively charged (fig. 3.10). This uneven balance of charges, or **polarity,** makes the water molecule an excellent solvent. That is, it can easily break the ionic bonds of other molecules and "shield" the ionic charge, thus maintaining the ions in solution and preventing them from interacting with other ions (Thinking in Depth 3.1). The polar nature of water is responsible for the shape of many biological molecules, as we will see in the next chapter.

## THE CHEMICAL MAKEUP OF CELLS

Water and the four types of biological molecules we have discussed are found in living things in various combinations. Table 3.3 shows the biological makeup of four representative organisms. The biological molecules listed make up almost 100 percent of these organisms.

What can we learn from the data given in table 3.3? It is immediately clear that water is the primary constituent of these organisms, ranging from more than 55 percent in the cow to 96 percent in jellyfish. Proteins are found in all organisms, while lipids vary from almost 25 percent in cows to just a trace in jellyfish.

| TABLE 3.3 | Molecules Composing Living Organisms as a Percent of Fresh Weight | | | |
|---|---|---|---|---|
| Compound | Cow (33 mos.) | Corn Plant (mature) | Sea Urchin Egg | Jellyfish |
| Carbohydrates | * | 17.5 | 1.36 | trace |
| Proteins | 16.73 | 1.8 | 15.18 | .67 |
| Lipids | 24.623 | .5 | 4.81 | trace |
| Water | 55.51 | 79.0 | 77.30 | 96.00 |

*Not available.*

Why does corn contain more carbohydrates than other organisms?

Living organisms contain high concentrations of hydrogen and oxygen. How do the data in table 3.3 support this fact?

We know that water and biological molecules make up all living things. Since we also know that living things are made up of cells, we can deduce that cells are made up of

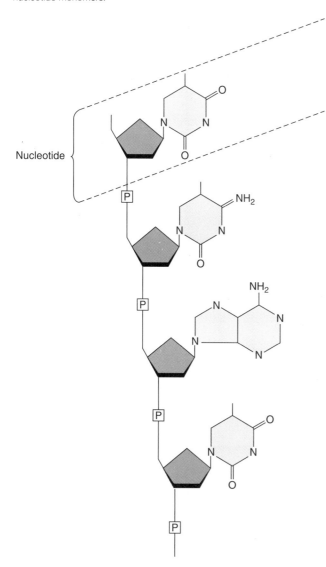

Nucleotide

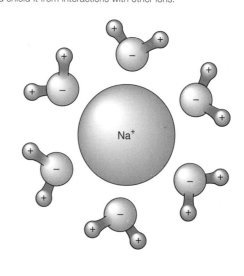

FIGURE 3.11    Membrane structure. In this proposed arrangement, lipid molecules form the membrane structure.

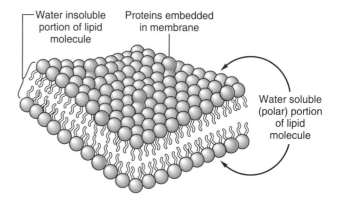

Water insoluble portion of lipid molecule

Proteins embedded in membrane

Water soluble (polar) portion of lipid molecule

biological molecules and water. What biological molecules are found in cells? Are organelles also made of the same biological molecules?

Using the technique of cell fractionation, which we discussed in chapter 2, we can separate cell organelles and analyze their composition. We get the following results.

Plasma membranes are composed of 50 to 75 percent lipids by weight, a very high percentage. The remainder is protein. Keep in mind that the plasma membrane forms a barrier regulating movement in and out of the cell. Since cells contain mostly water, and lipids are insoluble in water, the lipid molecules in plasma membranes can limit the movement of water (and other contents) in and out of cells. How do they do this? One end of the lipid molecules are polar, just as water molecules are. When the polar ends of

these molecules face the inside or outside environment, they interact with the polar water molecules. On the other hand, the rest of the lipid molecule is water insoluble. The most logical arrangement for these molecules is in a double layer with the polar groups in contact with water. The water insoluble portions would be in the interior of the double layer.

When the lipid molecules are arranged in a double layer with the polar groups facing the inner and outer environments, they form a wall that encloses the cell (fig. 3.11). The double layer formed by the lipids appears as relatively large, thin sheets in electron micrographs (fig. 2.6).

Many cell organelles, such as mitochondria and the Golgi complex, are composed of such thin sheets. It seems logical to guess that these organelles also have a high lipid

## WATER

The polar properties of the water molecule account for many of the unique characteristics of water. Because of their polarity, water molecules form hydrogen bonds with other water molecules (fig. 1). These hydrogen-bonded structures composed of many individual water molecules behave as much larger molecules. How does this hydrogen-bonding of water affect living organisms?

One such effect is to cool organisms when perspiration evaporates from their skin. A large amount of energy is required to break the hydrogen bonds in the water to convert liquid water (many hydrogen bonds) to water as a gas (no hydrogen bonds). This heat energy required to break the hydrogen bonds is carried away from the organism in the evaporated water molecules and the skin is cooled.

Another characteristic is the relatively large amount of energy required to heat water (one calorie per cubic centimeter), much larger than that required to heat other common liquids. As a result, water can absorb more heat with less change in temperature. In this way, water is an excellent heat stabilizer; it can maintain a stable temperature in an organism when the external temperature varies. Judging by the high water content of organisms (table 3.3), this temperature stabilization can be an important characteristic for many living things.

**FIGURE 1** Hydrogen bonding in water. Bonds form between positively charged hydrogen atoms and negatively charged oxygen atoms.

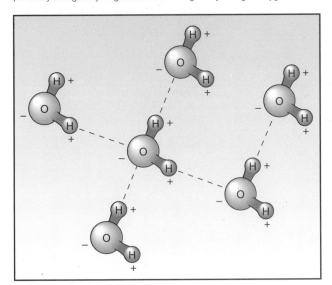

content. In fact, all the organelles that show extensive membrane structure in electron micrographs—mitochondria, Golgi complex, chloroplasts, and endoplasmic reticulum—contain 30 percent or more lipids. These organelles also contain a high percentage of protein. In the next chapter, we will see that much of this protein is enzymes, biological molecules that are essential to chemical reactions in the cell. Lysosomes also contain a high percentage of protein.

What other biological molecules make up cell organelles? Ribosomes and the nucleus contain nucleic acid as a primary component. Nucleic acid is necessary for synthesizing protein and transmitting genetic information. We will discuss these processes in chapter 12.

Chloroplasts, organelles in plant cells, contain a biological molecule called chlorophyll, which gives green plants their characteristic color. Cell walls in plants are composed almost entirely of cellulose.

As you can see, the molecular composition of organelles gives clues to their functions in the cell. In addition, the properties of the molecules indicate why organelles appear as they do in electron micrographs. Our knowledge of atoms and molecular structure can give us insight into how the cells of living organisms function.

## ENERGY

Another important characteristic of biological molecules is the **energy** contained in their chemical bonds. Energy is defined as the ability to do work; it is this energy that enables organisms to live and move. The energy in chemical bonds can take many forms. For example, when wood is burned, the cellulose in wood releases heat energy. How does this happen?

Cellulose is a carbohydrate. When it is burned, oxygen in the air combines with carbon in the carbohydrate to produce carbon dioxide and water. This chemical reaction can be written as

$$\text{oxygen} + \text{carbohydrate} \rightarrow \text{carbon dioxide} + \text{water} + \text{heat}$$

or

$$O_2 + \text{carbohydrate} \rightarrow CO_2 + H_2O + \text{heat}$$

The carbohydrate, a polymer, contains **chemical bond energy,** or energy that holds together the monomers. When the carbon dioxide and water are released, the chemical bonds are broken and bond energy becomes **heat energy.** Since energy is released, the chemical bonds in $CO_2$ and

## BOMB CALORIMETER

The energy in the chemical bonds of biological molecules can be measured using a bomb calorimeter (fig. 1). This instrument measures the amount of heat energy released when a weighed sample of material is burned (i.e., reacts with oxygen). As the figure shows, the sample to be burned is placed in the chamber at the center of the apparatus. Electrical wires are connected to the sample. To ignite the sample, an electrical current heats the wires and thus the sample. Oxygen is injected into the inner chamber before ignition so that sufficient oxygen will be present to support complete burning.

When the sample is burned, the water surrounding the chamber is heated by the heat released. The mechanical stirrer circulates the water so that heating is uniform. The increase in temperature is measured by the thermometer in the water.

For example, if one gram of carbohydrate is burned in the calorimeter and the water reservoir contains 1 liter (1,000 cubic centimeters) of water, then the temperature of the water would increase by 4.2 degrees centigrade. One calorie will raise the temperature of one cubic centimeter of water one degree centigrade; 4,200 calories are required to raise the temperature of 1,000 cubic centimeters of water 4.2 degrees.

Knowing the weight of the material (1 g), the increase in temperature of the water (4.2° C) and the volume of water (1,000 cu cm), the energy in 1 gram of material can be calculated (4,200 cal/g).

What would you be likely to observe if no insulation surrounded the water reservoir in the bomb calorimeter?

**FIGURE 1** Bomb calorimeter. A sample to be burned is placed in a central, sealed chamber and ignited with an electric current. Burning heats the water and the total heat produced can be measured. The sealed metal container and water both are shown.

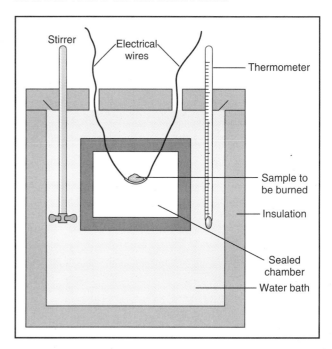

| TABLE 3.4 | Energy in Biological Molecules |
|-----------|-------------------------------|
| **Biological Molecule** | **Calories Released When One Gram Is Burned** |
| Carbohydrate | 4,200 |
| Protein | 4,300 |
| Lipid | 9,500 |

$H_2O$ have less energy than the bonds in the carbohydrate. Table 3.4 shows the amounts of energy released by different biological molecules when they are burned. You can see that lipids release more than twice the energy of carbohydrates and proteins.

Heat energy is measured in **calories.** One calorie is the amount of energy required to heat one gram of water one degree centigrade. We can precisely measure the bond energy of any biological material by burning it in a **bomb calorimeter** (Thinking in Depth 3.2).

In living organisms, large amounts of energy can be stored in the biological molecules. Under appropriate conditions, organisms can use that energy. For example, potato plants store energy in the form of starch in their tubers. When a new plant sprouts from the tuber, the stored starch provides it with energy. Humans store energy in the form of lipids, or fats, and carbohydrates. The fats and carbohydrates can be used as an energy source by the body. Knowing the way biological molecules store and use energy can help us understand how cells function.

Why is it an advantage for animals that migrate long distances to store energy as lipids rather than carbohydrates?

## SUMMARY

*Knowledge of atoms, molecules, and chemical bonds tells us a great deal about the composition of living organisms. At the most fundamental level, atoms and molecules indicate the ways in which living things are different from the nonliving world. Such knowledge will be used throughout this book to help explain how organisms and cells function. You should pay particular attention to the following points:*

1. All **matter** (material that occupies space and has mass) is composed of one or more of 88 elements. The smallest unit of an element is the **atom.**

2. Atoms are composed of **protons, electrons,** and **neutrons.** Protons and neutrons comprise the atomic **nucleus,** and electrons are arranged in **orbitals** around the nucleus.

3. Protons have a positive charge, electrons a negative charge, and neutrons no charge. Atoms have no **net** charge since the number of electrons and protons are equal.

4. Elements vary in their number of protons, electrons, and neutrons. The number of protons defines or is characteristic of an element.

5. Atoms are linked together by **chemical bonds** to form **molecules.** Covalent bonds are formed by sharing electrons. Ionic and hydrogen bonds are formed by electrical charge interaction between **ions** and polar groups.

6. Biological molecules can be grouped into four major classes.

   **Carbohydrates** are composed of carbon (C), hydrogen (H), and oxygen (O). This class includes sugars and **polymers,** such as starch and cellulose, that are composed of sugar subunits.
   **Proteins** are polymers composed of amino acids containing carbon, hydrogen, and oxygen. They include nitrogen (N) in their molecular structure.
   **Nucleic acids** are polymers of nucleotides. They are composed of carbon, hydrogen, oxygen, nitrogen, and phosphorus.

   **Lipids** are water insoluble molecules composed of carbon, hydrogen, oxygen, and sometimes phosphorous.

7. Water is an essential component of cells and organisms and a major constituent by weight of almost all organisms. The **polarity** of the water molecule strongly influences its interactions with biological molecules and structures.

8. The molecular composition of organelles can be measured by separating and isolating them and analyzing their molecular constituents.

9. Membrane structures (e.g., plasma membrane, mitochondria, Golgi complex, and endoplasmic reticulum) are made primarily of lipids. Because of the polar nature of lipid molecules and their interactions with polar water molecules, membrane structures can be formed that appear to separate the cell from its surroundings and parts of the cell from each other.

10. Protein is a primary constituent of almost all organelles. Protein functions as structural components and as enzymes in the cell cytoplasm and organelles.

11. Nucleic acid is found almost entirely in the nucleus and ribosomes and functions in the synthesis of protein and transmission of genetic information.

12. Energy is stored in the **chemical bonds** of molecules. Much of this energy is released as **heat energy** when biological molecules react with oxygen.

13. Carbohydrates and lipids in cells and organisms can store large amounts of energy; lipids contain more than twice as much energy as carbohydrates.

## EXERCISES AND QUESTIONS

1. A student claimed that the only similarity between cellulose and starch is that they are both carbohydrates. Which of the following is the best argument against the student's claim (page 28)?

   a. Both have nitrogen as part of their molecular structure.

   b. Only monomers of glucose make up both molecules.

   c. Starch contains phosphorous but cellulose does not.

   d. The two molecules are composed of carbon, hydrogen, and oxygen.

   e. Cellulose is a major component of cell walls.

2. Which of the following best explains why only one kind of cellulose can be formed (page 28)?

   a. Cellulose is composed of only carbon, hydrogen, and oxygen.

   b. The same kind of cellulose is found in all plant cells.

   c. Cellulose is a polymer formed from a large number of monomers.

   d. If monomers are rearranged in the molecule, the same polymer results.

   e. Many kinds of carbohydrates besides cellulose are known.

3. Which of the following best explains why more kinds of protein can be formed than kinds of nucleic acids (page 29)?

   a. Both have nitrogen as part of their molecular structure.

   b. Protein and nucleic acids are composed of different monomers.

   c. Nucleic acid is found in ribosomes and the nucleus, while protein is found throughout the cell.

*The Cellular and Chemical Basis of Life*

d. Phosphorous is part of nucleic acid structure but is not found in protein.

e. Proteins are comprised of twenty different monomers; nucleic acids are comprised of four different monomers.

4. Which of the following would you predict about an ion?

a. Electrons would be found as part of the atomic nucleus.

b. The nucleus would contain fewer than six protons.

c. The number of electrons and protons would not be equal.

d. No neutrons would be found as part of the atomic nucleus.

5. An analysis of a pure biological molecule indicated that the molecule contained nine different monomers. Based on this information, the molecule would most likely be a

a. carbohydrate.

b. lipid.

c. nucleic acid.

d. protein.

e. water molecule.

6. Regarding question 5, which of the following is the most likely prediction about the biological molecule if further analysis is done?

a. The molecule would be water insoluble.

b. Nitrogen would be found as part of the molecule.

c. The energy of the molecule would be greater than 8,000 calories per gram.

d. More than nine monomers would be found.

7. In table 3.3 (page 30), the amount of carbohydrate is greater in the corn plant than in any of the other organisms listed. Which of the following best explains this observation?

a. A low supply of glucose is available to plants.

b. Plants are relatively low in protein.

c. Although plants are high in carbohydrates, animals are relatively low.

d. Both cellulose and starch are polymers.

e. Plants are high in cellulose and starch.

8. Which of the following best explains why the data in table 3.3 indicate that hydrogen and oxygen are at high concentrations in living organisms (page 30)?

a. Water is at high concentrations.

b. Protein is found in all the listed organisms.

c. Some of the organisms contain almost no carbohydrates.

d. Carbohydrates are highest in plants.

9. The analysis of a biological molecule revealed the following characteristics:

a. The molecule was a polymer.

b. Nitrogen was part of the molecule.

c. Energy in the molecule was 4,300 calories per gram.

A student concluded that this molecule is a carbohydrate. Which of the characteristics contradicts the student's conclusion?

a. a only

b. b only

c. c only

d. a and c

e. None of the characteristics contradicts the student's conclusion.

10. If equal weights of the following materials were burned, which would produce the most heat?

a. dry wood

b. sugar

c. peanut oil

d. paper

11. What would you be likely to observe if no insulation surrounded the water reservoir in the bomb calorimeter (page 33)?

a. When the sample is burned, more heat would be generated by the reaction.

b. Oxygen would be lost from the chamber and the burning reaction would be incomplete.

c. The sample would not ignite when the electrical current was passed through the wire.

d. The sample would appear to generate less heat than was actually produced.

e. Water would leak from the reservoir and calculations on energy release would be in error.

12. Why is it advantageous for animals that migrate long distances to store energy as lipids rather than carbohydrates (page 33)?

a. Lipids are found in plants but in lower concentrations.

b. Lipids are not polymers, while carbohydrates are stored as polymers.

c. Some animals that move, such as a jellyfish, contain almost no lipids.

d. Lipids contain more energy per unit weight than do carbohydrates.

e. Lipids are not soluble in water.

# CHAPTER 4

# BIOCHEMICAL PATHWAYS AND ENZYMES

How are the unique, biological molecules discussed in chapter 3 synthesized (produced or created) by living cells? Why are some molecules synthesized or broken down by some organisms but not by others? For example, why is cellulose commonly synthesized by plants but not by animals? Why can some microorganisms break down or digest cellulose but other living organisms cannot? The key to answering these questions is found in understanding chemical reactions and the factors that affect these reactions.

What could you do to keep the reaction

$$A + B \rightleftharpoons C + D$$

from reaching equilibrium? That is, how could you shift the reaction to the right so that A and B are constantly decreased?

# CHEMICAL REACTIONS

The conversion of a molecule into a different molecule or molecules is called a **chemical reaction.** In chapter 3, we examined the chemical reaction involved in burning carbohydrates. If you recall, we summarized that reaction with a formula:

$$O_2 + \text{carbohydrate} \rightarrow CO_2 + H_2O + \text{heat}$$

In this reaction, the oxygen and carbohydrates are called **reactants;** they are the molecules being converted. The carbon dioxide and water are the **products** of the reaction.

Most chemical reactions are **reversible;** that is, the products can be changed back into the reactants. In the example above, carbon dioxide and water can be combined to make oxygen and a carbohydrate, although this may require energy input and rather special conditions. How does this happen? As two reactants, A and B, are combined, they form the products C and D. As this chemical reaction progresses, the amount and concentration of C and D increases. C and D are made of atoms, and we know that atoms are constantly in motion. As the concentration of C and D increases, their atoms are more likely to collide, because there are more of them. Under the right circumstances, when these collisions occur, C and D react with one another and form A and B. The two reactions can be written in one formula:

$$A + B \xrightleftharpoons{\text{reversible reaction}} C + D$$

The reversible nature of the reaction is indicated by the opposing arrows in the equation.

At some point, the concentrations of A and B and of C and D will be such that the speed of the forward reaction will equal the speed of the reverse reaction. The reactions will then be in **dynamic equilibrium.** We use the term "dynamic" because the reactants and products are constantly breaking down and re-forming; we use the term "equilibrium" because the concentrations of A and B and of C and D do not change with time.

# METABOLIC PATHWAYS

We have seen how molecules can change form in chemical reactions. In living organisms, many chemical reactions can be linked together, one after another, to create a **biochemical** or **metabolic pathway.** In the simplest form, a pathway could be thought of as an automobile assembly or disassembly line where something is added to or taken from the auto at each step. In the same way, a piece of the molecule may be added or removed at each step in a biochemical pathway. These pathways can branch or form loops, as shown in figure 4.1. Some pathways form more complex molecules as they progress, while others break down large molecules into smaller ones. In the following chapter, we will see many examples of metabolic pathways.

# ENZYMES

What makes chemical reactions link together into pathways? What drives the reactions to follow one pathway rather than another? The discovery of **enzymes** in living cells enabled biologists to answer these questions. As we discussed in chapter 3, enzymes are a class of proteins. The enzymes are able to **catalyze,** or increase the speed of, the chemical reactions. They force reactions, or a series of reactions, along certain paths by speeding them up.

It is important to recognize that an enzyme speeds up the rate at which a chemical reaction reaches equilibrium. It does not change the equilibrium point. If a reaction has not yet reached equilibrium, adding the right enzyme will help it do so. If, however, a reaction is already at equilibrium, adding an enzyme will have no effect.

Let us look at an experiment in which enzymes are used as a catalyst: the breakdown of sucrose into glucose and fructose. Sucrose, or table sugar, is composed of a glucose molecule and a fructose molecule bonded together. The reaction is written

$$\underset{\text{(reactant)}}{\text{sucrose}} \xrightarrow{\text{saccharase}} \underset{\text{(products)}}{\text{glucose} + \text{fructose}}$$
(enzyme)

FIGURE 4.1   Example of a metabolic pathway. The reactions convert substances in a series of steps. For example, A is converted to B, B is converted to X, B and G are converted to C and so on. The pathways can branch or form loops.

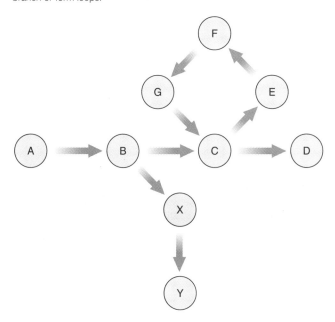

The speed of the reaction is measured by the disappearance of sucrose or by the appearance of glucose and fructose in the reaction mixture. Small amounts of an enzyme called yeast saccharase, which is extracted from yeast, are added to the mixture. The results of this experiment are shown in figure 4.2.

Note that as the concentration of enzyme increases, the conversion of sucrose into glucose and fructose also increases. Thus, the speed of the reaction increases as more enzyme is added. How can we interpret this result? The most likely interpretation is that the added enzyme is catalyzing the breakdown of sucrose. When more enzyme is present, more sucrose molecules are broken down in a given time. In other words, the more enzyme present, the faster the reaction.

A scientist conducted the experiment with sucrose and measured the reaction rate by measuring both the disappearance of sucrose and the appearance of glucose. What do you predict would be his observations?

In the experiment in figure 4.2, the reaction rate slows down after a period of time because the concentration of sucrose has decreased. Now we can observe another important facet of enzymes by conducting a variation of this experiment. If we add more sucrose to the mixture, we observe that the reaction rate returns to its original higher level, even

FIGURE 4.2   Enzyme-catalyzed conversion of sucrose to glucose and fructose. The enzyme (yeast saccharase) is extracted from yeast cells. The reaction rate is measured by the disappearance of sucrose from the reaction mixture or by the increase in glucose or fructose.   *Source: Data from E. Baldwin,* Dynamic Aspects of Biochemistry, *3rd edition, 1957, Cambridge University Press, London.*

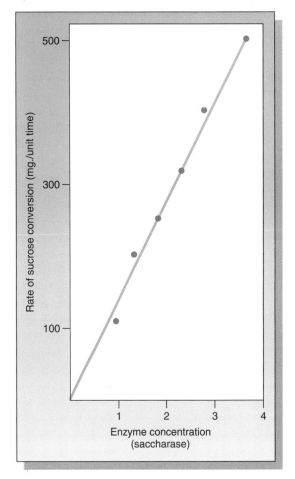

though we have not added any more enzyme. What does this mean? We interpret these results to mean that the enzyme is not destroyed or used up during the reaction. There is as much enzyme in the mixture when the reaction is over as there was when it began.

If the enzyme were used up in the sucrose reaction, what results would you expect when more sucrose is added to the mixture?

How do enzymes perform this role of catalyzing reactions? What special properties of these protein molecules enable them to affect other molecules? Perhaps the structure of the enzyme molecule will provide the key to its catalytic ability.

*Biochemical Pathways and Enzymes*

39

# STRUCTURE OF ENZYMES

Recall that in chapter 3 proteins were described as polymers of amino acids. The amino acids in enzymes join together to form a long, thin chain that can be folded into complex shapes, much as a chain of paper clips can be folded into many shapes.

The shapes formed by enzymes, however, are not quite as variable as the shapes formed by paper clip chains. The specific shape of an enzyme is determined by the sequence and properties of the amino acids that compose it. Some amino acids are able to make hydrogen bonds with other amino acids in the same enzyme chain and thus fold the enzyme, as shown in figure 4.3. Other amino acids are polar and interact with water. These are called **hydrophylic** amino acids. Still others are nonpolar and tend to avoid water. They are called **hydrophobic** and are usually found inside the enzyme structure away from contact with water.

How could you change the shape of an enzyme molecule?

How does the shape of an enzyme determine its ability to catalyze reactions? More than a century ago, scientists proposed that the shape of an enzyme "fits" the shape of a reactant much as a key fits a lock. The reactants, or **substrates,** as reactants in an enzyme-catalyzed reaction are called, are supposedly brought closer together and at the proper orientation for a reaction to occur. But why must substrates be brought together?

Without an enzyme, substrate molecules collide randomly. Only by chance do collisions occur at the appropriate orientation and energy and lead to chemical reactions, and those chances are slim. Left to themselves, substrates would react at a very slow rate, if at all. However, the reaction rate can be increased by higher temperatures. Heat causes the substrate molecules to move around more rapidly. At greater velocities, collisions would occur more frequently and at higher energy levels, increasing the probability that a reaction would occur. The level of energy at which substrates begin to react is called the **activation energy.** In organisms such as humans, however, body temperature is maintained at a constant level, so increasing the temperature is not an option. With enzymes, increased heat and molecular movement is not necessary. Thus enzymes can lower the activation energy of a reaction.

The proposed effect of enzyme shape can be stated as a hypothesis.

*Hypothesis* Enzymes catalyze reactions by providing a molecular shape that fits the shape of the substrate molecules and brings them into a position favoring a reaction.

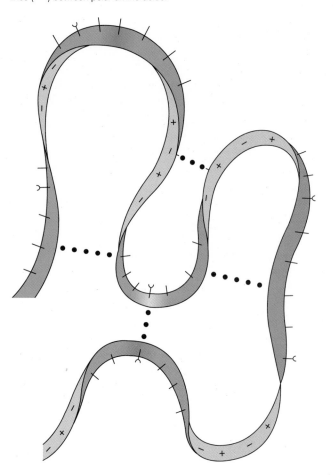

**FIGURE 4.3** Protein structure: Hydrogen bonds can form between some amino acids in a protein chain and help form a three-dimensional structure of the protein molecule. Hydrogen bonds are shown as dotted lines (•••) between polar amino acids.

Based on this "lock and key" hypothesis, it is logical to predict that when the shape of an enzyme is changed, its catalytic ability would be reduced. We can test this prediction by changing the molecular shape of enzymes and observing whether catalytic ability is, in fact, reduced.

## TEMPERATURE

Temperature is one factor that can change molecular shape. It does so by disrupting the hydrogen bonds in the protein structure. As we mentioned, hydrogen bonds between amino acids hold enzyme molecules in a folded, three-dimensional shape. As we increase the temperature, the relatively weak hydrogen bonds are broken by the increasing molecular motion and the enzyme shape changes, as shown in figure 4.4.

Using our knowledge about enzyme shape and temperature, we can predict that an increase in temperature will decrease the rate of a reaction catalyzed by an enzyme. That is, the increase in temperature will change the shape of the enzyme and render it ineffective. In an experiment to test this prediction, the reaction rate of amylase, an enzyme that

FIGURE 4.4 Disruption of enzyme shape. As temperature is increased, molecular motion can increase to the point where hydrogen bonds can be broken and change the molecule's shape.

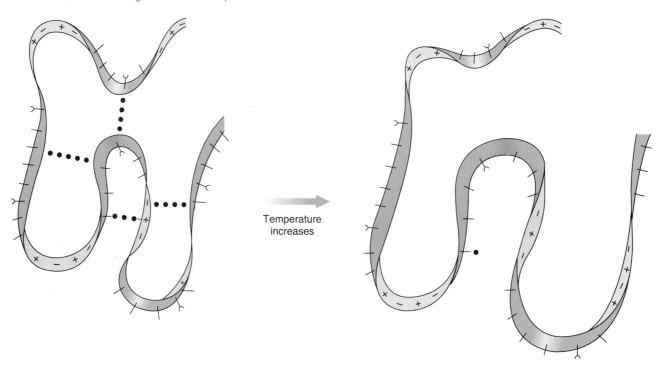

Temperature increases

breaks down starch into glucose, was measured as temperature was increased. The results are shown in figure 4.5.

You can see that initially the reaction rate increased as the temperature increased. At higher temperatures, however, the reaction rate began to decrease rapidly, reaching a maximum, or optimum, at about 50° C.

How can we interpret these results? The most likely interpretation is that at temperatures greater than 50° C, the enzyme structure is changed. This is consistent with the prediction and supports our hypothesis.

In figure 4.5, why does the reaction rate initially increase?
    Does the increase between 10° C and 40° C contradict the hypothesis?
Suppose the reaction mixture in figure 4.5. was cooled from 70° C to 50° C and the reaction rate remained low. How would you interpret this result?

## Acidity

Another variable that can affect enzyme shape is pH, which is a measure of the acidity of a solution. Acidity is determined by the concentration of hydrogen ions (H+) in a solution. A pH of 7 is neutral. Values of pH less than 7 are acid, and

**FIGURE 4.5** Enzyme reaction rate (activity) as a function of temperature. Reaction rate was measured at 10° C intervals. The enzyme used in this experiment (amylase) breaks down starch to glucose. *Source: Data from Gortner,* Outlines of Biochemistry, *3rd edition, 1949, John Wiley & Sons, New York, NY.*

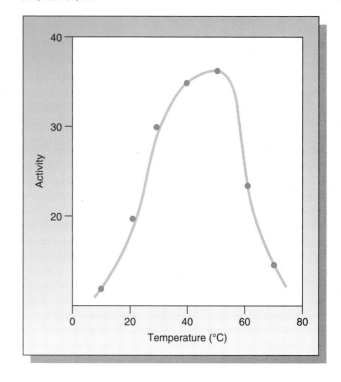

## THE pH FACTOR

The acidity of a solution is determined by the concentration of hydrogen ions. But where do hydrogen ions come from?

In pure water, some of the water molecules break apart to form hydrogen ions ($H^+$) and hydroxyl ions ($OH^-$).

$$H_2O \rightleftharpoons H^+ + OH^-$$

At equilibrium, water contains much more $H_2O$ than $H^+$ or $OH^-$. In natural waters such as lakes, ponds, or streams, or in any water exposed to air, carbon dioxide ($CO_2$) from the air quickly dissolves in the water. Dissolved $CO_2$ forms carbonic acid ($H_2CO_3$), which breaks down to form ions.

$$CO_2 + H_2O \longrightarrow H_2CO_3 \longrightarrow H^+ + HCO_3^- \text{ (bicarbonate ion)}$$

Since carbonic acid contributes a hydrogen ion to the solution, it is called an **acid.** Substances that remove hydrogen ions from solutions are called **bases.** Many biological molecules, such as amino acids, can contribute or remove hydrogen ions and affect pH. The full pH scale is shown in figure 1 with the pH of several biological substances indicated.

**FIGURE 1** The pH scale. A pH of 7 is neutral. As pH values decrease to one, they indicate more acid values. Increasing pH to 14 indicates more basic values. The pH value of several common substances are shown on the scale.

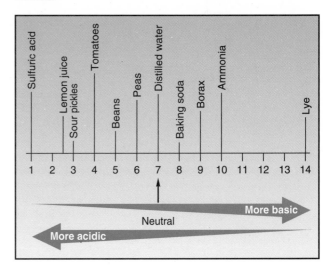

those greater than 7 are basic (Thinking in Depth 4.1). How does pH affect enzyme shape? Hydrogen ions can disrupt hydrogen bonds when they are attracted to the oppositely charged polar groups involved in the bond. The ions "screen" the charge on the polar group and break the bond. We can predict that if pH disrupts hydrogen bonds and thus affects enzyme shape, the rate of reactions catalyzed by enzymes should change as pH changes.

An experiment to test this prediction measured the activity of pepsin and trypsin as pH changes. Pepsin and trypsin are enzymes found in the human stomach that digest protein. In our experiment, we added these enzymes to a solution of gelatin (a protein) and measured the reaction rate. Pepsin and trypsin catalyze the breakage of the chemical bond between specific amino acids in the gelatin molecule. Figure 4.6 shows that the rate of reaction reached a maximum at about pH 2 for pepsin and pH 8 for trypsin. It decreased at higher and lower pH values. We can interpret these results to mean that the shape of the enzyme pepsin was disrupted at pH values greater than or less than 2 and the shape of trypsin was disrupted at pH values higher or lower than the optimum.

### Specificity

Another factor that affects the shape of enzymes is their **specificity.** If an enzyme must be a certain shape to fit a substrate, then it is logical that specific enzymes fit specific substrates. In other words, we cannot add just any enzyme to a substrate and expect it to catalyze a chemical reaction. If an enzyme has the wrong molecular shape, it will not fit the substrate. Thus we can predict that if we use the wrong substrate with an enzyme, or if the shape of the substrate is changed, a reaction will not be catalyzed.

Testing this prediction is relatively simple. We alter the chemical structure of a substrate and then compare its reaction rate with that of the unaltered substrate. What results do we observe? In general, the reaction rate of the altered substrate is changed. However, not all substrates react the way we might expect. Some enzymes exhibit **absolute specificity;** if the substrates to which they are added are changed in any way, the reaction will not be catalyzed. Other enzymes are not as specific. The substrate molecule can be changed slightly and the reaction will still occur, though the rate of reaction will be reduced.

In one experiment on specificity, an enzyme that acts on acetic acid was tested. The structural formula for acetic acid is

*The Cellular and Chemical Basis of Life*

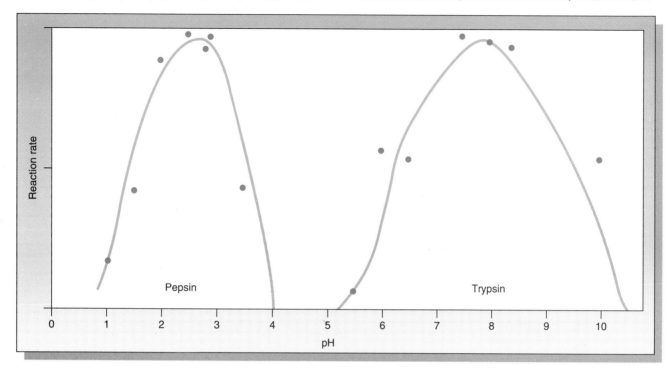

**FIGURE 4.6** Enzyme reaction rate as a function of pH. The graph shows the rate of gelatin (protein) digestion by the enzymes pepsin and trypsin as the pH is changed. *Source: Data from J. H. Northrop, M. Kunitz, and R. M. Herriott, Crystalline Enzymes, 1948, Columbia University Press, New York, NY.*

In the experiment, the structure of acetic acid was modified in two ways, as shown below.

1. A chlorine (Cl) is substituted for one of the hydrogens.

$$Cl - \underset{\underset{H}{|}}{\overset{\overset{H}{|}}{C}} - COOH$$

2. A fluorine (F) is substituted for a hydrogen.

$$F - \underset{\underset{H}{|}}{\overset{\overset{H}{|}}{C}} - COOH$$

When these molecules were exposed to the enzyme, no reaction occurred. When molecule 1 was mixed with normal acetic acid, the normal acetic acid underwent a reaction that was catalyzed by the enzyme. When molecule 2 was mixed with normal acetic acid, however, the reaction was inhibited. Not even the normal acetic acid would react.

How can we interpret these results? Figure 4.7 illustrates one interpretation based on enzyme specificity and shape. The normal, unaltered acetic acid fits the shape of the enzyme. The molecule with a chlorine substitute does not fit the enzyme and thus no reaction occurs. The molecule with fluorine substituted does fit the enzyme shape, but because the fluorine is present, no reaction takes place and the molecule remains attached to the enzyme. Since the unreacted molecule is not released by the enzyme, reactions with the unaltered acetic acid are blocked. This inhibition of the enzyme by blocking with a similarly shaped molecule is called **competitive inhibition.**

Each of our tests of the "lock and key" hypothesis appear to support the hypothesis. Temperature and pH affect the reaction rate, as does altering the substrate structure.

**FIGURE 4.7** Enzyme specificity. These diagrams interpret the experimental results when substrate molecule structure is modified. (*a*) Enzyme normally catalyzes reaction of acetic acid. (*b*) Chloride substitution does not allow the substrate to fit to the enzyme. (*c*) Fluoride substitution binds the substrate tightly to the enzyme.

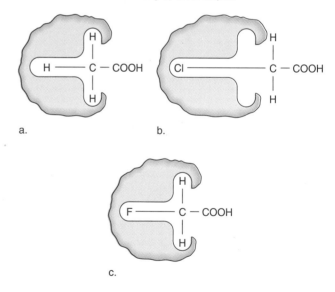

The complete picture of enzyme action drawn from these experiments is illustrated in figure 4.8. The substrate combines with the enzyme to form the **enzyme-substrate complex.** In the case shown in figure 4.8, the enzyme breaks the substrate into two parts, or products. After the reaction is completed, the products are released and the enzyme is ready to combine with another substrate molecule, continuing the reaction.

*Biochemical Pathways and Enzymes*

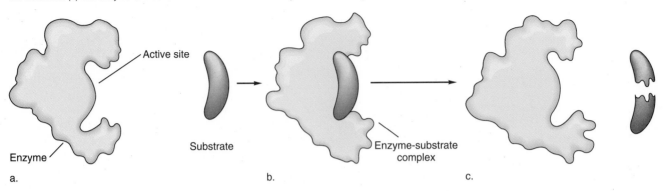

**FIGURE 4.8**　Enzyme action. (a) The enzyme and substrate are separate before reaction. The active site on the enzyme fits the shape of the substrate. (b) The enzyme and substrate combine to form an enzyme substrate complex. (c) When the substrate undergoes reaction (in this case by splitting into two products), products are released and the enzyme can combine with another substrate molecule.

Enzyme molecules work at an amazingly rapid pace. Table 4.1 lists four enzymes and the number of substrate molecules those enzymes can process per minute. Notice that the names listed in the right-hand column all end in "ase." This is a common ending for enzyme names.

Common sense suggests that, in spite of these impressive numbers, there must be a limit to how fast an enzyme can operate. Is it possible to measure the maximum speed of an enzyme? In one experiment, a small amount of enzyme was added to a substrate. Then the substrate concentration was increased. The results are shown in figure 4.9.

The graph shows that initially the reaction rate increased rapidly as the substrate concentration was increased. This is the effect we observed in figure 4.2. Eventually, however, the reaction reached a maximum rate at about 0.1 substrate concentration and did not increase any further, even though substrate continued to be added. The accepted interpretation of these results is that at 0.1 the enzyme was working as fast as it possibly could.

What would happen to the curve in figure 4.9 if more enzyme were added?

## ENZYMES IN METABOLIC PATHWAYS

Earlier in the chapter we learned that metabolic pathways are a series of chemical reactions. Enzymes catalyze each reaction, or each step in the pathway (fig. 4.10). In general, a different enzyme catalyzes each reaction. The overall movement of material along the pathway can be controlled by controlling the enzymes. For example, enzymes work at different rates, as we saw in table 4.1. If, say, the second enzyme in figure 4.10 ($E_2$) operates more slowly than all the other enzymes in the sequence, then material can move along the pathway no faster than the reaction at enzyme 2.

More precise control can be exercised by adjusting interactions between enzymes and products in the pathway. In figure 4.10, what would happen if $P_2$ could interact with $E_1$

| TABLE 4.1 | Enzyme Processing Rates |
|---|---|
| **Enzyme** | **Number of Substrate Molecules Reacting per Minute** |
| Amylase | 250,000 |
| Fumarase | 48,000 |
| Catalase | 2,400,000,000 |
| Acetylcholinesterase | 8,400,000 |

**FIGURE 4.9**　Reaction rate as a function of substrate concentration. At each of the indicated data points, reaction rate was measured for the substrate concentration provided. Rate becomes constant for all substrate concentrations above about 0.1.　*Source: Data from Ouellet, et al.,* Arch. Biochem. and Biophys., *39:41, 1952.*

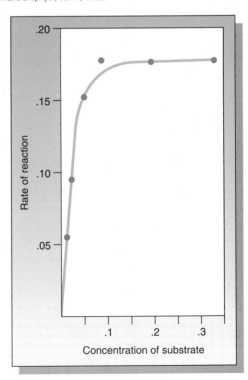

　*The Cellular and Chemical Basis of Life*

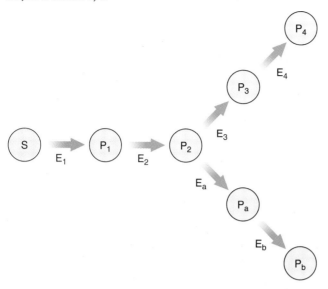

a.

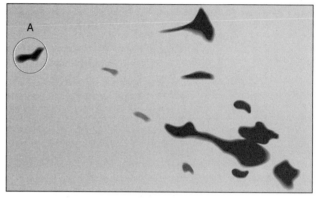

b.

to slow down (inhibit) enzyme action? With $E_1$ inhibited, less $P_1$ would be produced, and with less $P_1$ there would be less substrate ($P_1$) available for conversion to $P_2$. Thus, reactions all along the pathway would proceed more slowly and less $P_2$ would be produced.

If there is a decrease in $P_2$, there will be less $P_2$ to interact with $E_1$. The inhibition would be removed and the pathway would now speed up. This process is called **feedback inhibition** because a product at a later step in the pathway "feeds back" to an earlier step to regulate reaction rate.

What would happen if an enzyme were not present in a pathway? Occasionally this happens. Children can be born without the ability to produce an enzyme necessary for normal human metabolism. One well-known example is phenylketonuria (PKU). Children afflicted with PKU are born without the enzyme that breaks down the amino acid phenylalanine. When phenylalanine is not broken down, other products in the pathway accumulate and cause severe and permanent mental retardation. Newborn babies are routinely tested for this deficiency because the consequences are so serious. Fortunately, treatment is simple. Children are taught to eat a diet devoid of phenylalanine, so the enzyme that breaks it down is not necessary. The amino acid does not build up in the body and the child suffers no ill effects.

Metabolic pathways can be extremely complicated. Investigating the pathways and discovering the sequence of reactions can be a difficult task. One method researchers use is to label a substrate molecule with a radioactive atom. They allow chemical reactions to proceed for a short period of time and then destroy the organism or cells and separate the chemical components. Then they measure the separated components for radioactivity (Thinking in Depth 4.2). If they do this at various times, they can detect the appearance of radioactivity in new compounds and its disappearance from others. This establishes the sequence of reactions in the pathway (fig. 4.11).

---

In the metabolic pathway shown below, what would occur if the action of $E_3$ were inhibited?

$$S \xrightarrow{E_1} P_1 \xrightarrow{E_2} P_2 \begin{array}{c} \nearrow P_3 \xrightarrow{E_4} P_4 \\ E_3 \\ \searrow P_a \xrightarrow{E_b} P_b \\ E_a \end{array}$$

---

A scientist suspected that the substance at position A in figure 4.11 was the amino acid alanine. What could the scientist do to confirm that the substance was actually alanine?

In the experiment illustrated in figure 4.11, what was the experimental variable?

---

*Biochemical Pathways and Enzymes*

## TRACING METABOLIC PATHWAYS

If a molecule with one or more radioactive atoms is added to cells, the metabolic pathway followed when the radioactive atom is metabolized can be traced. For example, radioactive carbon dioxide or glucose can be added in place of the nonradioactive form. Scientists began using radioactive carbon in biological research in the mid-1940s. Research on atomic weapons and radiation in World War II yielded many advances in technology that enabled the production of radioactive elements such as radioactive carbon.

After the radioactive material has been added for a short period of time, the cells are killed and homogenized. The molecular compounds in this resulting homogenate can be separated by a variety of techniques and analyzed for radioactivity.

One commonly used separation technique, called autoradiography, is to add the homogenate to the edge of a material, allowing movement of the molecules as shown in figure 1. The edge of the sheet is placed in a solvent that moves up the sheet. Different molecules will be separated due to differences in solubility in the solvent and affinity to the sheet material. Then the sheet is rotated 90 degrees and again placed in the solvent. In this way, the molecular components are separated in a second dimension. At this point, the location of the separated materials is not usually visible. However, since the carbon is radioactive, the location of labeled compounds can be revealed by placing a sheet of photographic film over the sheet containing the separated material (fig. 1d). Radiation from the radioactive molecules will expose the film and indicate the location of the labeled molecule.

When this procedure is repeated at short intervals, the appearance and disappearance of compounds will indicate the relative positions of these compounds in a metabolic pathway. For example, in figure 4.11a, little substance appears at position A. At a later time, in figure 4.11b, more material appears at A. An obvious interpretation of these data is that the substance appearing at A is formed later in the metabolic pathway(s) than substances appearing at other positions in figure 4.11a.

**FIGURE 1**    Separation of metabolic compounds. (a) The homogenate is placed at a corner of the sheet. (b) The sheet is placed in solvent, which it absorbs. The compounds are separated. (c) The sheet is dried, rotated, and again placed in solvent to separate compounds in a second dimension. (d) A photographic film is placed on the sheet to locate the compound.

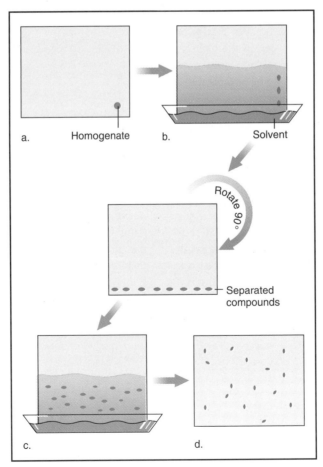

## SUMMARY

*Enzymes are absolutely essential in living organisms. Without them, chemical reactions in cells would occur so slowly that life could not exist. Understanding these amazing molecules and their mechanisms is necessary to understand other aspects of biology discussed throughout this book. Keep in mind the following key concepts as you read the rest of this book.*

1. **Chemical reactions** are described by showing the **reactants** and **products** as chemical symbols.

$$A + B \rightarrow C + D$$
$$\text{reactants} \quad \text{products}$$

2. Most reactions are **reversible;** when equilibrium is reached, the reaction is in **dynamic equilibrium.**

$$A + B \rightleftharpoons C + D$$

3. **Biochemical** or **metabolic pathways** are sequences of reactions in which molecules are modified in steps.

4. **Enzymes** are proteins that can **catalyze** (speed up) reactions without being used up or changed.

5. The shape of the enzyme molecule appears to be critical in catalyzing reactions. The "lock and key" hypothesis predicts that the **substrates,** or reactants, fit the enzyme molecule at an active site, the reaction is catalyzed, and the product is released.

6. Temperature and pH influence enzyme shape and also affect reaction rates.

7. Enzyme **specificity** is a characteristic of enzyme action. This means that one kind of enzyme will catalyze only a specific type of reaction.

8. Substrate molecules that have been structurally changed may still combine with an enzyme at the active site but not be released. In this way the enzyme is inactivated by **competitive inhibition.**

9. The reaction rate of enzymes can be extremely rapid, turning over more than 1,000,000 molecules a minute.

10. Enzymes catalyze almost all reactions in metabolic pathways. If an enzyme is missing, products can accumulate.

11. Control in metabolic pathways can be accomplished by **feedback inhibition** in which the concentration of a product in the metabolic pathway influences the action of an enzyme earlier in the pathway.

## EXERCISES AND QUESTIONS

1. Which of the following would cause A and B in the reaction shown below to constantly decrease (i.e., equilibrium would not be reached) (Page 38.)?

$$A + B \rightleftharpoons C + D$$

   a. Increase the concentration of C and D.
   b. Increase the temperature of the reaction mixture.
   c. Remove C and D from the reaction mixture.
   d. Add an enzyme to the reaction.

2. In the experiment in which sucrose breakdown is catalyzed by an enzyme (page 38), the resulting reaction rate is compared to which of the following?
   a. reaction rate at a higher temperature
   b. the decrease in sucrose concentration
   c. the decrease in enzyme concentration
   d. reaction rate with no enzyme present
   e. reaction rate when enzyme concentration equals sucrose concentration

3. A student examined the results shown in figure 4.2 and concluded that, even if the reaction continues for a long time, the reaction rate will remain unchanged. Which of the following was the student most likely assuming?
   a. A constant sucrose concentration is maintained.
   b. Enzyme is used up in the reaction.
   c. The temperature of the reaction mixture is decreasing.
   d. The enzyme is being denatured.

4. A scientist conducting the experiment shown in figure 4.2 measured the reaction rate at each enzyme concentration by measuring both the disappearance of sucrose and the appearance of glucose. When he compared these two measurements, which of the following observations would you predict (page 39)?
   a. Decrease in sucrose = two times the increase in glucose.
   b. Increase in glucose = increase in enzyme.
   c. Decrease in sucrose = increase in enzyme.
   d. Sucrose will continue to decrease when glucose remains constant.
   e. Decrease in sucrose = increase in glucose.

5. In the enzyme-catalyzed sucrose breakdown described on page 39, the measured reaction rate will slow down after a period of time. If additional sucrose is added to the reaction, and if enzyme had been used up in the reaction, which of the following would be the most likely prediction?
   a. Reaction rate would remain steady.
   b. Reaction rate would decrease rapidly.
   c. Reaction rate would increase well above original rate.

6. Which of the following would most likely change the shape of an enzyme molecule (page 40)?
   a. using up or removing substrate molecule(s)
   b. cooling the reaction mixture below the optimum reaction temperature

c. removing the product molecule(s)

d. changing the arrangement or order of the amino acids in the protein polymer

e. adding extra product molecules to the reaction

7. In the experimental results shown in figure 4.5, why does the reaction rate initially increase (page 41)?

a. The enzyme is denatured when the temperature is increased.

b. Increasing temperature increases collisions between enzyme and substrate.

c. The concentration of substrate is increasing at lower temperatures.

d. The concentration of product is decreasing at lower temperatures.

8. In the results shown in figure 4.5, the reaction rate at $70°$ C has decreased compared to the much higher rate at $50°$ C. The reaction mixture at $70°$ C is now cooled to $50°$ C, and it is observed that the reaction rate does not increase to the originally observed rate at $50°$ C. Which of the following is the best interpretation of this result (page 41)?

a. Other factors besides heat, such as pH, can change the shape of enzymes.

b. Enzyme shape is not the only factor affecting the catalytic function.

c. Fifty degrees centigrade is not the optimum temperature for enzyme function.

d. An enzyme molecule cannot return to its original shape once the shape is disrupted.

e. An enzyme catalyzes only one specific reaction.

9. What would happen to the curve in figure 4.9 if more enzyme were added (page 44)?

a. The reaction rate would reach a maximum of 0.2.

b. The curve would shift downward.

c. The curve would shift upward.

d. Changing substrate concentration would have no effect on the curve.

e. The curve would be unchanged.

10. In the metabolic pathway shown below, what would occur if the action of $E_3$ were inhibited (page 45)?

$$S \xrightarrow{E_1} P_1 \xrightarrow{E_2} P_2 \begin{array}{c} \nearrow P_3 \xrightarrow{E_4} P_4 \\ \searrow_{E_a} P_a \xrightarrow{E_b} P_b \end{array}$$

a. $P_3$ would increase.

b. $P_b$ would increase.

c. $E_2$ would be inhibited.

d. $P_2$ would increase.

e. $E_a$ would be inhibited.

11. A scientist suspected that the substance at position A in figure 4.11 was the amino acid alanine. Which of the following would the scientist do to confirm that the substance was actually alanine (page 45)?

a. Add alanine labeled with radioactive carbon and observe if it is metabolized by the cells.

b. Determine the structure of alanine to discover where radioactive carbon would be located.

c. Repeat the experiment several times and observe if material appears at position A.

d. Use only a sample of pure alanine in the procedure and observe if it appears at position A.

e. Repeat the experiment with other types of cells and observe if material appears at position A.

12. In the experiment in figure 4.11, what was the experimental variable (page 45)?

a. exposure of the chromatography sheet to photographic film

b. the amount of material placed on the chromatography sheet

c. the amount of radioactive material found at each position

d. the position at which A is found on the sheet

e. time of exposure of the cells to the added radioactive material

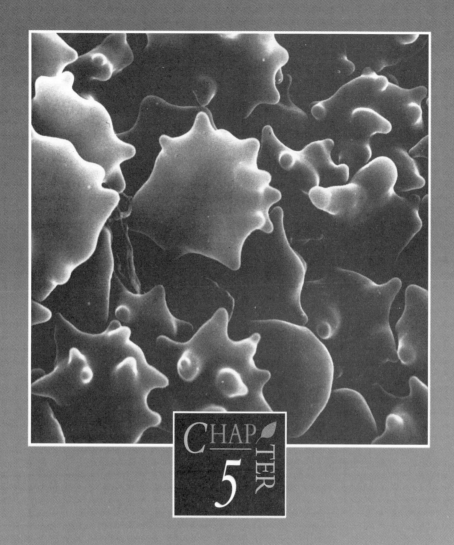

# MEMBRANE FUNCTION

he chemical reactions that make up metabolic pathways take place within the cells of living organisms. Where do cells obtain the molecules that serve as raw materials for these reactions? They must get them from their environment. We saw in chapter 2, however, that the plasma membrane surrounding the cell regulates the movement of material entering or leaving. Is it possible that the plasma membrane allows some materials to enter the cell but not others? If so, what factors affect this movement of materials through the plasma membrane? In other words, how does the plasma membrane function?

In nature, large amounts of material can move from one location to another by **bulk flow** (see chapter 8). The movement of a mass of material in one direction, such as water flowing in a river, is an example of bulk flow. In cells, bulk flow between the external environment and the cell interior is not possible because of the plasma membrane. Some other mechanism must be at work.

More than one hundred years ago, Robert Brown, an English botanist, made microscopic observations that provided a clue to molecular movement in cells. He observed that when very small particles (pollen grains, for example) are suspended in water, the particles vibrate in a random manner (fig. 5.1)

What could be causing the particles Brown observed to vibrate? Scientists have since interpreted Brown's observations as an indication that the water molecules surrounding the particles are in constant, random motion; they have named this movement "Brownian motion." The water molecules collide with the particles and cause them to randomly move too. In chapter 4 we learned that the random movement of molecules is increased by heat. If we heat the water molecules and increase their speed of movement, does the movement of the particles also increase? In fact, that is just what is observed.

Similar motions occur inside cells. Small particles or inclusions are observed to vibrate randomly. You can easily observe these vibrations with a microscope.

What does this random movement of cell particles have to do with moving molecules through the plasma membrane? Consider the situation shown in figure 5.2, where molecules are at a higher concentration in one area than they are in another. Since the molecules on both sides are in constant random motion and the concentration is greater on side A, more molecules will move from side A to side B than from B to A. The result will be a net gain of molecules for side B. This movement of molecules from A to B is called **diffusion.**

The molecules illustrated in figure 5.2 are moving from a higher concentration to a lower concentration. Thus they are said to be moving *down* the **concentration gradient.** In some cases, as we will see later, net movement can occur in the opposite direction, from a lower concentration to a higher concentration, but diffusion cannot account for this movement. Molecules are then said to be moving *against* the concentration gradient.

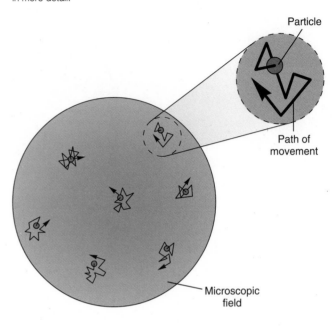

**FIGURE 5.1**  Microscopic view of particle vibrations. Particles are suspended in a fluid and are observed to vibrate in a random manner. The movements are shown by arrows and are somewhat exaggerated in this view. The expanded view at upper right shows the movement of a particle in more detail.

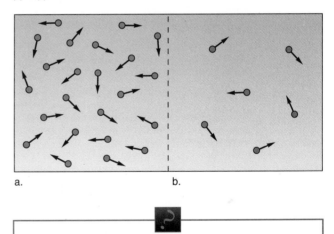

**FIGURE 5.2**  Net movement due to random motion. Since there are more particles in area (a), more particles will move from (a) to (b) than from (b) to (a).

In figure 5.2, what change in concentration would increase the movement of material from A to B? How would heating the solution affect diffusion?

Diffusion occurs the same way in living cells as in our example in figure 5.2. The constant molecular motion of the environment surrounding the cells and the cytoplasm inside the cells means that the plasma membrane is continually bombarded by molecules from without and within. A certain fraction of these collisions will result in molecules crossing the membrane to the opposite side. Several factors affect how much material crosses and which direction it travels.

FIGURE 5.3 Surface:volume ratio. For a large enclosed volume, the surface:volume ratio is very small. As the volume is divided into smaller and smaller pieces, the surface:volume ratio increases.

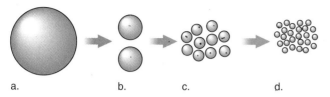

a.  b.  c.  d.

## SURFACE AREA OF CELLS

The larger the **surface area** of any object, the greater the number of molecules randomly colliding with it. This is true for cells as well. And if a greater number of molecules are colliding with the cell surface, more molecules are likely to enter the cell. But surface area alone does not determine the rate of diffusion. It is critical to examine surface area in relation to the volume of the cell. If the ratio of volume to surface area is too great, sufficient material will not diffuse into the cell to satisfy the needs of its metabolic pathways. A cell with a large volume has more chemical reactions and thus requires more molecular material. If the same cell has a large surface area, sufficient molecular material will diffuse into the cell to fuel its chemical reactions. Similarly, a large surface area will enable the cell to eliminate its waste products by diffusion across the plasma membrane more easily.

How can the ratio of surface area to volume be maximized? Consider what happens when volume is divided into smaller amounts, as in figure 5.3. Each time the volume is divided, the surface area increases, but the total volume remains the same. From this illustration you can see that many small cells have a greater surface area than fewer large cells. Most cells are around 50 micrometers (μm), or 1/1,000,000 of a meter, in diameter. If cells were any larger, they would most likely have an insufficient exchange of material across the plasma membrane to sustain life. This constraint on volume is probably the main reason cells are so small.

How could cell shape affect the ratio of surface area to volume?

In most mature plant cells, the vacuole fills the central part of the cell and the cytosol and organelles are pressed in a thin layer against the cell wall (see fig. 2.9). How do you think this arrangement aids in obtaining external material?

The surface area to volume ratio is also important in large, multicellular organisms that absorb or excrete material from or to the environment. The larger the surface area, the greater the exchange. Surface area will be considered in a variety of physiological processes throughout this book.

FIGURE 5.4 Permeability vs. lipid solubility. Each data point represents a different molecule. As lipid solubility increases, the permeability increases. *Source: Data from Collander,* Tran. Foraday Soc., *33:986, 1937.*

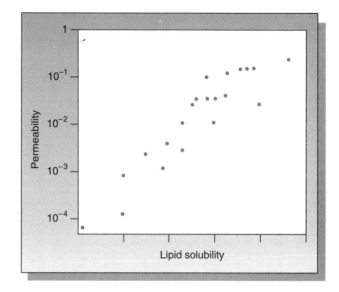

## PERMEABILITY

The ease with which a material moves across a membrane is a measure of the membrane's **permeability.** If a material easily moves across a membrane, the membrane is highly permeable to that material. If a material cannot move across the membrane, then the membrane is **impermeable** to that material. What factors affect membrane permeability?

### Lipid Solubility

Recall that lipid is a primary component of membrane structure (chapter 3). It is logical to predict that materials soluble in lipid will move more easily across membranes than materials insoluble in lipid.

What is the relationship between a substance's lipid solubility and its movement across membranes? See the results of the experiment in figure 5.4. You can observe that as lipid solubility increases, the movement across the membrane increases. The most likely interpretation is that the greater the lipid solubility, the more likely the molecule is to "dissolve" in the membrane and cross the membrane structure.

### Molecular Size

Another factor that might affect permeability is molecular size. Figure 5.5 indicates the relationship between the size of molecules and their movement across membranes. As molecular size increases, permeability decreases. In other words, large molecules have difficulty crossing membranes.

FIGURE 5.5 Permeability vs. molecular size. Each data point represents a different molecule. As molecular size increases, permeability decreases. *Source: Data from W. Ruhland and C. Hoffman,* Planta, *1:1, 1925.*

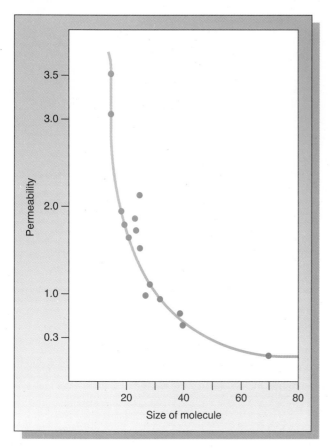

FIGURE 5.6   Facilitated diffusion. The rate of glucose entry into red blood cells as external glucose is increased. At low external glucose concentrations, entry increases rapidly. At higher concentrations, entry rate becomes constant. *Source: Data from W. D. Stein,* The Movement of Molecules Across Membranes, *1967, Academic Press, Orlando, FL.*

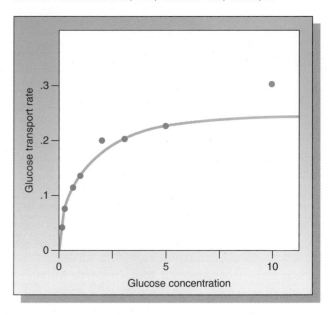

# FACILITATED DIFFUSION

Size and lipid solubility are just two of the factors that affect permeability. In fact, which molecules cross a membrane and how easily they cross is a complicated matter. Many kinds of molecules are apparently specifically selected to move across membranes, while other molecules that appear to be likely candidates are rejected. This selection, called **facilitated diffusion,** is so specific that scientists have proposed that receptors in the membrane aid the transport of certain molecules across the membrane. Such receptor and transport molecules would increase the membrane permeability for specific substances.

What evidence have researchers found for facilitated diffusion? One experiment measured the movement of glucose into red blood cells. The cells were bathed in a solution containing radioactive glucose (labeled with radioactive carbon; see Thinking in Depth 5.1) and the rate of glucose entry into the cells was observed. Figure 5.6 shows the results obtained when glucose entry was measured at different external glucose concentrations.

These results show that as glucose in the bathing solution increases, movement into the cells increases rapidly. At about 5 millimoles (m$M$), however, entry into the cells levels off. Further increases in glucose do not increase the uptake. Similar results have been observed for other sugars and amino acids.

If simple diffusion were operating in this experiment, we would expect glucose movement into the cell to continue increasing as the glucose in the bathing solution increased because, as we discussed earlier, as concentration increases, diffusion increases. What could explain the results of this experiment with glucose?

Researchers interpret these results as an indication that receptor molecules in the plasma membrane combine with the glucose molecules outside the cell, transport them across the membrane, and release them inside the cell. When the uptake rate levels off at about 5 m$M$, all the receptor molecules are transporting glucose as fast as possible. Increasing the glucose does not speed up the transport. (These results are similar in appearance to those shown in figure 4.9 in which the reaction rate remained constant above a certain level of substrate concentration.)

In the experiment with glucose, what is the experimental variable? How could you increase the maximum uptake of glucose above 5 m$M$?

As we saw in chapter 3, plasma membranes are made of a significant amount of protein. Biologists have observed that many of these proteins extend across the membrane and bind to specific molecules such as glucose. Some of these proteins are likely serving as the transport molecules in facilitated diffusion (fig. 5.7). That is, they are binding to glucose and bringing only glucose through the plasma membrane.

The advantages of facilitated diffusion are clear. Essential molecules and materials can move across the plasma membrane without allowing unrestricted entry of other materials.

*The Cellular and Chemical Basis of Life*

## RADIOACTIVE TRACERS

Radioactive tracers offer a convenient technique for measuring the movement of material across cell membranes. The molecule being investigated can be labeled and added to a solution bathing the cells. For example, a radioactive carbon can be added to glucose and the new glucose molecule can be added to a solution bathing red blood cells. Not every glucose molecule in the solution is radioactively labeled, however. Only enough labeled glucose is added to create a measurable amount of radioactive material in the cells. If we know the proportion of glucose labeled (called the **activity** of the glucose), we can calculate the total glucose entering the cells.

The glucose solution is gently mixed. After a period of time the red blood cells are removed and carefully washed with a nonradioactive solution. Next, the cells are placed in a radiation detector to "count" or measure the amount of radioactive material inside them (fig. 1). The amount of radioactive material tells us how much glucose has permeated the cell membranes.

Why were the red blood cells washed with a nonradioactive solution before the glucose molecules were counted? What error would result if the cells were not washed?

Radioactive tracing techniques can also be used to measure movement of molecules out of cells. In this case, the cells rather than the solution are "loaded" with radioactive materials. After loading, the cells are washed and placed in a nonradioactive solution. The bathing solution is periodically sampled and counted. Any radioactive molecules found in the solution have come from the cells, thus transport across the cell membrane can be measured.

**FIGURE 1**   Measurement of cell uptake. (*a*) Radioactive glucose added to red blood cells. After gentle mixing of the suspended cells for a period of time, the cells are centrifuged, (*b*) washed and recentrifuged, and then dried on a small plate. (*c*) Counting the radioactivity in the dried cells is a way to measure the glucose entering the cells.

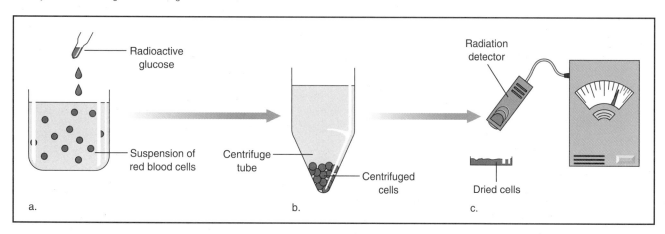

## ACTIVE TRANSPORT

We know that cells maintain internal concentrations of materials that are different from external concentrations. Otherwise there would be no difference between cells and their surroundings. If diffusion alone were acting on cells, an equilibrium would result in which internal and external concentrations were the same, but this does not happen. Why?

As we mentioned earlier in the chapter, it is possible for plasma membranes to move molecules *against* the concentration gradient. To do this, energy from the cells is required. Such movement against the concentration gradient

using cellular energy is called **active transport.** Active transport enables cells to "pump" materials across the membrane to maintain higher or lower concentrations than their exteriors. The generation and use of cellular energy is discussed in detail in chapter 6.

In an experiment to test for active transport, the energy-producing reactions of cells were "poisoned" with cyanide to eliminate any energy available for active transport. The movement of sodium ions ($Na^+$) across the membrane was observed. The concentration of sodium ions is normally much lower inside cells than in the external environment.

## FIGURE 5.7
**FIGURE 5.7** Membrane proteins. Many proteins are embedded in the cell membrane. Such proteins may serve as carriers to move specific molecules across the membrane.

**Exterior**

Transmembrane proteins

Peripheral membrane protein

Phospholipid bilayer

Peripheral membrane protein

Integral membrane proteins

**Cytosol**

We can hypothesize that sodium ions are actively transported out of the cell as they diffuse into the cell down the concentration gradient. This would maintain a lower internal Na$^+$ concentration inside cells. The hypothesis is illustrated in figure 5.8.

In the experiment, cells were placed in a solution containing radioactive Na$^+$. After a period of time, radioactive Na$^+$ had diffused into the cells. The cells were removed from the radioactive solution and placed in a bathing solution containing no radioactive Na$^+$. The rate of appearance of radioactive Na$^+$ in the bathing solution is a measure of the movement of Na$^+$ out of the cells.

After the movement of Na$^+$ was measured, the cells were placed in another bathing solution to which cyanide had been added. Cyanide stops the production of energy by cells. We can predict that if cells do actively transport Na$^+$, cells in the cyanide solution will stop active transport because they lack energy. In other words, movement of Na$^+$ out of the cell should be reduced. The results of this experiment are shown in figure 5.9.

We can see that cyanide does reduce the movement of Na$^+$ out of the cell. This supports our hypothesis that cells move materials against the concentration gradient by using energy for active transport.

If you examined the effects of cyanide on the movement of Na$^+$ into cells, what would you most likely find?

So far, we have discussed the movement of relatively small molecules across the plasma membrane. Is there any mechanism for large molecules to enter cells? Detailed observations of cell surfaces have revealed infoldings in the plasma membrane that enclose material outside the cell (fig. 5.10). The enclosed material appears to move into the cell and is later digested. Receptor proteins on the cell surface appear

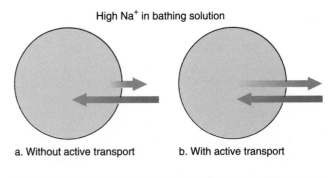

**FIGURE 5.8** Active transport. The length of the arrows represents the proposed rate of movement of Na$^+$ into and out of cells without active transport (*a*), and with active transport (*b*).

High Na$^+$ in bathing solution

a. Without active transport  b. With active transport

**FIGURE 5.9** Active transport. The movement of sodium ions (Na$^+$) out of the cell decreases when cyanide is added. Cyanide stops the production of energy by cells. This experiment indicates that active transport is responsible for the movement of Na$^+$. *Source: Data from W. D. Stein,* The Movement of Molecules Across Membranes, *1967, Academic Press, Orlando, FL.*

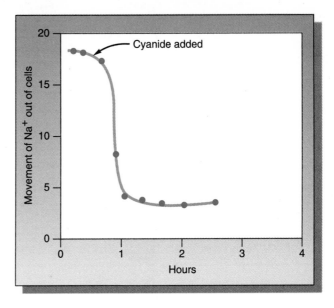

to bind to specific molecules outside the cell. This process, called **endocytosis,** can move large molecules into cells.

## WATER MOVEMENT

As we saw in chapter 3, cells are composed mostly of water, and most cells are bathed in a watery environment. Does water, like other kinds of molecules, move in and out of cells? Because the bulk of cells is made up of water, we can best answer this question by observing changes in cell volume when cells are placed in different water concentrations.

Let us look again at red blood cells. If we place red blood cells in pure water, the concentration of water outside the cells is greater than that inside the cells, because red blood cells do not contain pure water (fig. 5.11*a*). They contain a variety of dissolved materials that "dilute" the water, resulting in a lower water concentration. Once the blood cells are submerged, diffusion of water down the

**FIGURE 5.10** Endocytosis. (a) Infoldings form in the plasma membrane and contain material from the external environment. (b) These infoldings pinch off and move into the cell interior.

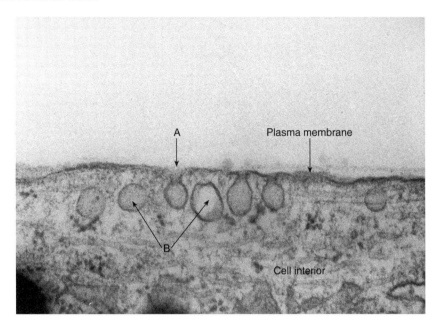

concentration gradient results in net water movement into the cells. Water will continue to diffuse into the cells until they burst (fig. 5.11b).

In another experiment, red blood cells are placed in a solution that has a lower water concentration than the cells themselves. As we would expect, water moves down the concentration gradient, leaving the cells and entering the solution. What happens to the cells as the water exits? Their volume decreases and the cells deflate (fig. 5.11c). Actually, the speed at which water moves in and out of a cell is impressive, much faster than almost any other substance. In other words, the permeability of the membrane to water is very high.

Water moves in and out of plant cells in a similar way, but there is an important difference. When water moves out of a plant cell, the cell contents shrink and pull away from the cell wall (fig. 5.12). Since the cell wall is rigid, the cell maintains its original shape rather than collapsing.

Similarly, when water moves into a plant cell, the cell does not burst as the red blood cells did. Why not? To answer this question, we can create an artificial system like the one shown in figure 5.13. We cover the larger end of a glass thistle tube with a membrane permeable to water but not other dissolved molecules such as sugar, and we place the tube in a solution of 99 percent water and 1 percent sugar. Since the solution inside the tube is 90 percent water and 10 percent sugar, the concentration gradient moves water into the thistle tube. As the volume in the tube increases, the water level in the tube rises. At a certain point, water stops entering the tube, but when we measure the two solutions, we discover that their water concentrations are not equal.

**FIGURE 5.11** Effects of water movement on cells. (a) Normal red blood cells have a characteristic disc-shaped appearance (b) when water moves into cells, the cells swell and burst. (c) When water moves out of cells, the cells collapse.

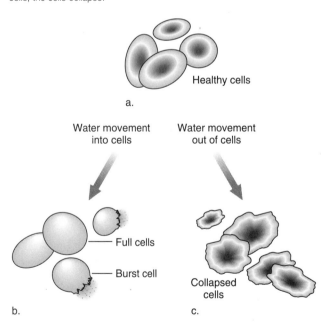

According to our measurements, water should continue moving down the concentration gradient into the tube. What has happened? As water moves into the thistle tube, the **pressure** in the tube increases because of the higher water level. The pressure forces water back out of the tube, balancing the movement of water into the tube. In the

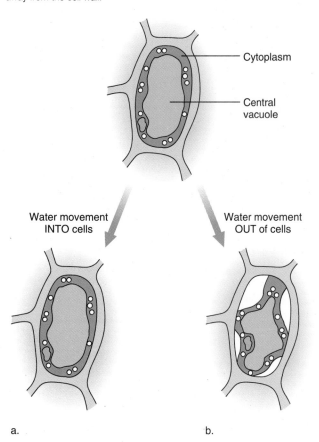

Cytoplasm

Central
vacuole

Water movement
INTO cells

Water movement
OUT of cells

a.

b.

plant cell, water moves into the cell until the cell wall exerts sufficient pressure to force water out of the cell as fast as it enters, establishing equilibrium. The movement of water caused by either pressure or the concentration gradient or both is called **osmosis.**

Suppose a bag made of a differentially permeable membrane was filled with a 5 percent sugar solution and sealed. What would happen if the bag was placed in pure water? In a 10 percent sugar solution?

The effect of water pressure in cells can also be demonstrated with thin potato slices. If a slice is placed in pure water, it becomes stiff and resistant to bending. Water has entered the slice, thus increasing its internal pressure. The slice is said to be **turgid.** If the slice is placed in a salt or sugar solution, it becomes much easier to bend, indicating that water has left the slice and the internal pressure has decreased.

FIGURE 5.13 Osmosis is the movement of water across a differentially permeable membrane due to water concentration and pressure gradients. The relative length of the arrows represents the relative amount of water flow.

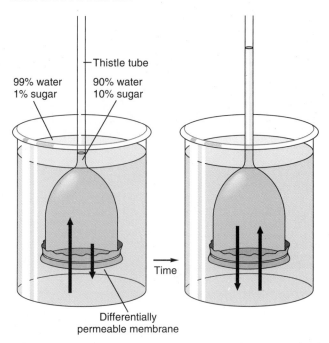

Thistle tube

99% water
1% sugar

90% water
10% sugar

Time

Differentially
permeable membrane

We have seen how the plasma membrane serves as a selective barrier for cells, separating the internal contents from the external environment, allowing the entry of some materials and actively transporting others. Other membranes in the cell serve similar functions for the organelles, separating them from other parts of the cell and isolating the metabolic pathways that occur within them. In other chapters of this book we will consider the functions of organelles and their membranes in more detail.

## MULTICELLULAR ORGANISMS

Many of the factors that move material in and out of cells also figure in the movement of substances in and out of multicellular organisms. Surface area, permeability, and osmosis all play a role in how living things function. For example, in most animals the digestive system has a large surface area. As we know from our discussion, a large surface area increases the rate of absorption, thus the digestive systems of animals allow them to absorb the maximum amount of nutrients. Similarly, the root systems of plants provide an enormous surface area for absorbing water and minerals from the soil. If, however, plant roots are exposed to salt water, which has a water concentration lower than that found in the roots, water will move down the concentration gradient from the roots to the soil and the plant will wilt. This is what happens when plants are fertilized too heavily, because fertilizers are salts. The factors that affect digestion and absorption will be discussed in chapter 9.

## SUMMARY

*The plasma membrane serves a vital purpose in the functioning of all living cells. Without it, cells would have no way to regulate material entering or leaving. Throughout this book we will refer to the important functions of membranes. As you read, keep in mind the following key concepts.*

1. Molecules and atoms are in constant, random motion. Such motion results in the net movement or **diffusion** of material from a higher to a lower concentration, referred to as down the **concentration gradient.**

2. The ratio of **surface area** to volume will determine whether diffusion of material into a cell can satisfy the cell's need for material. Smaller cell size increases the ratio.

3. **Permeability** indicates the ease of diffusion across a membrane. When a membrane is highly permeable to a substance, the substance diffuses easily across it.

4. Small molecular size and high lipid solubility aids the movement of molecules across membranes.

5. **Facilitated diffusion** is the enhanced movement of specific materials into cells down a concentration gradient. Membrane receptors are thought to bind to the molecules and effect transport across the membrane.

6. **Active transport** is the movement of material into or out of cells against a concentration gradient by utilizing cellular energy.

7. **Endocytosis** is the infolding and pinching off of part of the plasma membrane to move large molecules into the cell. The formation of membrane infoldings can be observed with the electron microscope.

8. **Osmosis** is the movement of water across a membrane due to diffusion down a concentration gradient or pressure differences. Volume changes in cells clearly indicate water movement.

9. Excess water movement into animal cells can cause the cells to burst. The walls of most plant cells, however, can resist the pressure resulting from water flow into the cells. Plant cells with a high internal water pressure become **turgid.**

## EXERCISES AND QUESTIONS

1. In figure 5.2, which of the following would increase the net movement of material from A to B (page 50)?
   a. Increase the concentration of material at A.
   b. Decrease the concentration of material at A.
   c. Increase the concentration of material at B.
   d. Decrease the concentration of material at both A and B.

2. How would heating the solution illustrated in figure 5.2 affect diffusion (page 50)?
   a. It would cause diffusion to move in only one direction.
   b. It would decrease molecular motion and slow down diffusion.
   c. It would increase molecular motion and speed up diffusion.
   d. It would cause net diffusion to move against the concentration gradient.

3. A sample of enzyme and a sample of amino acid were placed on a strip of agar as shown below. Agar is a jellylike substance allowing the diffusion of material.

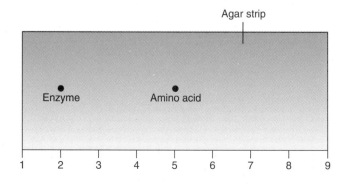

Agar strip

Enzyme    Amino acid

1   2   3   4   5   6   7   8   9

The enzyme catalyzes a reaction of the amino acid to form a yellow precipitate. At which position will the yellow precipitate first be observed? (*Note:* Large molecules diffuse more slowly than small molecules.)
   a. 1
   b. 2
   c. 3
   d. 4
   e. 5

4. Which of the following cell shapes would have the lowest surface area to volume ratio, i.e., the largest surface area in relation to volume (page 51)?
   a. a sphere
   b. a cube
   c. a fat cylinder
   d. a flat, ribbon shaped cell
   e. a rectangular shaped cell

5. In most mature plant cells, the vacuole fills the central part of the cell and the cytosol and organelles are pressed in a thin layer against the cell wall. How would this structure enhance the cell's ability to obtain sufficient material from its external medium (page 51)?
   a. Active transport would be necessary to move material from the external medium to the cytoplasm.
   b. External material need only move into the thin layer of cytoplasm near the wall.
   c. The plasma membrane will be located between the cytoplasm and the cell wall.
   d. Material from the cytoplasm could be lost to the large vacuole.
   e. The large surface to volume ratio of this cell would not favor sufficient uptake of material.

6. The results shown below were obtained in an experiment investigating the movement of a sugar into cells.

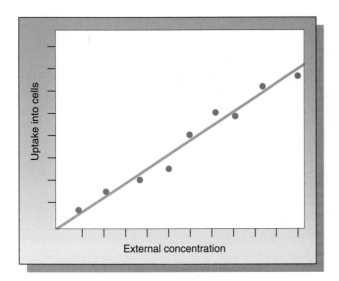

Which of the following would be the best statement of these results?
   a. The sugar is being transported by facilitated diffusion.
   b. The sugar is not being actively transported into these cells.
   c. Sugar is entering these cells by unaided diffusion.
   d. Further increasing the external sugar concentration would increase uptake.
   e. Sugar uptake is increasing as external concentration increases.

7. In the treatment diagrammed in Thinking in Depth 5.1, which of the following errors would most likely occur if the red blood cells were not washed before counting (page 53)?
   a. The cells would probably burst before counting is completed.
   b. The uptake of sugar would appear higher than it should be.
   c. The cell membrane would appear to be less permeable to the sugar than it is.
   d. The cells would lose excess water during the procedure.
   e. The uptake of sugar would appear lower than it should be.

8. If the procedure diagrammed in Thinking in Depth 5.1 is to give an accurate measure of sugar uptake, which of the following must be assumed?
   a. Red blood cells are impermeable to glucose.

   b. Glucose is transported into the cells by facilitated diffusion.
   c. The red blood cells are from humans.
   d. No internal sugar is lost during washing.

9. In the experiment on facilitated diffusion, the results of which are shown in figure 5.6, what was the measured experimental variable (page 52)?
   a. The amount of glucose entering the cells.
   b. The amount of time cells were exposed to radioactive glucose.
   c. The amount of glucose in the external solution.
   d. The amount of time the cells were washed in the nonradioactive solution.
   e. The time period used for counting the radioactivity.

10. In figure 5.6, which of the following would cause the greatest increase in the maximum uptake of glucose (page 52)?
   a. Increase the concentration of external glucose.
   b. Decrease the permeability of the membrane to glucose.
   c. Increase the concentration of internal glucose.
   d. Increase the amount of radioactivity in the external solution.
   e. Increase the number of transport proteins in the membrane.

11. In the experiment on the active transport of $Na^+$ (page 54), if you examined the effects of cyanide on $Na^+$ movement into cells, which of the following would you most likely find?
   a. $Na^+$ movement into the cell would increase.
   b. $Na^+$ movement before cyanide was added would be greater.
   c. Cyanide should have no effect on $Na^+$ movement.
   d. $Na^+$ movement out of the cell would increase.

12. Suppose a bag made of a differentially permeable membrane was filled with a 5 percent sugar solution and sealed. Which of the following would most likely occur if this bag was placed in a 10 percent sugar solution (page 56)?
   a. Water would diffuse into the bag down the concentration gradient.
   b. Water would flow out of the bag due to the pressure difference.
   c. Water would diffuse out of the bag down the concentration gradient.
   d. Water would flow into the bag due to the pressure difference.
   e. No net water movement would occur.

13. A plant cell is placed in pure water. Which of the diagrams in the figure most accurately indicates water movement across the cell's plasma membrane (the length of the arrows indicates the amount of water movement)?

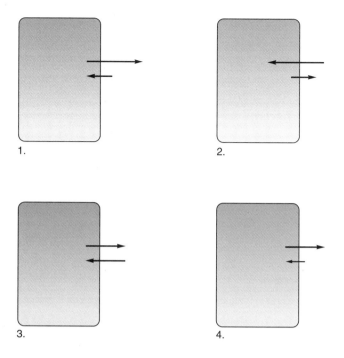

The exercises presented in this section provide an opportunity for you to explore relationships between biology and other academic disciplines such as law, literature, and psychology. There are no "right" answers to the questions posed in each exercise. Your response may be a short viewpoint or an in-depth research paper, depending on your teacher's instructions. References are provided with each exercise, should you want or need to refer to them as part of your assignment.

## Biology and Law
### SCIENCE AND THE U.S. LEGAL SYSTEM

In an article titled "Science and the Courts," *J. F. Ayala and Bert Black argued that the general public does not understand how the scientific process, including the formulation and testing of hypotheses, works. This has caused considerable confusion in U.S. courtrooms where "expert testimony" and "scientific evidence" are evaluated by standards different from those used by scientists.

For example, the following argument was used in a legal case claiming damage to human health by Agent Orange, a chemical used to kill plants: "The persons claiming damage complain of various medical problems; animals and workers exposed to extensive dosages of [ dioxin, a toxic contaminant in Agent Orange ] have suffered from related difficulties; therefore assuming nothing else caused the medical problems, Agent Orange must have caused them." In terms of the scientific process, what is wrong with this argument?

Should the legal system judge the reliability of scientific evidence on the basis of scientific procedures, or should it establish independent, legal rules?

*J. F. Ayala and Bert Black, 1993, "Science and the Courts," *American Scientist* 56, no. 3:230–39.

## *Biology and Sociology*
### OCCULT BELIEFS

Several studies have indicated that the American public's belief in the occult and the acceptance of occult beliefs have increased dramatically since the early 1960s. Phenomena Americans are coming to accept include UFOs, the Bermuda Triangle, spoon bending, and psychic healing.

Why do people accept beliefs that lack scientific validity? One study* has argued that the media plays an important role in public acceptance.

The public is chronically exposed to films, newspaper reports, "documentaries," and books extolling occult or pseudoscientific topics, with critical coverage largely absent. Our most trusted and prestigious sources of information are no exception. When the layperson hears about New Zealand UFOs from Walter Cronkite or reads about new frontiers in

psychic healing in a front-page article in the *Los Angeles Times,* he has prima facie reasons for believing such phenomena valid. Such uncritical coverage may be due in part to the inherent difficulties of science reporting in a popular medium. However, media coverage of the occult may sometimes be deceptive and monetarily motivated. It is our impression that the public is not aware of the extent to which First Amendment privileges permit such practices. The public seems to trust the printed work implicitly.

Are the media being irresponsible by promoting these unsubstantiated beliefs? Should media disclaimers indicating that these phenomena are "unproven" be stronger? Should scientists have a greater voice in media reports on science?

*B. Singer and V. A. Benassi, 1981, "Occult Beliefs," *American Scientist* 69: 49–50.

## *Biology and Literature*
### CATALYSTS

The poet T. S. Eliot* compared the mind of the poet to a catalyst that brings together ideas, feelings, and emotions. According to Eliot, while ideas and feelings may have a natural affinity for one another, they will not combine in any meaningful way without the help of the mind of the poet, which remains "inert, neutral and unchanged."

Does this analogy accurately describe the catalytic action of an enzyme? In what ways are the catalytic action of the poet's mind and the enzyme the same? In what ways are they different?

*T. S. Eliot, "Tradition and the Individual Talent," in *Selected Essays: 1917–1984* (New York: Harcourt, Brace, and Company, 1952).

## *Biology and Psychology*
### OBSERVATION OF CELLS

In the seventeenth century, scientists claimed that small human figures could be seen in egg and sperm cells observed under the microscope (see figure). These observations were presented as evidence for the preformation theory, which stated that the complete organism exists in its first cell. Later observations revealed that no such preformed organisms existed.

These scientists were careful observers. They published their observations and argued strongly for the preformation theory. Why would reputable scientists argue that they had observed something that did not exist? Were they biased in some way? If so, what was the source of their bias?

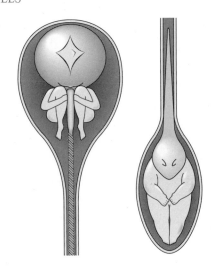

# PHYSIOLOGY OF PLANTS AND ANIMALS

CHAPTER
6

RESPIRATION

hy do living things need oxygen, and what do they use it for? What happens to oxygen once it enters the bodies of animals and plants? Living things obtain oxygen from their environment, and they must have a nearly constant supply in order to stay alive. If the human brain, for example, is deprived of oxygen for as short a time as four minutes, irreparable damage can occur. As we will discover in this chapter, the process through which living things obtain oxygen, called **respiration,** is a complicated but essential part of life on earth.

## OXYGEN UPTAKE

Let us look at some data showing how much oxygen enters the body. Figure 6.1 shows the concentration of oxygen, carbon dioxide, and nitrogen inhaled and exhaled by humans.

We can make several observations based on this data. First, the oxygen concentration in exhaled air is lower than that of inhaled air. The most logical interpretation of these data is that some oxygen is taken up, or used, by the human body.

Second, we observe that exhaled air contains more carbon dioxide than does inhaled air. The logical interpretation is that the body gives off carbon dioxide. We also observe that the concentration of nitrogen in inhaled and exhaled air does not change, suggesting that humans do not use nitrogen in the air they breathe. The data presented in this figure are fairly typical for all organisms that use oxygen, both plants and animals.

Many questions arise from our observations of the data in figure 6.1. How do organisms use the oxygen they take up? Where does the carbon dioxide in exhaled air come from? Why don't organisms use nitrogen? We will address these questions in this chapter as we explore the process of respiration.

We have established that animals and plants take in oxygen. Scientists can measure oxygen use in individual tissues with a Warburg respirometer, an instrument developed by German biologist Otto Heinrich Warburg (Thinking in Depth 6.1). Table 6.1 shows the oxygen use, or **rate of respiration,** for five different tissues. Notice that the animal tissues use much more oxygen than do the plant tissues. Researchers have measured the rate of respiration for an immense variety of plant and animal tissues and have found similar results.

Knowing that tissues use oxygen, we can reasonably guess that the individual cells that make up those tissues also use oxygen. When we use a respirometer to measure the respiration rate of cells, we find that cells do indeed take up oxygen. They also give off carbon dioxide. But what do cells do with the oxygen they take in? In chapter 2 we observed that cells are made up of various organelles. We were able to fractionate cells and examine each organelle separately. We can use the same process to separate organelles and test their respiration rates, thereby discovering what is happening to the oxygen taken up by cells.

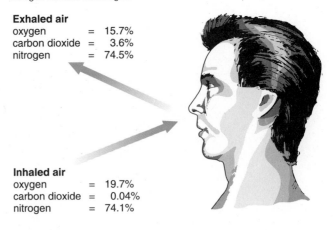

**FIGURE 6.1**    Gases entering and leaving the human body: exhaled air contains less oxygen and more carbon dioxide than inhaled air, but nitrogen remains unchanged.

**Exhaled air**
oxygen = 15.7%
carbon dioxide = 3.6%
nitrogen = 74.5%

**Inhaled air**
oxygen = 19.7%
carbon dioxide = 0.04%
nitrogen = 74.1%

| TABLE 6.1 | Respiration Rates in Plant and Animal Tissues |
|---|---|
| **Organism** | **Rate in Milliliters of Oxygen per Gram of Tissue Weight per Hour** |
| Plant leaf | 0.09 |
| Plant flower parts | 0.20 |
| Rat liver | 3.00 |
| Rat kidney | 7.50 |
| Rat muscle | 1.50 (at rest) |
| | 10.00 (active) |

When we measure the respiration rate of organelles, we find that the nuclei, endoplasmic reticulum, and plasma membrane use little, if any, oxygen. The mitochondria, however, use most of the oxygen taken up by the cell and give off most of the carbon dioxide that cells produce (fig. 6.2).

When we measured the oxygen use of mitochondria, why was it important to also measure the oxygen use of other organelles? How might we have misinterpreted the data if the respiration rate of other organelles were not measured?

## CELLULAR RESPIRATION

What are mitochondria doing with the oxygen they take up? Investigating this question has proved difficult and has involved a tremendous amount of work in biochemistry. Indeed, researchers conducted thousands of experiments over more than fifty years before they were able to understand

*Physiology of Plants and Animals*

## MEASURING RESPIRATION

We can measure the amount and rate of oxygen use with a respirometer of the type shown in figure 1. Tissue slices or individual cells are placed in the respiratory vessel, which is attached to the manometer, a long, thin tube. The manometer contains a dye that will rise or fall as gases are released or taken up by the tissue or cells in the respiratory vessel. For example, as cells take up oxygen, the volume of gas in the vessel decreases and the dye in the manometer falls. By observing the level of the dye at regular intervals, we can determine the rate of oxygen use.

The respiratory vessel has a well in the center that contains potassium hydroxide (KOH), which absorbs carbon dioxide. Why do you think it is important to absorb carbon dioxide in the respiratory vessel?

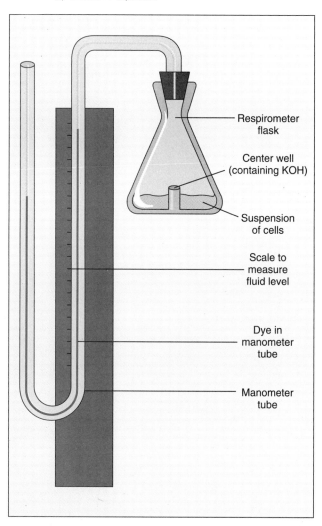

**FIGURE 1**    Warburg respirometer. As oxygen is consumed by cells in the vessel, the level of the dye in the manometer will fall and indicate how much oxygen is used. KOH is added to the center well of the vessel to absorb $CO_2$ produced in respiration.

Respirometer flask

Center well (containing KOH)

Suspension of cells

Scale to measure fluid level

Dye in manometer tube

Manometer tube

that biochemical pathways in mitochondria take up oxygen and give off carbon dioxide. Together these pathways are called **cellular respiration.** In the process of capturing and releasing energy that can be used to do mechanical work and synthesize other molecules, the mitochondria use oxygen. That is why oxygen is vital to living things.

The biochemical pathways in mitochondria that make up cellular respiration are diagrammed in figure 6.3. First we will consider the main features of these pathways. Later in the chapter we will examine the chemical reactions in more detail.

In the process outlined in figure 6.3, cells take in oxygen and break down nutrients to produce carbon dioxide and water and release energy. The chemical reactions are catalyzed by enzymes found in the mitochondria. We will look at one important product of these reactions: **adenosine triphosphate (ATP).**

Respiration in mitochondria begins when organic material (acetate) enters the pathway and is subsequently broken down. Oxygen enters the reaction only at the very end. It accepts electrons at the end of the biochemical pathway and produces water (Thinking in Depth 6.2). The reactions also

produce ATP, a compound that carries chemical energy and transfers this energy to other processes in the cell. Remember that biological molecules can contain large amounts of energy in their chemical bonds (chapter 3). The energy is released as heat when the molecules are burned (i.e., combined with oxygen) in a bomb calorimeter. Essentially the same process occurs in living cells, but much more slowly. The reaction can be represented by the formula

$$\text{organic molecules} + O_2 \rightarrow \text{energy} + CO_2 + H_2O$$
$$\text{(such as glucose)}$$

The energy given off is "captured" in ATP molecules and can be used by the cell.

ATP is formed from ADP (adenosine diphosphate) and $P_i$ (a free phosphate group) present in the cell. When the ADP and $P_i$ combine, they form a phosphate bond in ATP that picks up energy from other metabolic reactions in

FIGURE 6.3    Cellular respiration. At the start of the reaction, a two-carbon molecule (acetate) enters a cyclic series of reactions. $CO_2$ and electrons ($e^-$) are produced. The electrons then enter a series of electron transport molecules where the electrons are passed along, much like an electron "bucket brigade." At the end of the transport series, the electrons are accepted by oxygen to form water as a final product.

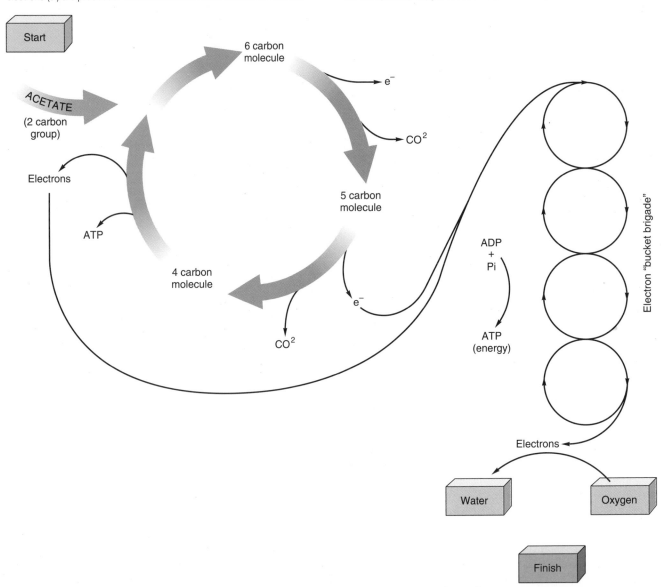

# OXIDATION AND REDUCTION

Some biological molecules can gain electrons from other molecules or lose them to other molecules. Which one they do depends on their structure; some molecules have a strong tendency to give up electrons while others have a strong tendency to accept electrons. The gain or loss of an electron often occurs with the gain or loss of a hydrogen atom.

Gaining an electron is called **reduction.** Losing an electron is called **oxidation.** Figure 1 illustrates an electron exchange in mitochondria in which a molecule of succinic acid gives up electrons and becomes fumarate. When the succinic acid molecule gives up electrons, we say that it is **oxidized.** The carrier molecule that receives the electrons is said to be **reduced.**

In the next step in the reaction, the carrier molecule transfers electrons to the first molecule in what is called the electron transport series. The electron transport series is a string of molecules that are alternately oxidized and reduced as an electron is passed from molecule to molecule (see fig. 6.3). The molecule that receives the electron is oxidized. The sequence of the molecules in the series depends on their tendency to gain or accept electrons. A molecule that accepts electrons would follow a molecule that gives up electrons, which would follow a molecule that accepts them, and so on.

The important feature of oxidation and reduction is that electrons enter the series with high energy and lose energy with each succeeding exchange in the series. Much of this energy is captured in the phosphate bond when ADP combines with $P_i$ to form ATP.

FIGURE 1

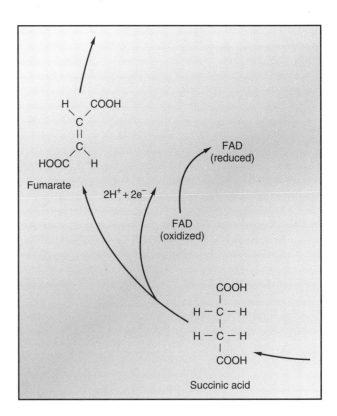

cellular respiration that are going on at the same time. The ATP molecule "carries" that energy to other places in the cell where it can be used. When the ATP is ready to release the energy, the phosphate bond is broken, forming ADP and $P_i$, which are recycled back into the chemical reactions that make up respiration.

We can now answer one of the questions we raised at the beginning of the chapter. If oxygen is not available to accept electrons at the end of the pathway as described in Thinking in Depth 6.2, the biochemical reactions will back up like cars in a traffic jam and the reactions in the mitochondria will not proceed. Energy (as ATP) will not be produced, and the cell will die.

What happens if cells lack organic material, or fuel, to begin the chemical reactions that make up cellular respiration? We can explore this question with an

experiment, the results of which are shown in figure 6.4. We place yeast cells in a respirometer but for 60 minutes provide them with no food or energy source. During this period the cells have no molecules to metabolize, so oxygen utilization is low. At 60 minutes, we add glucose to two respirometer vessels. To one vessel we add only glucose. To the other we add glucose plus a poison that blocks respiration at the point where electrons are given off. The figure shows that oxygen use in vessel 1 greatly

Why doesn't oxygen use increase in vessel 2?
Where would you find greater ATP production, in vessel 1 or vessel 2?

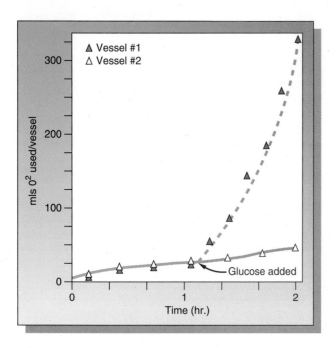

FIGURE 6.4    Oxygen utilization by yeast. Two vessels were prepared containing yeast cultures and the oxygen uptake measured from 0 to 60 minutes. At the arrow, glucose only was added to vessel 1, while glucose +, an inhibitor of electron transport, was added to vessel 2.    *Source: Data from Arthur Giese, Cell Physiology, 2nd edition, W.B. Saunders Co., Philadelphia, PA.*

increases, while oxygen use increases only slightly in vessel 2. The glucose provides fuel for cellular respiration in both vessels, but the poison blocks respiration in vessel 2.

## THE KREBS CYCLE AND ELECTRON TRANSPORT

The chemical reactions in the mitochondria that make up cellular respiration are so important in biological functions, and so interesting, we will take a closer look at them. The reactions can be viewed in two parts—the Krebs cycle and the electron transport system.

In the Krebs cycle, named after British biochemist Sir Hans Adolf Krebs, organic molecules are broken down to produce carbon dioxide, and electrons are transferred to electron carriers. These reactions, shown in figure 6.5, form a repeating circle.

Organic material enters the cycle as a two-carbon acetate molecule that is a product of the breakdown of carbohydrates, fats, and proteins (*a*). The acetate molecule combines with a four-carbon oxaloacetate molecule (*b*) to

form a six-carbon citrate molecule (*c*). In subsequent steps, splitting off a carbon dioxide atom (*d*) removes carbon from the reacting molecules. ATP is formed, and electrons are removed, reducing the electron carriers (*e*). After several more steps, a four-carbon oxaloacetate remains, picks up another acetate, and the cycle begins again. All $CO_2$ production occurs in the Krebs cycle, but further electron transfer reactions transfer the greatest amount of energy to ATP molecules. In other words, $CO_2$ is produced in one series of reactions (the Krebs cycle) while most energy capture occurs in another series of reactions (the electron transport series).

In the electron transport system, electrons removed from molecules in the Krebs cycle are passed along a chain of electron carriers. Electrons enter the chain with relatively high energy and gradually transfer this energy to form ATP molecules in coupled reactions. Coupled reactions are those in which two chemical reactions take place at the same time and energy is transferred. At the end of the chain, the electrons have little energy, but they are accepted from the electron carriers by oxygen, which has a high affinity for electrons. If oxygen did not accept the electrons, that is, remove them from the last carrier, all the carriers would become reduced and electron movement, as well as the Krebs cycle, would stop.

Some inhibitors can "uncouple" electron transport and the reactions that form ATP. As a result, electrons can be transported through the chain to oxygen without transferring energy to form ATP. Normally this would not occur. Electron transport and ATP formation must occur together. What would be the consequences to the cell if uncouplers were added?

Oxygen's essential role in cellular respiration is to accept electrons and thereby keep the metabolic machinery moving. Nitrogen, even though it is a far more common molecule in the atmosphere, has a much lower affinity for electrons and will not accept them from the transport chain. That explains why, as we saw in figure 6.1, the concentration of nitrogen in inhaled air and exhaled air is the same. Organisms do not use nitrogen for respiration.

The large amount of energy captured in ATP demonstrates how important mitochondria and cellular respiration are to the cell. We can study the relationship between ATP and the mitochondria in experiments using isolated mitochondria. As shown in figure 6.6, if succinic acid (a molecule that is part of the Krebs cycle) is added to isolated mitochondria, no oxygen is used unless ADP is also added. After a short period of time, oxygen use stops, but if we add more ADP, it increases again.

**FIGURE 6.5** Krebs cycle and electron transport. The reactions in the Krebs cycle and electron transport occur in the mitochondria. Acetate enters the cycle (a) and combines with a 4-carbon molecule (b) to form citric acid (c). In subsequent steps, carbon dioxide is split off and electrons are removed that enter electron transport.

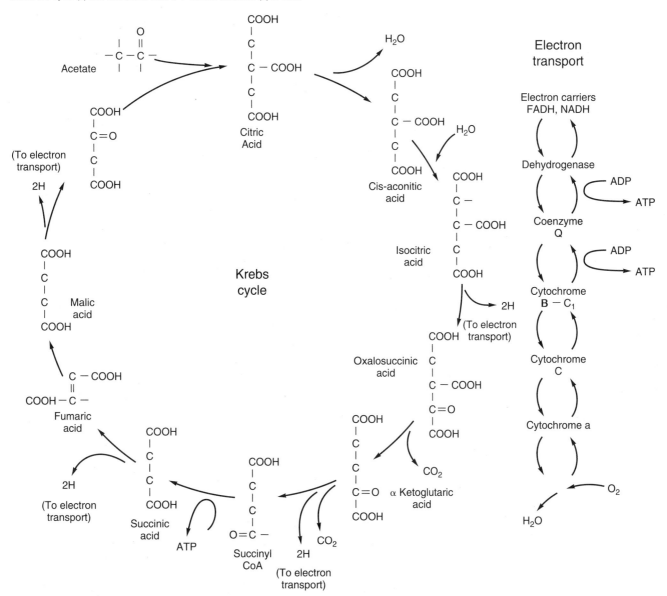

What compounds other than succinic acid could be added to the mitochondria to fuel respiration? Would the addition of ATP rather than ADP have increased oxygen use?

## GLYCOLYSIS AND FERMENTATION

The chemical reactions that occur in mitochondria are the main source of energy for cells. Additional energy transfer reactions occur in the cytoplasm through **alcoholic fermentation** and **lactate fermentation,** both of which metabolize sugars. The initial stage of this process is called **glycolysis.**

**FIGURE 6.6** Oxygen utilization by mitochondria. At time equals zero, a suspension of mitochondria and inorganic phosphate ($P_i$) was placed in a respirometer. At the indicated points, succinic acid or ADP was added.

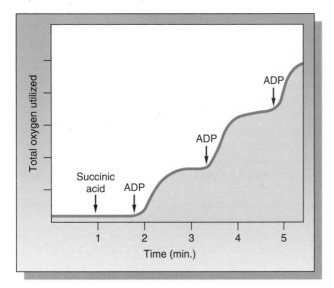

**FIGURE 6.7** Glycolysis. The reactions of glycolysis (glucose breakdown) occur in the cell cytoplasm. Glucose may be converted to alcohol or lactic acid in the absence of oxygen or to acetate in the presence of oxygen.

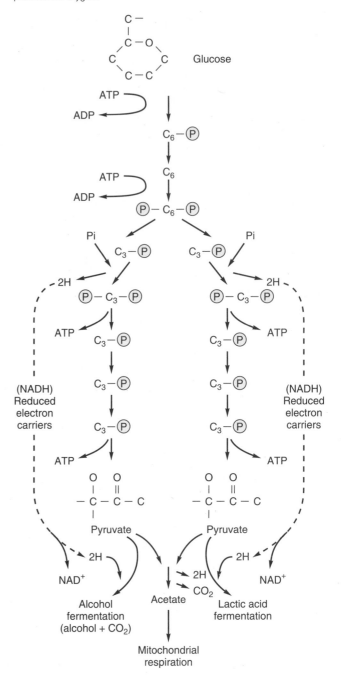

Sugar yields the greatest amount of energy when it is broken down by mitochondria, but reactions in the cytoplasm can transfer small amounts of energy by breaking down sugar even in the absence of oxygen. We know that this breakdown occurs in the cytoplasm and not in mitochondria because the enzymes that catalyze these reactions are found only in the cytoplasm. The chemical reactions of glycolysis that lead to alcoholic fermentation and lactate fermentation are shown in figure 6.7.

As you can see from the figure, the first stage of glycolysis uses ATP. When phosphate (P) is added, the sugar molecule (glucose) becomes less stable and the bonds are easily broken to release energy. The six-carbon sugar breaks into two three-carbon molecules, producing a net gain in ATP and reduced electron carriers that can enter the electron transport chain to produce additional ATP. When oxygen is added, the three-carbon molecules are converted to two-carbon acetate molecules and enter cellular respiration, as we saw in the previous section.

Some cells can carry out alcoholic fermentation when oxygen is not present. Yeast and many plant cells can reduce the three-carbon pyruvate in the last stages of glycolysis (fig. 6.7) to yield alcohol and carbon dioxide. The carbon dioxide produced by yeast is what makes bread dough rise, and the alcohol makes alcoholic beverages.

In lactate fermentation, the reduction of pyruvate leads, by another biochemical pathway, to the formation of lactic acid. This reaction occurs in human muscle cells when oxygen concentration is low. The buildup of lactic acids leads to soreness in the muscles until the acid is metabolized. Your muscles may become sore after exertion because the muscle cells lack oxygen.

Suppose the electron carrier (NAD) reduced in glycolysis did not later give up electrons in fermentation. What would happen to the fermentation process?

| TABLE 6.2 | A Comparison of ATP Production in Metabolic Pathways | |
|---|---|---|
| Metabolic Pathways | Glucose Breakdown | Electron Transport |
| Glycolysis (including acetate formation) | 2 ATP | 10 ATP |
| Krebs cycle | 2 ATP | 22 ATP |
| Total ATP | 4 ATP | 32 ATP |
| | 4 ATP + 32 ATP = 36 ATP | |

| TABLE 6.3 | Respiration Rates in Cold-blooded and Warm-blooded Animals |
|---|---|
| Organism | Rate in Milliliters of Oxygen per Gram of Tissue Weight per Hour |
| Mussel | 0.02 |
| Crayfish | 0.047 |
| Mouse | 2.50 |
| Human | 0.20 |

While fermentation does allow glycolysis to proceed rapidly in the absence of oxygen, the energy yield is low. As indicated in table 6.2, fermentation of one molecule of sugar yields only four ATP while the complete respiration of sugar yields thirty-six ATP.

Each molecule of ATP carriers 7.3 kilocalories per mole (kcal/mole) in the phosphate bond. Thus fermentation captures 14.6 kcal/mole while complete oxidation transfers 262.8 kcal/mole to ATP. Recall that when glucose is burned in a bomb calorimeter, 686 kcal of heat energy are released. If 262.8 kcal are transferred to ATP in cellular respiration, that means about 40 percent of the energy in glucose is captured. Compare that ratio to gasoline, in which about 25 percent of the energy can be used to power an automobile. Living cells are far more efficient than mechanical devices at transferring energy from chemical bonds to do useful work.

# HEAT ENERGY

If 40 percent of the energy in chemical bonds is captured in respiration, what happens to the other 60 percent? That energy is given off as heat. Just as heat is produced when molecules are burned in a bomb calorimeter, heat is given off in cellular respiration. In living cells, however, the reactions proceed much more slowly so heat is produced more slowly. Organisms can increase the amount of heat they give off by increasing their metabolic rate. When you exercise, for example, you raise your rate of cellular respiration and thus increase your body heat.

In so-called warm-blooded animals, including mammals and birds, the heat energy produced by cellular respiration maintains the body's heat at a relatively constant temperature, which is higher than that of the surrounding environment. Since the body is continually losing heat to the cooler environment, cellular respiration must produce heat to maintain a constant temperature.

So-called cold-blooded organisms, such as reptiles, fish, and plants, also obtain oxygen by respiration, but the body temperature of these organisms is approximately equal to that of their environment. If these organisms are producing heat energy through respiration, why are their bodies not heated? One explanation is that the heat they produce is lost to the environment as rapidly as it is generated. The amount of heat produced is insufficient to raise body temperature. If this explanation is correct, then we can predict that cold-blooded organisms use oxygen at a lower rate than warm-blooded organisms, because their cellular respiration is slower.

Table 6.3 demonstrates that our prediction is correct. The warm-blooded mouse and human use oxygen at a much higher rate than do the cold-blooded mussel and crayfish. We can conclude that the higher rate of respiration (and thus heat production) in mammals and birds maintains their higher body temperature.

If warm-blooded organisms require higher rates of respiration to survive, why be warm-blooded? Most animals are cold-blooded and produce sufficient energy to live. Are there advantages to being warm-blooded?

Apparently there are. The great majority of large land animals in all parts of the world are warm-blooded. The most widely accepted explanation is that higher body temperature allows greater work output of muscles and body organs and that a higher optimum (and constant) temperature can be maintained for enzyme function. (Remember

Warm-blooded animals require ten to thirty times the amount of food that cold-blooded animals require. Why do you think this is so?

from chapter 2 that enzyme function is strongly influenced by temperature.) Warm-blooded animals can work harder, endure longer, move faster, and obtain food more efficiently than cold-blooded organisms.

## SUMMARY

*Respiration is a critical process in the life of cells, allowing organisms to use oxygen to obtain the energy they need to grow and survive. Without adequate oxygen, living things and their cells quickly die. While some energy can be obtained in glycolysis and fermentation, the majority of an organism's energy comes from cellular respiration in the mitochondria, where the presence of oxygen opens up the biochemical pathways in the Krebs cycle and the electron transport system and allows greater energy extraction.*

*Respiration is an important part of the biological processes we are examining in this book. As you read, keep in mind the following key points.*

1. Oxygen that enters the bodies of organisms is utilized by cells. Inside cells, oxygen combines chemically with other substances in the mitochondria.
2. **Cellular respiration** takes place in the mitochondria. Enzymes catalyze each of the chemical reactions that make up respiration, and the end result is the production of carbon dioxide, water, and energy.
3. Most energy is released as heat but some is captured in the chemical bonds of **adenosine triphosphate.** ATP carries energy to a great many energy-requiring processes in the cell.
4. Organic material enters the mitochondria and pathways of cellular respiration as acetate, a two-carbon compound that is a product of carbohydrate, protein, and lipid breakdown.
5. Reactions in the mitochondria can be viewed as two parts: the **Krebs cycle** and the **electron transport system.**
   a. In the Krebs cycle, the acetate combines with other compounds, carbon is removed and forms $CO_2$, and electrons are removed and enter the electron transport system.
   b. In the electron transport system, electrons gradually lose energy as they are passed down a chain of enzymes. This energy is captured in ATP. At the end of the chain the electrons are accepted by oxygen and water is produced.
6. In the absence of oxygen, enzymes in the cytoplasm can catalyze **alcoholic fermentation** or **lactate fermentation.** These processes transfer some energy to ATP, though far less than is captured in cellular respiration.
7. Some heat is produced in cellular respiration. In almost all plants and cold-blooded animals this heat is quickly lost and the organisms' temperature remains the same as the surrounding environment. Warm-blooded animals, however, produce much more heat and can maintain a body temperature higher than their surroundings.

## EXERCISES AND QUESTIONS

1. Experimental measurements and comparisons of oxygen uptake in cell organelles indicate a high rate of uptake by mitochondria but little if any uptake by other cell organelles such as nuclei and endoplasmic reticulum. Why was it important to measure the uptake in organelles other than mitochondria in this experiment in order to conclude that mitochondria are the primary oxygen users in the cell (page 64)?
   a. The organelles placed in the respirometer may not be fully isolated and separated from other organelles.
   b. When oxygen enters a cell, all organelles will be exposed to the oxygen, not just mitochondria.
   c. Carbon dioxide may be produced by the other organelles.
   d. Other organelles may also have indicated a high oxygen uptake.
   e. In most cells, there will be many more mitochondria present than nuclei.
2. In the experimental results shown in figure 6.4 (page 67), why does oxygen uptake not increase in vessel 2?
   a. Oxygen was not available in vessel 2.
   b. Electron movement down the transport chain was prevented.
   c. Oxygen has a high affinity for electrons and will combine with electrons before the transport chain.
   d. The inhibitor added to vessel 2 could have a high affinity for electrons.
3. In figure 6.4, where would ATP production be the greatest (page 67)?
   a. In the first sixty minutes in vessels 1 and 2.
   b. In vessel 2, after glucose and the inhibitor are added.
   c. In vessel 1, after glucose is added.
   d. ATP production should be the same in all vessels at all times in this experiment.
4. Some chemicals can "uncouple" electron transport from the ATP-forming reactions so that electrons are transported through the chain to oxygen without forming ATP. (Usually this would not occur; electron transport and ATP formation occur together.) Which of the following would most likely occur if a person consumed an "uncoupling" chemical?
   a. The level of ATP in the body would increase while ADP would decrease.
   b. Electrons in the electron transport chain would not be accepted by oxygen.

c. The amount of oxygen used by the body would decrease.

d. Fewer electrons would be removed from molecules in the Krebs cycle.

e. Less of the energy in the food consumed would be converted to energy in the body.

5. In the figure in Thinking in Depth 6.2 (page 67), which of the following, instead of succinic acid, could be added to the mitochondria to fuel the observed mitochondrial respiration?

a. oxygen

b. electron carriers such as NAD

c. $P_i$

d. $CO_2$

e. citrate

6. In the experiment shown in figure 6.6 (page 70), which of the following would most likely occur if ATP rather than ADP were added?

a. No oxygen would be used by the mitochondria.

b. Oxygen would be used even if succinic acid were not added.

c. Oxygen would be used more rapidly by the mitochondria.

7. If the mitochondria in the experiment in figure 6.6 were "uncoupled," which of the following would occur?

a. The addition of both ATP and ADP would be required to increase oxygen use.

b. Oxygen use would increase without adding succinic acid.

c. Water would not be formed when oxygen is used.

d. Oxygen use would increase without adding ADP.

8. A student observed that liver cells contain many more mitochondria than skin cells. The student then concluded that liver cells use much more energy than skin cells. Which of the following was the student most likely assuming?

a. Liver cells contain more mitochondria than skin cells.

b. Mitochondria in skin cells use more oxygen than mitochondria in liver cells.

c. The metabolic rate in all mitochondria is the same.

d. Liver cells use more energy than skin cells.

e. The skin cells and liver cells came from the same individual.

9. Suppose the electron carrier (NAD) reduced in glycolysis did not later give up electrons in fermentation. Which of the following would most likely occur (page 70)?

a. The process of fermentation would stop.

b. Cellular respiration would proceed more rapidly than fermentation.

c. Alcoholic fermentation would occur rather than lactic acid fermentation.

d. Additional electrons would be provided to the electron transport chain.

10. Which of the following most likely explains why warm-blooded animals require ten to thirty times as much food as cold-blooded animals (page 71)?

a. Warm-blooded animals can move faster for longer periods of time.

b. The higher rate of metabolism requires more chemical body energy.

c. Most land vertebrates are warm-blooded.

d. Warm-blooded animals are probably more effective at obtaining food.

11. In table 6.3 (page 71), the oxygen consumption per unit body weight of a mouse is over ten times that of a human, even though the body temperatures are approximately the same. Why do you think this is so?

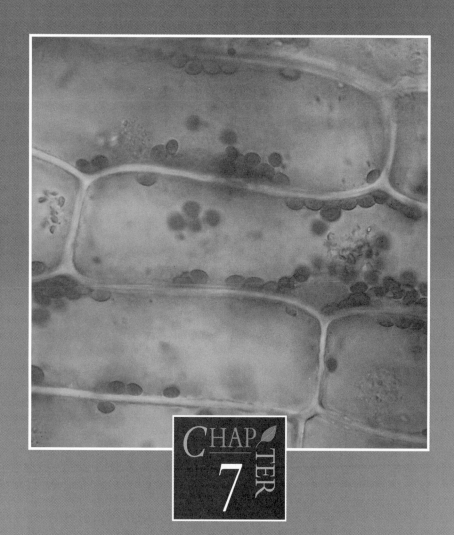

# PHOTOSYNTHESIS

e have accepted that chemical bonds contain energy, but where does this energy come from? What is its source? We have also taken for granted that the molecules used by cells come from somewhere outside the cell, but where? In this chapter we will explore the process through which living things on earth capture the energy in sunlight and manufacture food.

Organisms on earth can be broken into two types, depending on how they obtain energy (fig. 7.1). **Heterotrophs** must obtain energy and food from organic material by consuming other organisms or organic remains. Examples are animals, most bacteria, and fungi. **Autotrophs** obtain energy from the nonliving environment. Green plants are the primary example, but some bacteria also obtain energy this way. Almost all autotrophs use **photosynthesis** to capture light energy from the sun and transform carbon dioxide and water into organic molecules. It is here that the process of life on earth begins.

The process of photosynthesis is so well known as to be taken for granted. You probably studied photosynthesis early in your school career and have some idea how it works. The basic process can be expressed in the formula

$$CO_2 + H_2O \xrightarrow[\text{chlorophyll}]{\text{light}} \text{organic} + O_2$$

(from the atmosphere)     molecules (oxygen)

Although the basic facts of photosynthesis were understood more than two hundred years ago, details of the process have been revealed only in the past forty years. A great many experiments and sophisticated techniques have been required to discover the complex steps in the process.

In this chapter, three questions will guide our study of photosynthesis and our examination of the experiments undertaken to explain it.

- How did early experiments reveal the processes of photosynthesis?
- How do light and chlorophyll interact to capture energy?
- What metabolic pathways convert carbon dioxide and water into organic molecules and oxygen?

## EARLY EXPERIMENTS

The basic features of photosynthesis may seem obvious to us because they have been explained so many times and accepted as fact for so long. Two hundred years ago, however, the process was a mystery, and many ingenious experiments were necessary to establish these facts.

The first question biologists considered was "Where does a growing plant obtain material to increase its size?" We have all observed this phenomenon. A tree, for example, may be a foot tall when planted as a seedling but quickly grow to

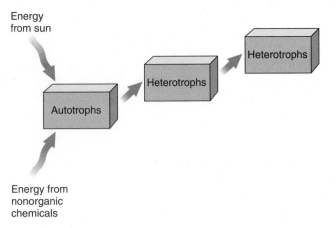

**FIGURE 7.1** Autotrophs obtain energy and material from the nonliving environment. Heterotrophs, on the other hand, obtain energy and most material from other living organisms—either autotrophs, other heterotrophs, or the remains of these organisms.

ten times that height. What makes the tree grow bigger? We know that plants grow in soil and that they need sufficient water to survive. These facts lead to two hypotheses.

*Hypothesis 1*    The plant obtains material from the soil.
*Hypothesis 2*    The plant obtains material from water.

A Flemish physician and chemist named Jan Baptista van Helmont performed an experiment in the early seventeenth century to test these hypotheses. He planted a tree seedling in a measured amount of soil. He watered the tree but added nothing else to the soil. After the tree had grown for five years, he measured the weight of the tree and the weight of the soil. The results of his experiment, shown in figure 7.2, seem to support hypothesis 2. The biologist concluded that water was the raw material used to form the body of the tree.

What did van Helmont assume when he drew his conclusion that water provided the raw material for the tree?

After van Helmont conducted his experiment, biologists discovered through chemical analysis that plant material contains a large amount of carbon (see tables 3.1 and 3.3). Since carbon is not part of the water molecule ($H_2O$), the carbon in the tree must have come from some other source. But where?

Stephen Hales, an English clergyman working in the early 1700s, offered a clue to what this source might be. He placed a plant under an inverted glass chamber with water covering the open end of the glass (fig. 7.3). He observed that after several days water was drawn up in the glass chamber. Water in another inverted glass chamber that did not contain a plant rose very little. This change in water level indicated that, in the chamber with the plant, the volume of air was decreasing.

Tree weight = 5 lb
Earth weight = 200 lb

Tree weight = 169 lb 3 oz
Earth weight = 199 lb 14 oz

5 years
watered regularly

Why did Stephen Hales use a control in his experiment?
What does the control show?

Could it be possible that trees obtain carbon from the air? Detailed studies of air reveal a composition of primarily carbon dioxide, oxygen, and nitrogen. Biologists have interpreted Hales's experiment to mean that plants remove carbon dioxide from the air and use it to form organic molecules.

Would you predict that the plant in figure 7.3 will continue to grow?

**FIGURE 7.3** Hales's experiments. A plant was placed under a glass chamber so that the open end of the chamber was covered by water. After several days Hales observed that water had risen in the chamber. In a similar chamber containing no plant, the water level did not rise as high.

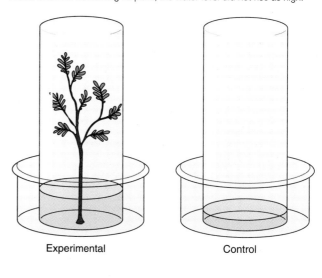

Experimental                    Control

**FIGURE 7.4** Ingenhousz's experiment. In separate water-filled chambers, (a) green stems, (b) green leaves, or (c) nongreen plant parts were exposed to sunlight. The gases produced by the three materials were analyzed.

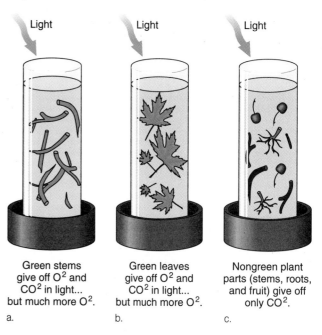

Light          Light          Light

Green stems give off $O_2$ and $CO_2$ in light... but much more $O_2$.
a.

Green leaves give off $O_2$ and $CO_2$ in light... but much more $O_2$.
b.

Nongreen plant parts (stems, roots, and fruit) give off only $CO_2$.
c.

Another experiment in the mid-1700s, this one done by Jan Ingenhousz, a Dutch physician, suggested that carbon dioxide is not the only atmospheric gas involved in the manufacture of energy by plants. When Ingenhousz immersed plants or parts of plants in water and exposed them to sunlight, he discovered that they released oxygen and a little carbon dioxide (fig. 7.4). Parts of plants that were not green, however, such as roots and fruits, gave off only carbon dioxide, not oxygen.

(*a*) Location of chloroplasts in leaf cells, (*b*) Electron micrograph of chloroplasts clearly indicate grana where chlorophyll is localized.

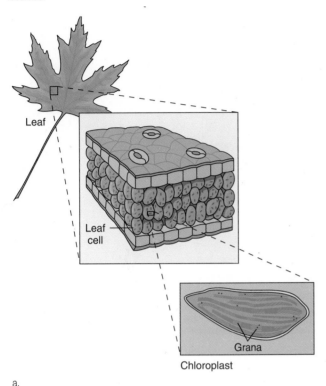

a.

In figure 7.4*c*, what was the source of the carbon dioxide produced by the nongreen plant parts?

Figure 7.4 shows no control used in Ingenhousz's experiment. What would be an appropriate control? What observations would you predict?

Several interpretations can be drawn from Ingenhousz's experiment. Since oxygen is given off, it is most likely a product of plant growth. And, since only green plant parts produce oxygen, the green pigment in plants, called **chlorophyll,** must be necessary for oxygen production.

These and many other experiments clearly established the equation of photosynthesis given at the beginning of this chapter. This equation, however, indicates only the first and last steps in the process. Biologists have conducted further research in recent years to discover the precise mechanisms of photosynthesis. Their questions include, Why is light necessary? What is the role of chlorophyll? and How is carbon dioxide used? We will answer these questions in the following section.

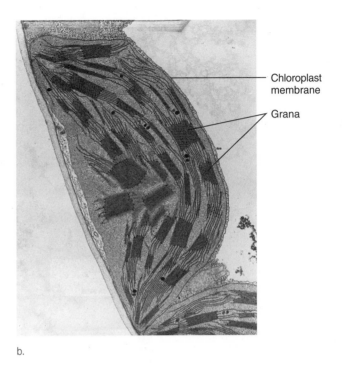

b.

## LIGHT AND CHLOROPHYLL

When we observe plant cells we see that chlorophyll is localized in the **chloroplasts** (fig. 7.5). More detailed examination reveals that chloroplast membranes are organized into collections of flattened sacs, called **thylakoids,** that are arranged in stacks, called **grana.** The chlorophyll molecule is composed of a complex ring structure called **porphyrin** made of alternating single and double chemical bonds (fig. 7.6). Attached to the porphyrin is a **phytol** tail of carbon atoms linked mostly by single bonds.

Why is chlorophyll green? The answer has to do with the color of light. As shown in figure 7.7, the color of light varies with its **wavelength.** A ray of light travels as a wave with a specific wavelength, and different wavelengths specify different colors of light. Visible light lies within the wavelength range of 400 nanometers (blue) to 700 nanometers (red).

Different molecules will absorb only specific wavelengths of light. Which wavelengths a molecule absorbs depends on its molecular structure. That is because absorption occurs when electrons in the molecule are "excited" to higher energy levels. Since every molecule has a unique configuration of electrons, every molecule absorbs specific wavelengths of light. The mechanism of absorption is explained in Thinking in Depth 7.1.

*Physiology of Plants and Animals*

## FIGURE 7.6    Structure of the chlorophyll molecule.

Porphyrin

Phytol

If a molecule's electrons are not excited by a particular wavelength, the molecule will reflect or transmit that wavelength instead of absorbing it. The wavelength that is reflected is the color that the object appears to be. Chlorophyll absorbs red and blue wavelengths but reflects green and yellow wavelengths. Thus green is the perceived color of plants.

The range of wavelengths absorbed by an object is called its **absorption spectrum.** Figure 7.7 shows the absorption spectrum for several plant pigments, including two kinds of chlorophyll that differ only slightly in their

## FIGURE 7.7    Absorption spectrums for several plant pigments. Chlorophyll a is the molecule shown in figure 7.6. Absorption spectra measure the amount of light absorbed at different wavelengths of light.

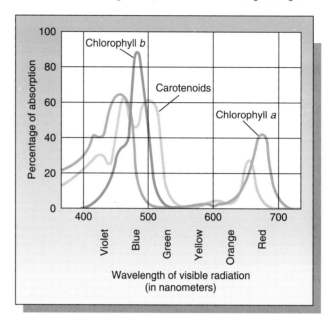

molecular structures. The structure of the chlorophyll molecule that we observed in figure 7.6 determines its absorption spectrum.

Chlorophyll's green pigment is certainly an interesting feature of plants, but does it have anything to do with how plants obtain energy from sunlight? We can begin answering this question by comparing chlorophyll's absorption spectrum to its **action spectrum.** The action spectrum shows the physiological responses of a plant, such as oxygen production or plant growth, at different wavelengths of light. An action spectrum for a green plant is shown in figure 7.8. Compare it to the absorption spectrum shown in figure 7.7. The close correspondence between the two curves strongly suggests that the sunlight absorbed by chlorophyll is used to produce oxygen and create plant growth.

When we compared the absorption spectrum of a green plant to its action spectrum, what hypothesis were we testing?

The evidence so far shows a relation between light absorption, chlorophyll, and physiological response (i.e., growth), but the question we considered earlier in the chapter remains: How do plants use light energy to convert carbon dioxide to organic molecules?

## LIGHT ABSORPTION

**FIGURE 1**    Light absorption. When light of a wavelength and energy "matches" the energy to excite an electron from the ground level to a higher level, the light is absorbed. Light is not absorbed when an electron is not excited.

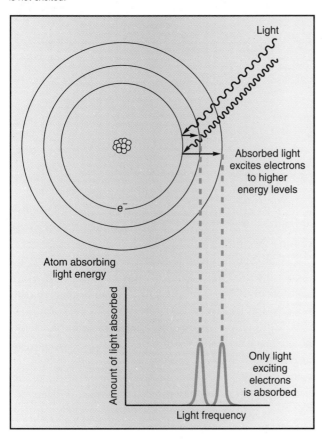

**FIGURE 2**    Spectrophotometer. A light source (*a*) shines through a slit to produce a narrow beam of light falling on a prism (*b*) splitting the light into different wavelengths. (*c*) A second slit allows only a narrow band of wavelengths to pass. This light passes through a sample and a photomultiplier tube (*d*) measures the amount of light passing through the sample.

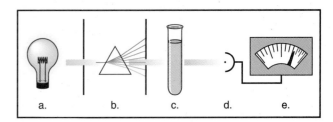

Why do plants appear green, oceans blue, and cardinals red? The colors of these objects are determined by how their molecules interact with light.

We know that electrons occupy orbitals around the nuclei of atoms. The specific orbitals of these electrons is determined by their energy levels. Scientists have learned that the energy levels of electrons can be affected by light.

Light can "excite" the electrons of an atom or molecule to a higher energy level. When this happens, the light is absorbed and we do not see it. Light comes in different wavelengths, which are measured in nanometers. The energy level of the wavelength must match the energy level of the electron in order for the light to be absorbed (fig. 1).

Since molecular composition and chemical bonding are the major factors that determine energy levels of electrons, the wavelengths of light absorbed can be used to identify molecules. The absorption spectrum for any molecule can be measured with a **spectrophotometer,** shown in figure 2.

In the spectrophotometer, light is split into narrow wavelength bands using a prism or grating. A specific wavelength is then passed through a sample of material. The amount of light passing through the material is detected with a phototube connected to a meter that gives a reading of the output from the phototube. The process is repeated for each wavelength until an absorption spectrum is constructed.

Figure 7.7 indicates absorption spectrums for three plant pigments. What does a comparison of these absorption spectrums and the action spectrum shown in figure 7.8 reveal about the use of absorbed light in photosynthesis?

One of the organic molecules that make up plants is carbohydrates. All the sugars making up the carbohydrates contain one carbon atom and one oxygen atom for every two hydrogen atoms: $CH_2O$. Sugars all have the fundamental ratio of 1 carbon : 2 hydrogen : 1 oxygen. A simple answer to our question might be that the energy from sunlight splits the $CO_2$ molecule, releases the oxygen, and combines the carbon atom with a water molecule. That reaction could be written

$$CO_2 + H_2O \longrightarrow O_2 + CH_2O$$

**FIGURE 7.8** Action spectrum for photosynthesis. When light of one wavelength illuminates a green plant, the response or "action" such as the production of oxygen can be measured. If this measurement is separately made at many different wavelengths, the data points can be used to construct an action spectrum.

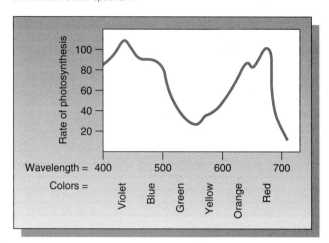

**FIGURE 7.9** Fluorescence. When an electron is excited to a higher energy level by the absorption of light, the electron will often drop back to the original energy level and re-emit light as fluorescence.

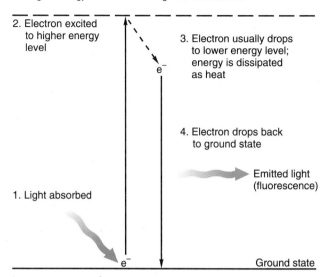

Further experiments have revealed, however, that this explanation is misleading. We know, for example, that certain types of bacteria can use $CO_2$ to make carbohydrates without releasing oxygen. There must be another explanation.

In one of the experiments designed to discover how light and chlorophyll interact, chlorophyll was extracted from cells and placed in a solution. The experiment showed that chlorophyll could chemically reduce (i.e., add electrons to) other molecules in the solution when it was illuminated. Apparently, the light was causing the chlorophyll to release electrons, which were accepted by the molecules being reduced. Additional investigations have indicated that isolated chloroplasts can exhibit the same kind of reaction. In chloroplasts, however, NAD and ADP are reduced to form NADPH and ATP. Recall that production of these "high energy" molecules is achieved by energetic electrons flowing through a series of oxidation-reduction reactions (see chapter 6). Perhaps light absorbed by chlorophyll in the chloroplast produces an energetic electron that is accepted by cytochromes (see fig. 6.5). Sequential oxidation-reduction of the cytochromes may be coupled with ATP and NADPH production (see fig. 6.5) as occurs with electron transport in mitochondria.

Additional information was provided by experiments on the fluorescence of chlorophyll molecules. **Fluorescence** is the reemission of absorbed light, generally at a longer wavelength than the light that was absorbed. As we have learned, when light is absorbed by an atom or molecule, an electron is excited to a higher energy level. This excited electron often will spontaneously drop back to a lower energy state and transmit the excess energy as light, as illustrated in figure 7.9.

The light that is emitted is called fluorescent light. The wavelength that is emitted is longer (i.e., has less energy) and thus is a different color than the absorbed light.

Extracted chlorophyll in solution will fluoresce brightly when illuminated with blue or red light. However, when cells or whole leaves are illuminated, there is very little fluorescence. These observations indicate that, in the chloroplast, electrons excited by light absorption are transferred to some acceptor of the electron's energy.

The fate of excited chlorophyll electrons has been revealed by ingenious and complex experiments, each making a small contribution to the overall picture. Many of these experiments used a spectrophotometer (see Thinking in Depth 7.1) to detect the oxidation and reduction of molecules in the electron pathway. The entire process of photosynthesis is diagrammed in figure 7.10.

How could a spectrophotometer detect the oxidation and reduction of molecules?

Through these experiments and many others too numerous to describe in this chapter, biologists discovered that two light reactions, not one, occur in photosynthesis. In photosystem II, light excites an electron from chlorophyll. This electron is accepted by a primary electron acceptor and passed down a series of electron transport molecules to form ATP. An electron is obtained from a water molecule in exchange

FIGURE 7.10 Light reactions in photosynthesis. Two different light reactions occur. In both, light absorption produces an excited electron. Movement of the electron through a series of transport molecules transfers energy to form ATP and NADPH.

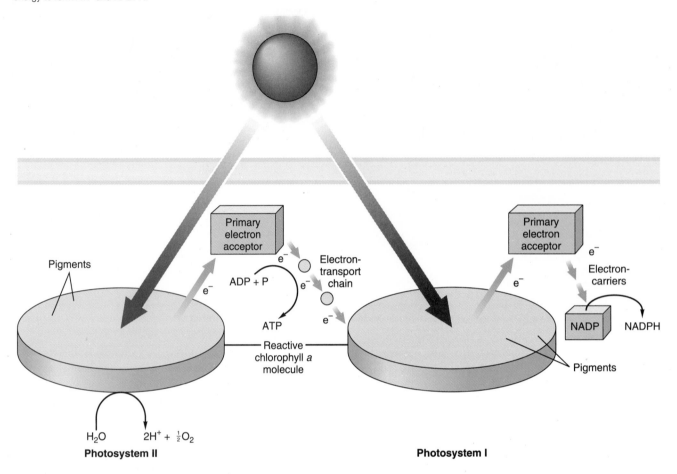

for the one excited by the absorbed light. In the process, oxygen is formed. A similar reaction occurs in photosystem I where NADP accepts excited electrons and NADPH is formed. The vacancy in the chlorophyll molecule left by the excited electron is filled by the electron from photosystem II. The main features in this scheme are those described in the investigations earlier in this chapter and in chapter 6—the production of ATP and NADPH by electron transport, and light absorption by chlorophyll.

Is light's only role to produce energetic electrons and thus generate ATP and NADPH? If so, $CO_2$ could be synthesized into organic molecules using only these energetic compounds in a series of "dark" reactions. The only "link" between the light reactions in photosynthesis and the dark reactions utilizing $CO_2$ would be ATP and NADPH.

Biologist Daniel Arnon investigated this question with a series of experiments in the 1950s. He hypothesized that ATP and NADPH manufactured in light reactions were the only products necessary to "fix" $CO_2$ in dark reactions. Based on this hypothesis, he predicted that when $CO_2$ was *not* supplied to isolated chloroplasts, ATP and NADPH

would accumulate in large amounts. He did indeed observe this accumulation in his experiments. However, while these results were consistent with the hypothesis, they were not considered to be strong support for it.

Why do you think the results of Arnon's experiment were not considered strong support for his hypothesis? Why did ATP accumulate?

In a second experiment, Arnon extracted from chloroplasts the enzymes for $CO_2$ assimilation. He discarded the green part of the chloroplasts, including the chlorophyll. Next he added ATP and NADPH, both of which were manufactured during the light reaction, to the enzymes, and he added $CO_2$. When he added these compounds, $CO_2$ was incorporated into the same molecules synthesized in whole chloroplasts. In other words, *only* ATP, NADPH, and $CO_2$ were necessary to manufacture the molecules. This finding

*Physiology of Plants and Animals*

was much stronger support for his hypothesis since any possible link with the light reaction (other than ATP and NADPH) had been eliminated with removal of the chlorophyll.

Arnon performed a third experiment in which ATP and NADPH from animal cells were added to the extracted enzymes from chloroplasts. $CO_2$ was again incorporated into molecules in the dark, further confirming that ATP and NADPH are the only necessary contributions of the light reactions.

Why did Arnon do the third experiment? What possible interpretation did it eliminate?

## $CO_2$ INCORPORATION

Earlier investigations of metabolic pathways in a wide variety of nonphotosynthetic cells, such as liver cells, showed that $CO_2$ could be incorporated into biomolecules *if* sufficient ATP and NADPH were supplied. This process appeared to be a reverse of respiration as we described it in chapter 6. Could the same process be taking place in the chloroplast using the energy from the light reaction?

Green plant cells were exposed to $CO_2$ that contained a radioactive carbon. The radioactivity was added so that the path of carbon in the photosynthetic process could be traced. At various times after the plant cells were exposed, they were quickly killed and the molecular constituents were separated using the chromatographic techniques described in chapter 4. Within seconds after the initial exposure to radioactive $CO_2$, an almost bewildering variety of biomolecules were labeled with radioactive carbon. Establishing the sequence of reactions and products was a painstaking process that required a great many experiments over many years.

The metabolic pathways that were ultimately revealed indicated that although the process was not a simple reversal of respiration, many of the intermediate biomolecules were identical in the two processes. In the first step, $CO_2$ is added to an existing five-carbon compound (ribulose) as shown in figure 7.11. The resulting six-carbon compound splits to form two three-carbon compounds (phosphoglycerate, or

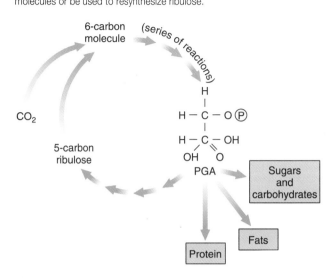

FIGURE 7.11    $CO_2$ incorporation. $CO_2$ inside the chloroplast is chemically bonded to an existing ribulose molecule. PGAL is produced by a series of reactions. The PGA can be converted to a variety of biological molecules or be used to resynthesize ribulose.

PGA). After a series of reactions utilizing the ATP and NADPH from the light reaction, PGA is converted to phosphoglyceraldehyde (PGAL), another three-carbon compound. At this point, the PGAL can be shunted to a number of different metabolic pathways.

Most of the PGAL is ultimately used to re-form ribulose, but about 20 percent will be utilized to synthesize carbohydrates, fats, and protein. Some will enter the respiratory cycle to produce energy as ATP and NADPH, as described in chapter 6.

After biologists discovered the basic scheme of $CO_2$ incorporation described here, other slightly different pathways were revealed. However, the addition of $CO_2$ to ribulose and the production of PGAL is by far the most common process in green plants.

Photosynthesis, then, adds one carbon atom at a time (as $CO_2$) to existing molecules in the chloroplast. By a series of complex metabolic pathways, this carbon is used to synthesize the multitude of biological molecules produced by green plants. Living organisms are totally dependent on this process for their carbon source, either directly as photosynthetic plants or indirectly as heterotrophs that consume other living organisms.

## SUMMARY

*The complicated process of photosynthesis has been revealed through a great many experiments over many years. Through photosynthesis, plants are able to transform light energy from the sun into food for all organisms on earth. As you read about other biological processes in this book, keep in mind the following important points about photosynthesis.*

1. **Autotrophs** obtain energy and material from the nonliving environment. **Heterotrophs** obtain energy and material from other living organisms.

2. **Photosynthesis** can be summarized in the formula

$$CO_2 + H_2O \xrightarrow[\text{chlorophyll}]{\text{light}} \text{organic molecules} + O_2$$

3. Experiments conducted more than two hundred years ago established the general characteristics of photosynthesis, i.e., that green plants (with chlorophyll) remove $CO_2$ from the atmosphere and in the presence of light produce organic molecules and release oxygen to the atmosphere.

4. **Chlorophyll** is localized in **chloroplasts,** membrane-bounded organelles with an extensive internal membrane system of flattened sacs called **thylakoids** arranged in stacks called **grana.**

5. Chlorophyll absorbs blue and red wavelengths of light and reflects green and yellow wavelengths. The amount of absorption of each wavelength of light is used to construct an **absorption spectrum.** When light is absorbed, an electron in the absorbing molecule is excited to a higher energy level.

6. An **action spectrum** shows the relationship between light absorbed and a physiological response such as oxygen production.

7. **Fluorescence** is the reemission of absorbed light by a molecule, usually at a longer wavelength.

8. Numerous experiments suggested that excited electrons from chlorophyll produced NADPH and ADP by oxidation-reduction reactions similar to those in respiration. More detailed investigations revealed two photosystems:
    a. Photosystem II produces ATP by passing an excited electron through a cytochrome chain. An electron from water replaces the one lost in chlorophyll and oxygen is produced.
    b. Photosystem I produces NADPH using an excited electron from chlorophyll.

9. Incorporation of $CO_2$ in the dark reactions in photosynthesis utilizes *only* the ATP and NADPH from the light reactions to produce organic molecules.

10. $CO_2$ first combines in the chloroplast with a ribulose molecule to form a six-carbon molecule. Through a series of reactions, the six-carbon molecules are converted to three-carbon molecules of phosphoglyceraldehyde (PGAL).

11. Most of the PGAL regenerates ribulose, but about 20 percent is used to synthesize the biomolecules produced by green plants.

## REFERENCES

Arnon, Daniel I. November 1960. "The Role of Light in Photosynthesis." *Scientific American* 203, no. 5: 104–9.

Bassham, J. A. June 1962. "The Path of Carbon in Photosynthesis." *Scientific American* 206, no. 6: 88–100.

Govinjee, Rajni. December 1974. "The Absorption of Light in Photosynthesis." *Scientific American* 231, no. 6: 68–82.

## EXERCISES AND QUESTIONS

1. In van Helmont's experiment, what did he assume when he concluded that water was the substance converted into new plant material (page 76)?
   a. Water was added as needed to the pot holding the tree.
   b. The tree grew over the five years of the experiment.
   c. The weight of the soil remained approximately constant for the duration of the experiment.
   d. Water and soil were the only material that could be converted into new plant material.

2. In Stephen Hales's experiment, why was a control necessary (page 77)?
   a. To maintain a constant water level in all chambers.
   b. To detect any volume change in the glass chamber with no plant.
   c. To measure the amount of water used by the plant.
   d. To detect any effects of water on the shape of the glass chamber.

3. In figure 7.3, would you predict that the plant under the glass chamber will continue to grow (page 77)?
   a. Yes, because the plant is well lighted.
   b. No, because no soil is available to the plant.
   c. Yes, because adequate oxygen is produced by the plant and is available for respiration.
   d. No, because the plant will use up $CO_2$ in the chamber.
   e. Yes, because enough water is available to supply the needs of a growing plant.

4. What most likely produced the $CO_2$ generated in figure 7.4c (page 78)?
   a. respiration
   b. the effect of light on water
   c. photosynthesis
   d. the decrease in water level in the chamber
   e. dissolved $CO_2$ in the water

5. In Ingenhousz's experiment, illustrated in figure 7.4, what would have been an appropriate control (page 78)?
   a. a chamber with both green stems and green leaves
   b. a chamber with both green and nongreen plant parts
   c. a chamber containing no plant parts
   d. a chamber with more green leaves than the chamber shown in figure 7.4b
   e. a chamber with green stems, leaves, and nongreen plant parts

6.  Which of the following hypotheses is being tested by comparing the action spectrum of chlorophyll (fig. 7.8) with its absorption spectrum (fig. 7.7) (page 79)?
    a.  Green light is a wavelength absorbed by chlorophyll.
    b.  Light absorbed by chlorophyll is used in photosynthesis.
    c.  Chlorophyll absorbs light most strongly in the blue and red wavelengths.
    d.  Light is essential for photosynthesis.
    e.  Chlorophyll is a green pigment found in plants.

7.  Which of the following could most likely be concluded from a comparison of the absorption spectrums in figure 7.7 and the action spectrum in figure 7.8 (page 80)?
    a.  Chlorophyll *a* and chlorophyll *b* are not found in the same plant.
    b.  Some plant pigments assist with $O_2$ production while others assist with $CO_2$ incorporation.
    c.  Not all plant pigments absorb light used in photosynthesis.
    d.  When green light is absorbed, photosynthesis is inhibited.
    e.  Light absorbed by several pigments is used in photosynthesis.

8.  The data shown below were collected from a large tank of green algae. The oxygen concentration of the water in the tank was measured at the times indicated. Which of the following is the best explanation for the changes in $O_2$ concentration?
    a.  No ATP is being formed in the dark as it is when the algae are illuminated.
    b.  Oxygen is combining with ATP and NADPH in both the light and dark.
    c.  Photosynthesis requires that no respiration occurs to consume the oxygen produced.
    d.  Respiration of the algae removes $O_2$, but photosynthesis produces excess $O_2$ in the light.

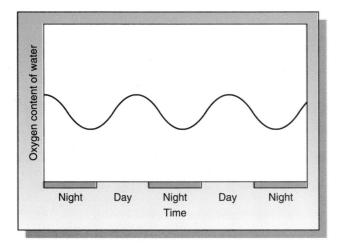

9.  Which of the following would be most necessary in order for the oxidation and reduction of molecules to be detected with a spectrophotometer (page 81)?
    a.  The oxidized and reduced forms of the molecule must be chemically separated and isolated.
    b.  An action spectrum must be generated for each of the different forms (oxidized or reduced).
    c.  Oxidized and reduced forms must absorb different wavelengths or amounts of light.
    d.  The spectrophotometer must be able to detect wavelengths shorter than visible light.
    e.  The spectrophotometer must be able to detect reflected as well as absorbed light.

10. Why was the experiment showing accumulation of ATP and NADPH in the absence of $CO_2$ *not* strong support for the hypothesis that ATP and NADPH are the only components necessary for $CO_2$ incorporation (page 82)?
    a.  The lack of $CO_2$ may have had other effects on the cell, causing ATP and NADPH accumulation not associated with photosynthesis.
    b.  The light reactions are known to produce ATP and NADPH that can be used for a variety of biochemical reactions.
    c.  Strong support for the hypothesis would require that $CO_2$, ATP, and NADPH be present in the experiment.
    d.  If $CO_2$ is not provided, there will be no carbon source for the synthesis of organic molecules.

11. If the hypothesis that ATP and NADPH are the *only* products of the light reaction necessary to incorporate $CO_2$ into organic molecules is valid, why would ATP be expected to accumulate when $CO_2$ is absent (page 82)?
    a.  ATP energy could be used for purposes other than photosynthesis.
    b.  The production of ATP from ADP requires a source of external energy such as glucose or sunlight.
    c.  Without $CO_2$ present, the metabolic pathways using ATP would not proceed.
    d.  The hypothesis should have considered the possibility that ATP could be produced by sources other than the light reactions.

12. Why did Daniel Arnon do the third experiment in which ATP and NADPH from animal cells were added to the enzymes from chloroplasts (page 83)?
    a.  To eliminate the possibility that some unknown substance essential for photosynthesis was present other than ATP and NADPH.
    b.  To check for the possibility that metabolic pathways similar to photosynthesis were present in animal cells.
    c.  To eliminate the possibility that ATP and NADPH from animal cells and from plant cells were different.
    d.  To check for the possibility that only ATP *or* NADPH were necessary for photosynthesis.

13. In an experiment, green plants were exposed to flashes of light separated by short dark periods. The amount of photosynthesis was measured as oxygen production or plant growth. As shown in the graph to the right, the amount of photosynthesis increased as the length of the dark period increased. Which of the following is the best interpretation of these data?

a. The use of $CO_2$ to synthesize organic molecules can inhibit events in the light reaction.

b. Accumulated ATP and NADPH can inhibit the light reaction but are used during the dark periods.

c. If the flashes of light were longer, then a dark period would have little or no effect.

d. ATP and NADPH from the light reactions can be used by metabolic pathways other than $CO_2$ incorporation.

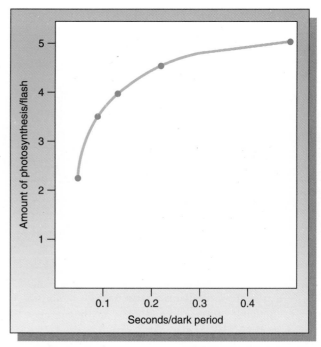

Source: Data from R. Emerson and W. Arnold, J. General Physiol., 15:391, 1932.

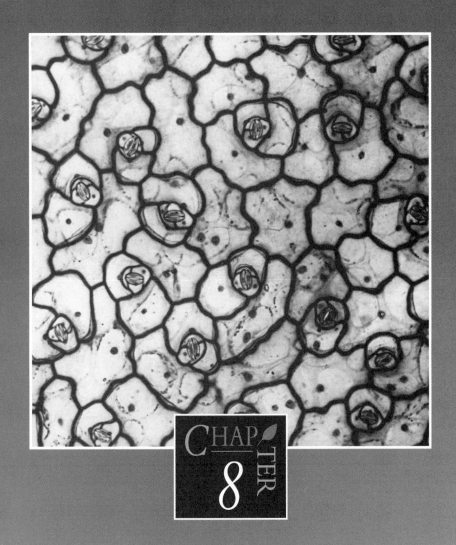

# GAS EXCHANGE

e know from previous chapters that plants require carbon dioxide and that both plants and animals require oxygen. We have seen how cells in plants and animals take in and give off these atmospheric gases through respiration. Diffusion across the plasma membranes of animal cells, for example, allows them to obtain sufficient oxygen and eliminate carbon dioxide.

Larger organisms, however, cannot obtain sufficient oxygen by diffusion alone. They must make use of specialized organs. In this chapter we will look at the organs in plants and animals that bear the primary responsibility for **gas exchange,** or movement of gases into and out of the body. What factors influence the structures of these organs, and how do they function?

The most important function of gas exchange organs is to permit sufficient oxygen and carbon dioxide exchange with the external environment for the organism to survive. This is more difficult than it might sound, because diffusion is a relatively slow process if distances between the cell and its source of gas are too great. A typical cell is about .001 centimeters in diameter. Material can diffuse across this distance in a mere .01 seconds (fig. 8.1). As distance increases, however, diffusion time slows considerably. Table 8.1 shows the amount of time materials need to diffuse over increasing distances.

You can see that at longer distances material diffuses much more slowly. Over a distance of only 1 centimeter, diffusion time is almost three hours! In larger organisms, many cells are at least a centimeter, if not more, from the environment outside the body. How do these cells obtain gases necessary for their growth and survival? Without specialized organs to exchange gases between cells inside the body and the environment outside the body, the cells, and thus the organism, would die.

You probably know that in your own body your lungs are responsible for gas exchange. There are many types of gas exchange organs in addition to lungs, and their diversity is impressive. In this chapter we will examine three of the most common gas exchange organs and how they function.

## FACTORS CONTROLLING GAS EXCHANGE

Several factors determine how gas exchange organs are structured and how efficiently they work. The first is **surface area.** As we saw in chapter 5, the surface area of a structure, such as the plasma membrane, has a major effect on how much gas it can exchange with its external environment. The greater the surface area, the greater the exchange of gas.

**Permeability** is a second important factor. The greater the permeability, the more easily material can be exchanged and therefore the greater the amount of material that can diffuse across the membrane.

FIGURE 8.1   Diffusion times. (a) Diffusion across an average cell diameter of 0.001 cm occurs in about 0.01 second. (b) At longer distances, diffusion is much slower.

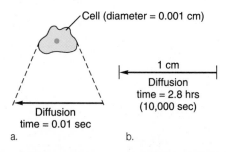

| TABLE 8.1 | Diffusion Times over Increasing Distances |
| --- | --- |

| Distance (cm) | Time of Diffusion (sec) |
| --- | --- |
| .001 | .01 |
| .01 | 1.0 |
| .1 | 100 |
| 1.0 | 10,000 (2.8 hrs) |

The third factor affecting gas exchange organs is **diffusion distance.** As we just saw, the smaller the diffusion distance, the more rapid the exchange. Gas exchange organs must decrease the distance across which gases move.

All three of these factors are important in gas exchange, but it just so happens that all three factors also contribute to water loss for many organisms, particularly land plants and animals. How do organisms balance their need for sufficient gas exchange against their need for sufficient water? As we discuss three types of gas exchange organs in the following section, we will also consider the problem of water conservation.

## LUNGS

The gas exchange organs used by most terrestrial animals are the **lungs,** which can be thought of as elastic bags within the body cavity (fig. 8.2). Air enters the lungs through the **trachea,** a conducting passageway that branches into **bronchi** and then into increasingly smaller tubes. These tubes end in small chambers called **alveoli.**

Air moves into the lungs because of a pressure difference between the outside of the body and the interior of the lungs. The body achieves this pressure difference by expanding the body cavity, which lowers the internal pressure. Humans expand their body cavities by contracting our **diaphragms** and rib muscles. This increases the space in our bodies, and air rushes in (fig. 8.3). Some animals, such as frogs, have lungs but no diaphragm. They achieve lung expansion by contracting only their rib muscles.

When we relax our rib muscles and diaphragm, the volume of the chest cavity decreases and air is exhaled. This

**FIGURE 8.2** The structure of the lung. (a) Air enters the lungs through the trachea and bronchi. Successive branching of the air passageways leads to the alveoli. (b) The enlarged view shows alveoli as small saclike structures in close contact with blood vessels.

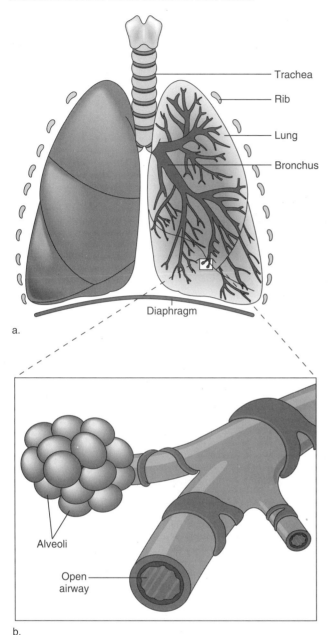

**FIGURE 8.3** Breathing. (a) When the diaphragm and rib muscles contract, the chest cavity expands and air flows into the lungs. (b) When these muscles relax, the chest cavity volume decreases and air flows out of the lungs.

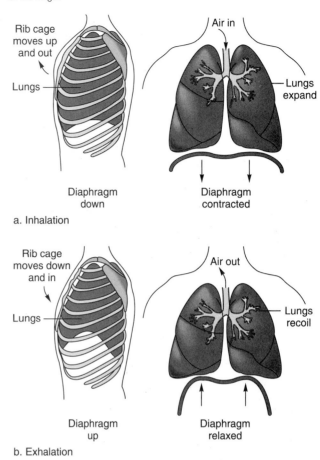

a. Inhalation

b. Exhalation

**TABLE 8.2 Gas Exchange in Human Respiration (in percentages)**

|  | Atmospheric Air | Alveolar Air | Exhaled Air |
|---|---|---|---|
| $N_2$ | 74.09 | 74.9 | 74.5 |
| $O_2$ | 19.67 | 13.6 | 15.7 |
| $CO_2$ | 0.04 | 5.3 | 3.6 |
| $H_2O$ | 0.50 | 6.2 | 6.2 |

Would it be possible for the oxygen concentration in the alveoli to decrease to near zero?

process of inhaling and exhaling is called **breathing.** Most of the time it is involuntary—we breathe without planning to—but we can take in greater or lesser amounts of air by breathing consciously.

Recall from chapter 6 that breathing is the method we use to obtain oxygen for respiration. Our bodies use oxygen from the air we breathe and give off carbon dioxide. Table 8.2 shows the percent of various gases in the air before we breathe, after we breathe, and in our lungs. Notice that in the alveoli, the oxygen concentration decreases to two-thirds of the amount in atmospheric air, while carbon dioxide increases to 120 times the amount in atmospheric air!

We can see from table 8.2 that oxygen in exhaled air is *higher* than that in the alveoli. If our bodies use oxygen, shouldn't the oxygen content of exhaled air be lower? We can explain this observation by recognizing that in the trachea and bronchi, very little gas is exchanged. When a breath is exhaled, alveolar air mixes with air from the trachea, resulting in a slightly higher oxygen content.

**TABLE 8.3** Water Loss in Humans

| Method of Water Loss | Water Loss per Day in Milliliters |
|---|---|
| Lungs | 350 |
| Urine | 1,400 |
| Sweat | 100 |

Once air has entered the alveoli, oxygen diffuses into the blood vessels, where the blood carries it to all parts of the body. The membrane separating the alveolar air space from the blood vessels is highly permeable to both oxygen and carbon dioxide and is only about 0.0004 centimeters thick. Rapid and adequate gas exchange occurs over this short distance.

The surface area of lungs is impressively large. Total alveolar area for the human lung is about 100 square meters, approximately fifty times the surface area of human skin. Clearly the specialized structure of our lungs amplifies the surface area available for gas exchange. If gas exchange occurred through our skin, oxygen and carbon dioxide would diffuse much too slowly to support life. Some smaller animals, however, such as frogs and salamanders, do exchange significant amounts of gases through their skin.

## Water Loss from Lungs

Inhaled air quickly becomes saturated with moisture in the lungs. When we exhale this air, our bodies lose that moisture. Table 8.3 shows the amounts of water the human body at normal body temperature loses per day from the lungs and through urine and sweat.

Would water loss from the lungs change when a person is exercising vigorously?

Under most conditions, water loss by breathing does not pose a danger to organisms. Animals living in very dry environments, however, may experience difficulties from too much water loss. Most animals can regulate the amount of moisture lost through urine and sweating, but loss of water from the lungs is more difficult to control since animals must breathe all the time.

## GILLS

Almost all aquatic organisms use some form of **gills** to exchange gases. Gill structures can be as simple as finger-like projections on the surface of a starfish or as elaborate as the finely divided sheets of cells that make up the gills of fish (fig. 8.4).

How could a starfish increase gas exchange with its external environment?

As water passes over the gills, oxygen from the water diffuses to blood vessels in the gill structure, much like oxygen diffuses from alveoli to blood vessels in the lungs of land animals. The platelike structure of the gills creates a very large surface area, which promotes gas exchange. Fish can increase gas exchange by pumping water over the gills with their mouths or swimming with their mouths open.

Organisms that obtain oxygen from water face two problems not encountered by air-breathing animals. First, the oxygen content of water is much lower than that of air. Water contains at most 0.005 percent oxygen, while air contains 21 percent. Recall that the amount of a substance diffusing across a membrane depends partially on the concentration of that substance (chapter 5). Because the oxygen content of water is so low, less oxygen is available for aquatic organisms to use.

The second problem is that diffusion occurs much more slowly in water than it does in air. This is to be expected since the molecules in water are much closer together than those in air. When air molecules move randomly, they travel for greater distances before colliding with another molecule. This creates a rate of diffusion in air one thousand times faster than the rate of diffusion in water.

Water presents one additional problem for aquatic animals. Since it is a liquid and much more dense than air, it is more difficult to move. Some aquatic animals expend as much as 20 percent of their metabolic energy just moving water over their respiratory organs. Moreover, the resistance created when water flows over an object forms a layer of "unmixed" liquid on the surface of gills. This layer, once its oxygen has been taken up by the gills, increases the distance over which diffusion must take place (Thinking in Depth 8.1). Especially in water, where diffusion times are long, this unmixed layer can seriously limit gas exchange.

Air flows two ways in lungs (in and out the alveoli and trachea) but only one way in gills (through the organism's body). Why is the latter more advantageous than the former?

Given the difficulties of gas exchange faced by organisms that obtain their oxygen from water, it is not surprising that they are invariably cold-blooded. As you will recall from chapter 6, cold-blooded animals require less oxygen than warm-blooded animals because they maintain lower body temperatures. Water-breathing animals can get only so much

*Physiology of Plants and Animals*

## BOUNDARY LAYERS

When either air or liquid flows over an object, a stationary or "unmixed" layer forms at the surface. This **boundary layer** is caused by friction between the object and the material flowing over it. Air has very little friction, so its boundary layer is thin. Water is more dense than air and creates greater friction. Thus the boundary layer formed when water moves over an object is thicker.

Boundary layers can cause problems when it comes to diffusion because they increase the distance between membranes and gases. For example, when water flows over the gills of a fish (fig. 1), the water close to the surface of the gills moves more slowly than the rest of the water. A thin layer of water immediately next to the surface does not move at all. The more rapidly the water moves, the thinner the boundary layer is. Oxygen is depleted quickly from this layer. Once it is gone, oxygen must diffuse from the moving water through the boundary layer and across the gill membrane before it reaches the gill interior. The distance added by the boundary layer makes diffusion of oxygen in water difficult. The same problem exists for the diffusion of carbon dioxide, but in the opposite direction.

FIGURE 1    Boundary layers. (a) Water flowing over the gill surface moves more slowly near the gill due to friction between the water and gill. (b) Water immediately adjacent to the gill is stationary, forming an additional barrier to diffusion.

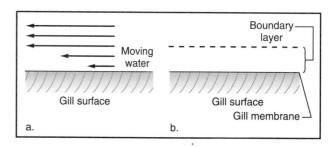

FIGURE 8.4    Gills. (a) In the starfish, gills are fingerlike projections from the surface. (b) Fish gills are located just behind the head and protected by a flap of skin. Water enters through the mouth and flows over and between the flattened, platelike structures of the gill.

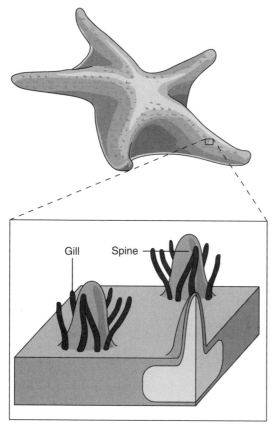

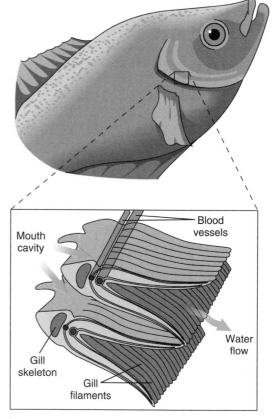

a.

b.

oxygen through gas exchange, so they are limited to being cold-blooded. The few aquatic animals that are warm-blooded, such as whales and otters, breathe air instead of water.

## Water Loss in Aquatic Animals

Water loss by fish may seem like a contradiction, but for ocean-dwelling fish, water loss is a constant problem. Because sea water contains dissolved salts, its water concentration is actually less than that found in the bodies of fish and other aquatic organisms. Thus these organisms are continually losing water through osmosis. The constant flow of water over the thin membranes of the gills accounts for much of this water loss. Freshwater fish face the opposite problem. Osmosis constantly moves water into the body of the fish. Maintaining a relatively constant water balance requires physiological adaptations that we will discuss in chapter 9.

# TRACHEAE

As you have probably observed, most insects are much smaller than mammals or fish. Their size is still too great, however, for sufficient gas exchange through their skin. How do they exchange oxygen and carbon dioxide?

Most insects have neither lungs nor gills. If you examine an insect carefully, you will observe tiny openings on its body called **spiracles** (fig. 8.5). Insects can block these openings by closing **valves** controlled by muscles in the body wall. Spiracles are the entry and exit points to a system of branched, air-filled tubes called **tracheae** (not to be confused with the trachea in animals). The tracheae provide a pathway for oxygen and carbon dioxide movement throughout the body. The tracheal system is so extensive that almost every cell in an insect's body lies near a branch.

The tracheae provide a large surface for the exchange of oxygen and carbon dioxide. Also, insects breathe air, which diffuses more rapidly than water. However, the diffusion distances in tracheae are longer than those of alveoli in the lungs, limiting the size of insects. Oxygen will diffuse 1 centimeter in 10 seconds; a tracheal tube of that length and a diffusion time of 10 seconds is sufficient to supply most insects' needs for oxygen. If distances were much longer than that, insects would not receive enough oxygen to live. Diffusion distance is probably the main limitation on the size of insects. Some larger insects, such as grasshoppers, can move air more rapidly through their tracheal systems by flexing their abdomen. This squeezes and expands the tracheal tubes, increasing the air flow.

Could insects obtain sufficient oxygen from water?

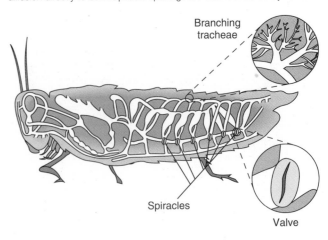

**FIGURE 8.5** The tracheal system of insects. Tracheal tubes form a system of branching tubes extending throughout the body and allowing diffusion directly to cells. Spiracle openings are often controlled by valves.

Branching tracheae

Spiracles

Valve

## Water Loss in Tracheae

Because the tracheal system lies within the bodies of insects, we might expect to see considerable water loss from this respiratory system. Let's examine an experiment in which the actual water loss from insects was measured under three conditions: (*a*) when the spiracles were completely open (the valves controlling the spiracles were prevented from closing), (*b*) when the spiracles were artificially blocked, and (*c*) when conditions were normal and the spiracles were untreated. The results are shown in figure 8.6.

When the spiracles were kept open, water loss was ten times greater than when they were kept closed. When the valves were operating normally, water loss was greater than for the blocked spiracles, but still minimal. We can interpret these observations to mean that water can indeed be lost through the tracheal system but that insects are able to control water loss by opening and closing the tracheal valves.

What hypothesis was being tested in the experiment shown in figure 8.6? What other interpretations can you draw from the data presented?

# PLANTS

Green plants absorb carbon dioxide and release oxygen by photosynthesis, reversing the order of respiration in animals. Do the same principles of gas exchange that apply to animals apply to plants? Essentially, yes. In this section we will discuss the nature of gas exchange between plants and their environment.

FIGURE 8.6 Insect water loss. Loss of water was measured under three conditions. (*a*) Spiracles completely open, (*b*) spiracles untreated and (*c*) spiracles completely blocked. Water loss was measured at varying levels of relative humidity. *Source: Data from E. Bursell,* Proc. R. Ent. Soc. Lond. (A), *32:21, 1957.*

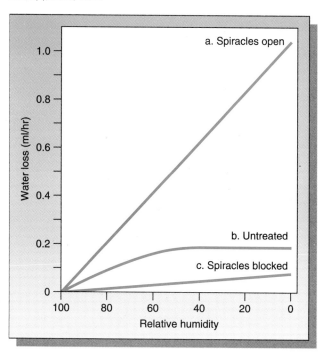

FIGURE 8.7   Stomata. The surface of leaves are covered with tiny structures, called stomata, that open and permit gas exchange between the leaf interior and the external atmosphere.

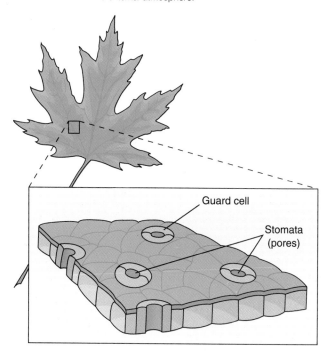

The "organ" through which plants exchange gases is their leaves. As we saw at the beginning of this chapter, surface area, permeability, and diffusion distance are all factors that affect gas exchange. Leaves and fronds, which are thin and flat, rank high in all three factors. Their structure provides adequate surface area and permeability for exchange and decreases diffusion distance.

How does carbon dioxide reach the photosynthetic cells inside leaves? If you observe a leaf closely, you will see on its surface minute openings called **stomata** (fig. 8.7). Stomata open to expose the leaf interior to the external atmosphere, allowing rapid gas exchange.

Stomata are composed of an opening bounded by two elongated cells called **guard cells.** The guard cells can change shape and thereby control the size of the opening (fig. 8.8). How are the guard cells able to change shape? When water flows into them, turgor causes them to bend, which opens the stoma.

How are stomata able to close and decrease water loss when the plant's external atmosphere is dry?

What other factors affect the opening and closing of stomata? Let's look at an experiment in which stomata were observed at different times of the day and night. The data from this experiment are shown in figure 8.9. As you can see, the stomata are generally open during the day and closed at night. We can propose two hypotheses to explain these observations.

*Hypothesis 1*   The guard cells detect light and for some reason take in water, opening the stomata.

*Hypothesis 2*   Light causes photosynthesis, lowering the concentration of carbon dioxide. The carbon dioxide concentration somehow controls the guard cells' uptake of water.

We can test these hypotheses by keeping one variable (light) constant, changing the other variable ($CO_2$ concentration), and observing the stomata. Doing so, we obtain the results shown in figure 8.10. We can see from the figure that when $CO_2$ is absent, stomata are open; when $CO_2$ is present, stomata are closed. These data support hypothesis 2 and are inconsistent with hypothesis 1.

Why do you think stomata close at night? What observations would you predict if carbon dioxide concentration were varied in the dark?

What results would you expect to observe in the experiment shown in figure 8.10 if light, rather than carbon dioxide, had been the controlling factor?

FIGURE 8.8   Stoma opening. (a) Stomate closed. (b) When water moves into guard cells, the increasing turgidity causes the cells to change shape and increase the stomatal opening.

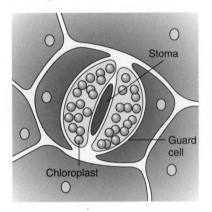

a.

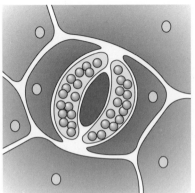

b.

FIGURE 8.9   Stomatal opening vs. time of day. The diameter of the stomatal opening was measured at different times of the day and night. Opening was greatest during the daylight hours, but stomata were closed at night.   *Source: Data from Frank B. Salisbury and Cleon Ross,* Plant Physiology, *1969, Wadsworth Publishing Co., Belmont, CA.*

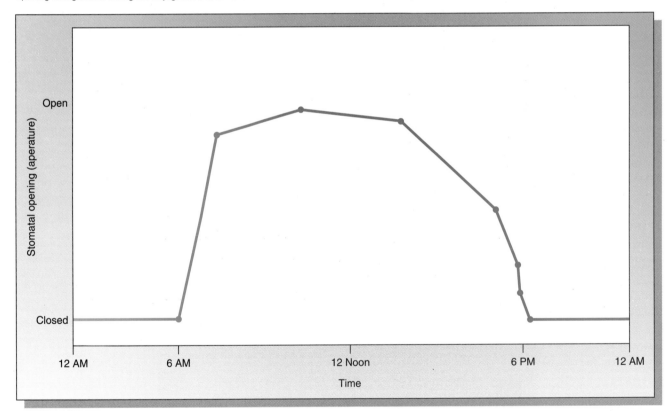

*Physiology of Plants and Animals*

FIGURE 8.10    Effects of $CO_2$ on stomatal opening. Stomatas were exposed either to $CO_2$ or to $CO_2$-free air. In both cases, light intensity was constant. Observations of stomatal opening indicate that stomatas were closed when $CO_2$ concentration was high but open when no $CO_2$ was present.    *Source: Data from C. K. Pallaghy, "Stomatal Movement and Potassium Transport in Epidermal Strips of Zea Mays" in* Planta (Berl), *101, 287-295, 1941.*

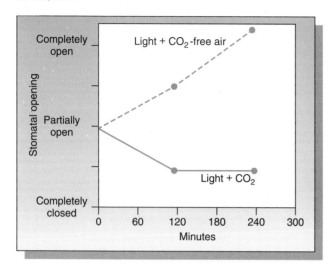

## Water Loss in Plants

The structure of leaves is ideal for gas exchange, but it also favors rapid water loss. This is not a problem for aquatic plants, but terrestrial plants can wither and die in a short time if they lack sufficient water. Unlike animals, which can seek out sources of water, plants are limited to the moisture available at their roots. For plants, water conservation is a life and death matter.

Most leaves are coated with a waxy covering that impedes water loss. In other words, the permeability of the leaf surface to water is decreased. In addition, the stomata provide protection by regulating the amount of water lost from the leaves. This protection is sufficient for plants in a moist environment, but how do plants in the desert, which is extremely dry, survive? If their stomata are open during the day, water loss will be extreme. If they are closed, however, the plants will not be able to obtain carbon dioxide for photosynthesis. Desert plants have adapted to their extreme environment with unusual biochemical pathways that are able to store carbon as organic acids. During the night, the stomata take in carbon dioxide, which is converted to organic acids. The following day, while stomata are closed and prevent water loss, these organic acids are used in photosynthesis. Desert plants have adapted this unique way of exchanging gas through their leaves while minimizing water loss.

## SUMMARY

*Living organisms have effective mechanisms to ensure adequate exchange of oxygen and carbon dioxide with their external environments and, at the same time, minimize water loss. The structure of the specialized gas exchange organs, which include lungs, gills, tracheae, and leaves, varies depending on the organism and its environment. As you read more about biological processes, keep in mind the key points of this chapter on gas exchange.*

1. Molecules diffuse rapidly over small distances, but as diffusion distances increase, the rate of diffusion increases exponentially. Materials diffuse a thousand times more rapidly in air than they do in water.

2. The three most important factors affecting gas exchange in respiratory organs are **surface area, permeability,** and **diffusion distance.**

3. Organisms can lose a significant amount of water through their gas exchange organs, presenting a disadvantage for terrestrial plants and animals.

4. **Lungs** serve as gas exchange organs for most air-breathing animals. Air enters the lungs when air pressure in the lungs is lower than atmospheric pressure.

5. Air enters the lungs through **trachea** that branch into increasingly smaller tubes. These tubes end in small, air-filled sacs called **alveoli** where primary gas exchange occurs.

6. The surface area of human lungs is about 100 square meters, fifty times the area of human skin.

7. **Gills** are projections or platelike structures that exchange oxygen and carbon dioxide with water.

8. Two disadvantages of obtaining oxygen from water are the low diffusion rate of gases in water and the low solubility of oxygen in water.

9. Insects exchange oxygen and carbon dioxide through their **tracheae,** a system of branching, air-filled tubes extending throughout their bodies. Air enters the tracheal system through openings in the body wall called **spiracles.**

10. Plants exchange gases through openings called **stomata.** The stomata are opened and closed by **guard cells** that change shape when they take in or lose water.

11. Carbon dioxide concentration appears to control the opening and closing of stomata.

## REFERENCES

Comroe, J., Jr. February 1966. "The Lung." *Scientific American* 214: 56–68.

Feder, M. E., and W. W. Burggren. November 1985. "Skin Breathing in Vertebrates." *Scientific American* 253: 126–43.

Salisbury, Frank B., and Cleon Ross. 1969. *Plant Physiology* (Belmont, Calif.: Wadsworth Publishing).

## EXERCISES AND QUESTIONS

1. Which of the following best explains why the oxygen concentration in alveoli would not decrease to zero (page 89)?
   a. Exhaled air will have a higher oxygen concentration than will air in the alveoli.
   b. The large surface area of the lungs allows exchange of a large amount of oxygen.
   c. Carbon dioxide increases in exhaled air.
   d. Oxygen will equilibrate between alveoli and blood.
   e. The size of the alveoli is very small.

2. In the lung, diffusion is important in net oxygen movement in which of the following locations?
   a. trachea only
   b. bronchi only
   c. alveoli only
   d. trachea and alveoli
   e. bronchi and alveoli

3. Holding your breath before exhaling would most likely have which of the following effects on gases in exhaled air?
   a. $CO_2$ would be higher.
   b. $O_2$ would be higher.
   c. Less water vapor would be lost.
   d. The total volume of exhaled air would be lower.

4. Vigorous exercise would most likely result in which of the following (page 90)?
   a. decreased water loss
   b. no effect on water loss
   c. increased water uptake by the body
   d. increased water loss

5. Which of the following would most likely increase the gas exchange between a starfish and the external environment (page 90)?
   a. Decrease the number of gill projections.
   b. Place the starfish in deeper water.
   c. Increase the length of the gill projections.
   d. Turn the starfish upside down.
   e. Increase the length of the starfish's arms.

6. An experiment was performed to investigate the effect of flowing water velocity on photosynthetic rate in aquatic plants. A plant was placed in a clear plastic tube. Water was forced through the tube and around the plant at different speeds. The rate of photosynthesis was measured by the rate of oxygen production. The results of this experiment are shown in the graph.

Effects of water flow velocity on photosynthetic rate. The rate of photosynthesis (oxygen production) by an aquatic plant as the velocity of water flow over the surface of the plant is varied.

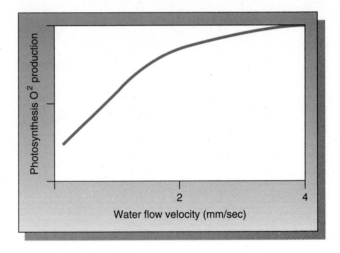

Which of the following is the best interpretation of these results?
   a. Carbon dioxide diffuses faster than oxygen in water.
   b. Boundary layer thickness decreases as water flow velocity increases.
   c. Oxygen is more soluble in flowing water than in still water.
   d. Boundary layers are thicker near aquatic plant leaves than near the walls of the plastic chamber.
   e. Carbon dioxide is more soluble in flowing water than in still water.

7. Which of the following would most effectively prevent an insect from obtaining sufficient oxygen (page 92)?
   a. spraying the insect with water
   b. filling the tracheal system with water
   c. a boundary layer of air around the insect
   d. preventing the insect from moving
   e. placing the insect in a dry atmosphere

8. Which of the following is most likely the hypothesis being tested in the experiment shown in figure 8.6 (page 92)?
   a. Not all insects have valves that can minimize water loss.
   b. Tracheae extend throughout the insect's body.
   c. Water loss in insects can exceed the amount of water intake.

d. The greater the diffusion distance in tracheae, the lower the water loss.

e. Most water loss in insects occurs via the respiratory system.

9. How are stomata able to close and decrease water loss when the plant's external atmosphere is dry (page 93)?

a. When stomata are open, excess water is lost.

b. The cuticle of the leaf minimizes water loss when stomata are closed.

c. When stomata are closed, relatively little water is lost from the leaf.

d. The two guard cells around each stoma control the stomatal opening.

e. Water loss from guard cells will cause the stomata to close.

10. Based on the results in figure 8.10, why would you expect stomata to close at night (page 93)?

a. Carbon dioxide will be high in the leaf due to respiration.

b. The absence of light will be a controlling factor.

c. Water loss from the leaf is drastically decreased when the stomata are closed.

d. Carbon dioxide is normally present in the atmosphere.

e. The size of the stomatal opening is controlled by the guard cells.

11. What observations of the stomatal opening would you predict if $CO_2$ concentration were varied in the dark (page 93)?

a. Stomata would remain constantly open.

b. Stomata would close at low $CO_2$ concentrations.

c. Stomata would remain constantly closed.

d. The same as observed in light.

e. Stomata would open at high $CO_2$ concentrations.

12. What results would you expect to observe in figure 8.10 if light rather than carbon dioxide had been the factor controlling stomatal opening (page 93)?

a. $CO_2$ would vary under natural conditions and control stomatal opening.

b. Stomatal opening size would remain constant as $CO_2$ is varied.

c. Some stomata would open and some would close.

d. Light would increase $CO_2$ concentration and close stomata.

e. Light would be varied in the experimental design.

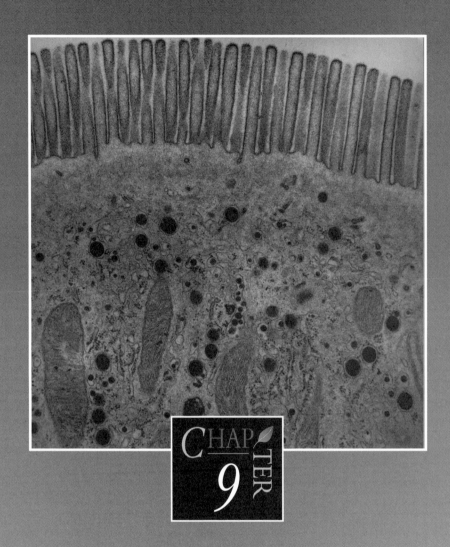

# DIGESTION AND EXCRETION

e saw in earlier chapters how cells take in organic molecules and water. We also saw that cells cannot take in large molecules such as proteins and starch, yet these molecules are essential for the health of many organisms. How do molecules, especially large ones, enter and leave the body? Organisms must have ways of breaking down these larger molecules into particles they can use. Indeed, both plants and animals have specialized ways of taking in nutrients.

If you are like most people, you have probably eaten a meal within the last several hours. Perhaps you are snacking on something while you read. Like most animals, humans spend a good part of their time consuming food, transforming it to obtain useful energy, and eliminating the wastes that the body cannot use. These processes are essential for life and are the focus of this chapter.

## DIGESTION IN ANIMALS

The process through which animals take in food and reduce large molecules to small ones is called **digestion.** In most animals, the **digestive system** is essentially a tube that runs through the body and consists of several compartments (fig. 9.1). Animals take in food through an opening at one end, break it down into small molecular fragments, and excrete the remainder as waste through an opening at the other end. Once the material is broken down into small molecules it is absorbed by the blood and transported throughout the body where individual cells use the nutrients.

How could you argue that materials inside an animal's digestive system are actually outside the animal's body?

How does the digestive system break down food into useful materials? Recall our discussion in chapter 4 of how enzymes catalyze chemical reactions. Among those reactions are ones that break the chemical bonds between sugar monomers in carbohydrates and the peptide bonds between amino acids in proteins. Animals can absorb the relatively small sugars and amino acids into their cells much more easily than they can absorb carbohydrates and proteins.

As you might have guessed, the enzymes that break down carbohydrates and proteins are present in the digestive system. Table 9.1 shows these enzymes, the molecules they break down, and the digestive organs where they are active.

We can make three observations based on the information given in table 9.1: Enzymes occur in different locations; enzymes are produced by a variety of sources; and most enzymes are found in the small intestine. The data also show that the human digestive system contains enzymes that can act on all the major biological molecules: proteins, carbohydrates, fats, and nucleic acids.

**FIGURE 9.1**   The digestive system in most animals is basically a tube extending through the body. The system is divided into several compartments where different stages of digestion occur.

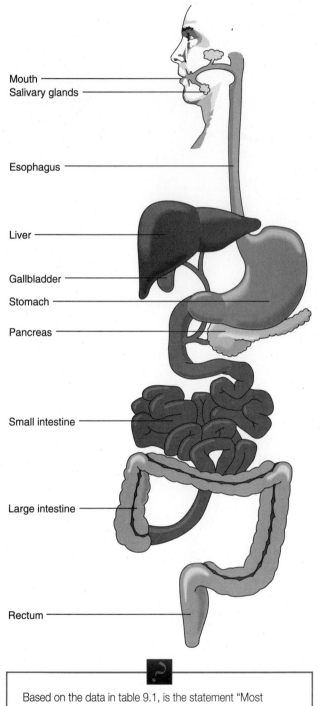

Mouth
Salivary glands
Esophagus
Liver
Gallbladder
Stomach
Pancreas
Small intestine
Large intestine
Rectum

Based on the data in table 9.1, is the statement "Most digestion occurs in the small intestine" an observation or an interpretation?

The digestive system in most animals begins with the **mouth,** where food is taken in, chewed, and swallowed. **Salivary glands** in the mouth produce enzymes that begin to break down the food. In birds, which often consume seeds whole, the food enters an organ called the **crop** that contains small stones or grit. The grinding action of the crop breaks the food into smaller pieces.

*Physiology of Plants and Animals*

TABLE 9.1  Enzymes in the Human Digestive System

| Enzyme | Source | Substrate | Site of Action |
|---|---|---|---|
| Salivary amylase | Salivary glands | Starches | Mouth |
| Pepsin | Stomach lining | Proteins | Stomach |
| Pancreatic amylase | Pancreas | Starches | Small intestine |
| Lipase | Pancreas | Fats | Small intestine |
| Chymotrypsin | Pancreas | Polypeptides | Small intestine |
| Deoxyribonuclease | Pancreas | DNA | Small intestine |
| Aminopeptidase | Small intestine | Polypeptides | Small intestine |
| Maltase | Small intestine | Maltose | Small intestine |
| Lactase | Small intestine | Lactose | Small intestine |
| Sucrase | Small intestine | Sucrose | Small intestine |
| Phosphatases | Small intestine | Nucleotides | Small intestine |

FIGURE 9.2   Stomach lining. (a) Lining as observed with scanning electron microscope. (b) Cross section of the stomach wall. Hydrochloric acid and enzymes are secreted by the gastric glands and enter the stomach through the gastric pits.

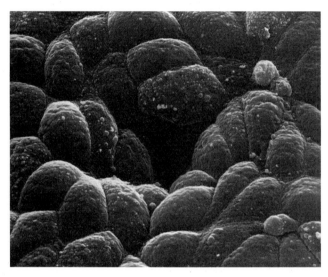

a.

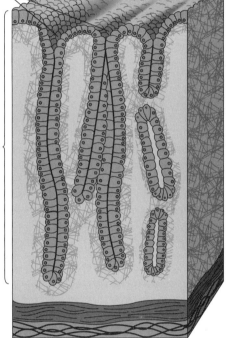

Gastric pits

Gastric glands

b.

Like the cells of the stomach lining, skin cells are replaced rapidly. Why do you think this is so?
Given the structure of the stomach lining shown in figure 9.2, what would you predict about the absorption of material across the stomach wall into the blood vessels?

From the mouth food moves down the **esophagus** and enters the **stomach** where enzymes continue to break down carbohydrates and sugars. In addition, the stomach contains hydrochloric acid (HCl), a highly corrosive acid deadly to cells. If HCl is corrosive, why doesn't it damage the stomach wall? Why doesn't the stomach digest itself? Biologists have discovered that the cells lining the stomach are closely joined to one another by tight junctions that prevent the hydrochloric acid or enzymes from penetrating. In addition, mucus coats the cells and appears to provide some protection. Figure 9.2 illustrates the relatively smooth lining of the stomach.

Biologists have also learned that stomach cells flake off from the lining at a rapid rate and are quickly replaced by new cells. In fact, the surface layer of cells in the stomach is completely replaced every three days! The same sort of rapid cell replacement occurs farther down the digestive system in the small and large intestines, where the movement of food material constantly abrades the living cells.

From the stomach, food travels into the **small intestine,** where a variety of enzymes are active. Many of these enzymes are produced by glands in the small intestine wall. Others are produced by a large gland called the **pancreas.** The **liver** produces **bile,** a mixture of salts that break up fats

*Digestion and Excretion*

**101**

into small droplets that can be more easily attacked by **lipases,** which are fat-digesting enzymes.

Other enzymes in the small intestine include amylases, which break starch into sugars, and the enzymes that act on proteins. Some of these enzymes attack the protein molecule at the ends, while others cleave the molecule between specific amino acids and break it into smaller protein fragments called **polypeptides.** These are further digested by other enzymes.

In most people these enzymes work efficiently to digest a wide variety of foods, but some people have difficulty digesting certain foods because they lack certain enzymes. One example of an enzyme deficiency is **lactose intolerance.** Lactose is a sugar composed of two monomers, glucose and galactose, chemically bonded together. It is found in milk and is commonly called milk sugar. Many people produce very low concentrations of **lactase,** the enzyme that digests lactose. When they consume milk products, they suffer from cramps and diarrhea.

Physicians can detect lactose intolerance by removing material from a person's small intestine and testing it for lactase activity. There is an easier method, however. Figure 9.3 shows the results when researchers tested someone for lactose intolerance. The individual was fed lactose. Afterward, her blood glucose level was measured to monitor any change. Next, she was fed sucrose and researchers measured her blood glucose levels again.

Keep in mind that lactose is composed of glucose and galactose and sucrose is composed of glucose and fructose. When the individual was fed lactose, researchers found no increase in blood glucose, indicating that the lactose was not being digested.

What result would you expect to see in figure 9.3 if lactase were present in the individual's small intestine?

Why did researchers have the person consume sucrose in addition to lactose?

What would you expect to observe if a lactose intolerant person consumed lactose and sucrose at the same time?

The pH in different parts of the digestive system can exert a strong effect on digestion. We saw earlier that the stomach is highly acidic. The small intestine, by contrast, is neutral to alkaline. Remember from chapter 4 that enzyme activity is highly sensitive to pH balance. In the stomach, the enzyme pepsin is most active at a pH of about 2, while in the small intestine enzymes are most active at a pH of about 7 to 8. These observations indicate how enzyme activity, and thus digestion, are influenced in different parts of the digestive system.

Most of the nutrients taken in by the digestive system are absorbed in the small intestine, thus you would

FIGURE 9.3  Lactose intolerance. When a lactose intolerant individual is fed lactose, blood glucose does not increase (dotted line). When fed sucrose, however, blood glucose increases rapidly (solid line).  *Source: Data from T. M. Bayless and N. S. Rosenweig, "A Racial Difference in Incidence of Lactose Deficiency" in Journal of the American Medical Association, 197, #12:968-972, September 1966.*

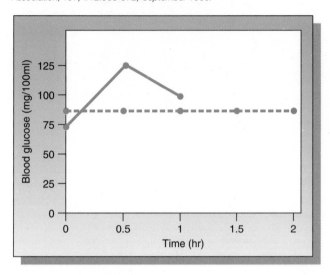

expect the surface area of the small intestine to be quite large. The human small intestine is about 6 meters long. If the lining were smooth, the surface area would be less than 1 square meter. But, in fact, the total surface area of the lining of the small intestine is about 300 square meters. How can this be so?

Detailed examination has revealed fingerlike projections on the wall of the small intestine called **villi** (fig. 9.4). Cells on the surface of the villi have microscopic projections called **microvilli**. Together, the villi and microvilli greatly increase the surface area of the lining and thus increase the absorption of material.

We have seen how enzymes digest the variety of food material that travels through the small intestine. Some molecules, however, cannot be broken down by enzymes made in the body. For example, cellulose, the primary structural molecule of plants, passes through the human digestive system virtually unaffected. Humans produce no enzymes capable of breaking the chemical bonds between the sugar monomers in cellulose. Even herbivores, which consume only plants, produce no cellulose-digesting enzymes. This may seem strange, since herbivores consume large amounts of cellulose. The answer lies in the bacteria and other microorganisms that live in the digestive systems of herbivores. They produce

A student proposed that humans could consume enzymes to increase the digestion of food, or even allow digestion of material that could not normally be broken down. How could this proposal be contradicted?

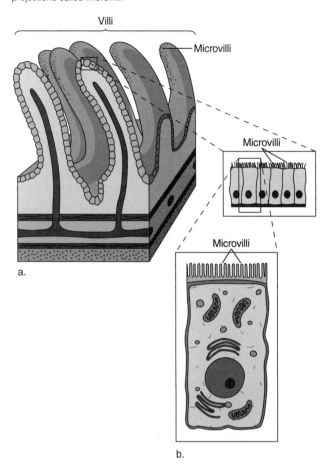

a.

b.

the enzymes to break down cellulose, forming a partnership with the herbivore. Such microorganisms do not, however, live in the human digestive system, so humans cannot benefit from consuming cellulose.

From the small intestine, material moves into the **large intestine** where water and sodium ions are absorbed and the material becomes more solid. Here, large populations of bacteria live. In humans, these bacteria appear to digest some previously undigested food and suppress the growth of harmful bacteria. Intestinal bacteria can be upset by sudden changes in diet or other factors, all of which may cause diarrhea or other intestinal disorders. The digestive system ends with the **rectum,** the opening where wastes are eliminated.

A student was skeptical that bacteria could live in the large intestine. He argued that as waste material is eliminated from the body, bacteria would also be eliminated, leaving no microorganisms. What was this student most likely assuming?

As we mentioned earlier, the entire digestive system in most animals forms a tube. Figure 9.5*a* and *b* show this tube-like structure in worms and insects. Not all organisms have tube-shaped digestive systems, however. Some, such as bread mold, shown in figure 9.5*c*, digest food outside their bodies. Bread mold releases digestive enzymes to the environment in order to break starch into sugars or other molecular fragments that can diffuse into the cells of the mold. Other organisms such as single-celled animals, specialized cells, and small multicellular animals may take in nourishment through large food particles that directly enter their cells. Recall from chapter 5 our discussion of **endocytosis,** the engulfing of large molecules and particles by folding of the plasma membrane. The hydra, shown in figure 9.5*d*, digests food this way. **Phagocytes** in the human circulatory system and single-celled amoebae can consume bacteria and foreign particles by engulfing material.

## UPTAKE BY PLANTS

Plants take in material for food much differently than do animals. As we saw in chapter 7, plants manufacture their organic molecules through photosynthesis. They must absorb, however, water and a small amount of essential minerals. These minerals are not digested in the sense that digestion occurs in animals, but they must come from the external environment and be transported throughout the plants.

In terrestrial plants, the primary organ of water and mineral absorption is the **root.** Roots are made of progressively branching structures like the ones shown in figure 9.6. This branching results in an amazingly large root surface. In one experiment, a single young rye grass plant was planted in a box of soil measuring 12 inches square by 22 inches deep. After four months of growth, researchers carefully removed the roots from the soil and measured their length. The total root length was 387 miles; the surface area was 2,554 square feet.

The surface area of roots is increased even more by projections from the root surface cells called **root hairs** (fig. 9.6*b*). In the experiment just mentioned, the root hairs increased the root length to about 7,000 miles and the surface area to about 7,000 square feet. We know from previous chapters that surface area is a critical factor in how well an organ absorbs material from the environment. The greater the surface area, the greater the absorption.

Both digestion and plant uptake of water and minerals are processes that absorb small molecules into the body. The digestive system lining and the root system provide large surface

Root hairs are analogous to what structures in the human digestive system?

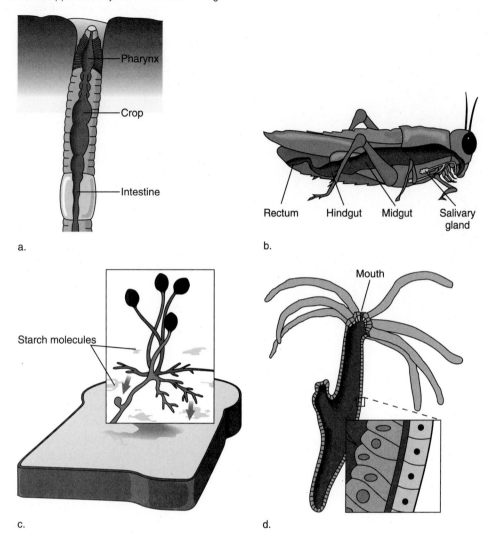

a.

b.

c.

d.

areas for absorption so material can be moved from the external environment to the organism's interior.

# EXCRETION

As food travels through the digestive system of an animal, enzymes break down materials, but not all these materials are used by the organism. Some are of no use to the animal and are eliminated from the body as wastes. This process of elimination is called **excretion.**

The excretory system in animals is extremely important. If animals were unable to rid their bodies of waste materials, they would die. Moreover, some of the wastes produced by digestion are toxic and will poison other systems in the body if they are not removed.

The excretory system performs one other important function that we will look at momentarily, and that is the regulation of water and salts in the body. Because most terrestrial animals have limited access to water, they must retain as much water as possible. The excretory system enables them to do so.

We saw in chapter 8 how the most common waste material in animals, carbon dioxide, is eliminated through gas exchange. Most other wastes in animals are eliminated in the form of liquids or solids. One example is ammonia ($NH_3$), a nitrogen product formed by the breakdown of proteins and nucleic acids. Ammonia is highly toxic; at very low concentrations it will damage cells. Surprisingly, many animals do not eliminate ammonia directly. Instead, they first convert it to other nitrogen compounds. Table 9.2 lists three types of nitrogenous wastes and the animals that produce each type.

**FIGURE 9.6** Roots. (a) Repeated branching of roots results in a large root surface area. (b) Root hairs at root tips further increase the root surface area.

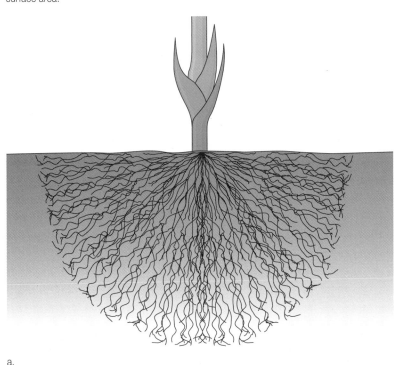

a.

b.

| **TABLE 9.2** Nitrogenous Wastes in Animals | |
|---|---|
| **Animals** | **Waste** |
| Many aquatic animals | Ammonia ($NH_3$) |
| Birds, reptiles, insects | Uric acid |
| Mammals | Urea ($NH_2$-C-$NH_2$) |

Uric acid and urea are more complex compounds than ammonia. In other words, energy is required to turn ammonia into uric acid or urea. Why would animals expend energy to form these compounds when they will just be excreting them anyway? Why not simply excrete the nitrogen waste as ammonia?

Table 9.2 provides a clue. Notice that aquatic organisms excrete ammonia and terrestrial organisms excrete uric acid and urea. Aquatic animals have unlimited supplies of water while land animals do not. These observations lead to a hypothesis: Land animals convert ammonia to uric acid and urea in order to conserve water.

Analysis reveals that ammonia is far more poisonous than uric acid or urea. This means that animals must get rid of ammonia faster and in smaller quantities than they must excrete uric acid and urea. Aquatic animals can do this because they can take in large amounts of water and excrete ammonia in small, diluted quantities. Land animals, which must conserve water, convert ammonia to uric acid and urea and are able to store greater amounts of waste and excrete it in concentrated forms. Eliminating wastes infrequently

**FIGURE 9.7** Malpighian tubules. Closed tubes extend from the intestine wall. As fluids circulate throughout the body, waste material diffuses into the tubules and is concentrated. This material moves down the tubule and into the gut and is eliminated.

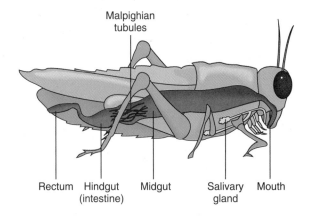

conserves water. Uric acid and urea are less toxic than ammonia, but they must eventually be excreted.

Just as there are organs that do the work of digestion, there are organs that do the work of excretion. These organs take waste products from the blood or body fluid of the animal and pass them to the outside. The excretory organs lie in close contact with the organism's blood system so that waste products and water can diffuse into the organs. In insects, water and waste products diffuse into closed tubules called **Malpighian tubules** (fig. 9.7).

FIGURE 9.8 Kidney. (a) The kidney is a fist-shaped organ containing about 1 million nephrons. (b) Small arteries contact the nephron at Bowman's capsule. Filtrate moves from the glomerulus into the capsule and proximal tubule. Fluid continues through the Loop of Henle and distal tubule before entering the collecting duct.

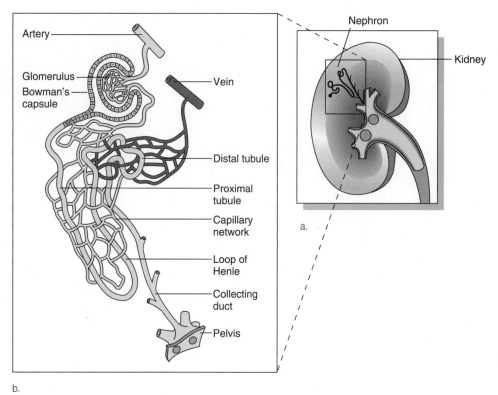

a.

b.

In animals with backbones, called **vertebrates,** the primary organ of excretion is the **kidney.** Humans have two of these fist-sized organs (fig. 9.8). The kidneys contain small blood vessels called capillaries that contact the kidney at sites called **nephrons** (fig. 9.8b). In humans, each kidney contains about 1 million nephrons. The point at which the blood vessel and the nephron connect is called the **glomerulus.** The glomerulus is surrounded by a cuplike structure called **Bowman's capsule.** Fluid from the blood is forced into the Bowman's capsule. As it is forced, some molecules are filtered out. Thus the fluid that enters the nephron is called **filtrate.** Table 9.3 compares the concentrations of certain molecules in blood to the concentrations in the filtrate entering Bowman's capsule.

As you can see, concentrations of ions and smaller molecules in the blood and filtrate are equal, but the filtrate has no protein or blood cells. The blood vessel walls filter these larger molecules and cells from the fluid entering Bowman's capsule.

When a person consumes a dye or pigment, sometimes that dye later appears in the urine. What can you conclude from this observation?

**TABLE 9.3   Concentrations of Materials in Blood and Filtrate**

| Material | Blood Concentration (mM/1) | Filtrate Concentration (mM/1) |
|---|---|---|
| Urea | 26 | 26 |
| Glucose | Equal* | Equal* |
| K+ | 5 | 5 |
| Na+ | 142 | 142 |
| Cl- | 103 | 103 |
| Protein | High* | 0 |
| Blood cells | High | 0 |

*Concentrations may vary depending on diet.

Approximately 125 milliliters of filtrate enter the kidneys each minute. This filtrate passes through the nephrons and is gradually modified to become urine. As the fluid passes through each portion of the nephron, it decreases in volume. Measurements of fluid flow in the nephron have produced the data shown in figure 9.9. The amount of fluid flowing through the nephron structures consistently decreases, falling from 125 milliliters per minute at Bowman's capsule to 1 milliliter per minute leaving the collecting duct. Apparently almost all the water in the filtrate is reabsorbed before leaving the kidney.

*Physiology of Plants and Animals*

FIGURE 9.9 Fluid flow at different points in the nephron. The amount of fluid flow consistently decreases as fluid moves farther through the tubule.

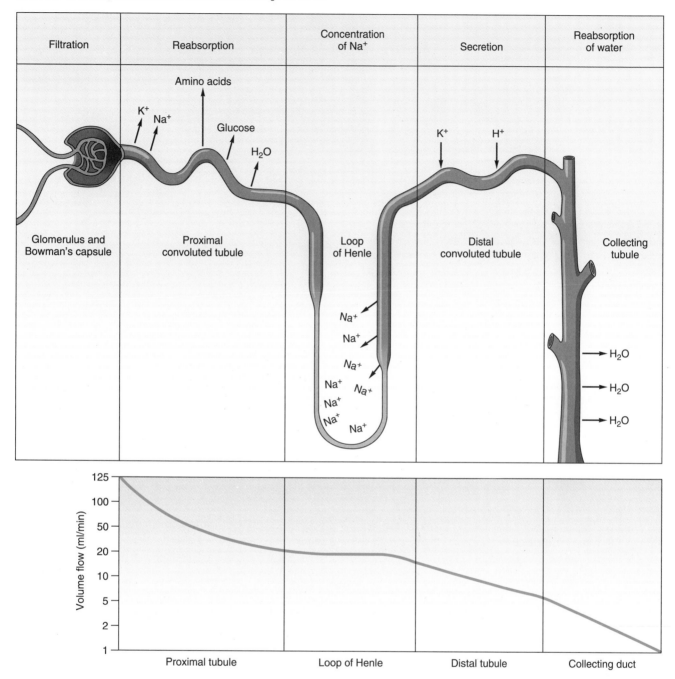

How is the water reabsorbed? There is no known mechanism of active water transport, so how is water reabsorption accomplished? Table 9.4 provides a clue.

Several substances are removed as fluid moves through the tubular system. This is most obvious with glucose and sodium, although substantial reabsorption into the blood capillaries near the nephron occurs for other substances as well. As the solute concentration within the tubules decreases, water concentration increases. As a result, water moves out of the kidney tubules

**TABLE 9.4  Concentrations of Materials in Filtrate and Urine**

| Materials | Filtrate Concentration (mM/1) | Urine Concentration (mM/1) |
|---|---|---|
| Urea | 26 | 1,820 |
| Glucose | Equal to blood | 0 |
| K$^+$ | 5 | 60 |
| Na$^+$ | 142 | 128 |
| Cl$^-$ | 103 | 134 |

## HEMODIALYSIS THERAPY

Millions of people in the United States and Canada have kidneys that do not function because of injury or disease. Without some way of eliminating wastes from their bloodstreams, these people could not survive. Fortunately, most kidney patients have access to machines that perform a procedure called **hemodialysis.** *Hemo* denotes blood and *dialysis* indicates the use of a differentially permeable membrane to separate large molecules and cells from smaller molecules and ions that can diffuse across the membrane.

The kidney machine operates by pumping blood from a patient's artery through a dialyzing tube (fig. 1). The tube is bathed by a dialyzing solution in which a low concentration of urea is maintained. Urea diffuses out of the blood into the solution. Ion concentrations in the blood can be adjusted by maintaining constant ion concentrations in the dialyzing solution. After it passes through the tube, the cleansed blood is returned to a vein. Within a few hours, all the body's blood can circulate through the machine and be cleansed of urea.

What would happen if low urea concentrations in the dialysis solution were not maintained?

What would happen to a person's blood glucose if higher glucose were present in the dialysis fluid?

FIGURE 1   Artificial kidney machine. Blood is pumped from an artery through an extended dialysis tube constructed of a dialyzing membrane. The tube is immersed in a dialyzing fluid. After passing through the tube, blood is returned to the body via a vein.

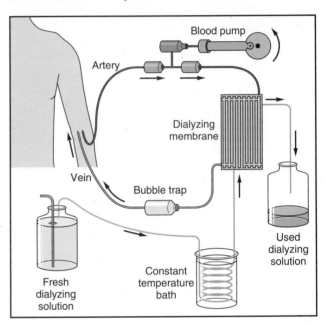

and into the blood by osmosis (chapter 5). Water passively follows (by osmosis) the active transport of other materials. This reabsorption into the blood allows the conservation of water and essential substances. Some materials, such as glucose, are completely reabsorbed at normal concentrations. Urea, on the other hand, becomes progressively more concentrated as water is reabsorbed. In this way the type and amount of material excreted can be finely regulated by the kidney.

Sometimes kidney function can be reduced due to injury or disease and waste material cannot be excreted. When this occurs, urea will accumulate to poisonous levels in the body. Unless the urea is removed (Thinking in Depth 9.1), the body will die.

## SUMMARY

*In this chapter we have examined how material moves into and out of an organism's body. This movement requires specialized organs that digest and absorb nutrients and eliminate and excrete waste material. Keep in mind the following points as you study these biological processes further.*

1. The digestive system of animals is basically a tube, usually consisting of a series of compartments, extending through the body of the organism.
2. Some cells and smaller multicellular animals can ingest large molecules and particles by **endocytosis.**
3. Large particles of food can be reduced in size by chewing, while enzymatic digestion reduces large molecules to sizes that can be absorbed by cells.

4. In the human digestive system, food and water enter the **mouth** and pass through the **esophagus** into the **stomach,** the **small intestine,** and the **large intestine.**
5. Each part of the digestive system varies in pH levels and enzymes. The stomach, where protein digestion begins, is highly acidic.
6. Most enzymatic digestion occurs in the small intestine. In the large intestine, water is reabsorbed and waste material is concentrated.

7. Not all material entering the digestive system can be broken down. If appropriate enzymes are not present, material such as cellulose will pass through the digestive system of humans unaffected.

8. The intestines provide a large surface area that absorbs material after digestion. Projections from the small intestine lining called the **villi** and **microvilli** significantly increase the surface area of the small intestine.

9. Plants take up water and minerals by diffusion via a root system.

10. Roots provide a large surface area for absorption through extensive branching and **root hairs.**

11. Some by-products of digestion, such as ammonia, must be excreted since they can have poisonous effects on the body. In some cases, ammonia is converted to urea or uric acid before excretion.

12. In insects, waste from body fluids diffuses into **Malpighian tubules** and moves into the digestive tube.

13. In many animals, the **kidney** concentrates waste material before excretion.

14. Fluid enters **Bowman's capsule** from the blood as **filtrate.** Molecules and ions in the filtrate are at the same concentration as in the blood.

15. As fluid passes through the **nephron,** many materials are reabsorbed by active transport to the blood. Water is reabsorbed, while urea and other waste products become concentrated to form urine.

## EXERCISES AND QUESTIONS

1. Which of the following statements provides the strongest support for the argument that material inside the human digestive system is actually outside the body (page 100)?
   a. Food should be chewed thoroughly before it is swallowed and enters the digestive system.
   b. The pH in different parts of the digestive system can vary significantly.
   c. The intestinal wall provides a large surface area for absorption of material into blood vessels.
   d. An object could pass completely through the digestive system without crossing a membrane.
   e. Many different enzymes that are released by specialized glands are found in the digestive system.

2. Is the statement "Most digestion occurs in the small intestine" an observation or an interpretation of the data in table 9.1 (page 100)?
   a. An observation, because many enzymes are present in the small intestine.
   b. An interpretation, because there is no direct evidence that food molecules are broken down in the small intestine.
   c. An observation, because more enzymes are found in the small intestine than in other parts of the digestive system.
   d. An interpretation, because the presence of enzymes was established by measuring intestine contents.
   e. An observation, because data were also collected on the presence of enzymes in other parts of the digestive system.

3. Which of the following best explains why skin cells must be replaced rapidly (page 101)?
   a. The skin is subjected to abrasion by objects in the external environment.
   b. The digestive system provides a much larger surface area than does the skin.

   c. Even though it has a large surface area, the skin has a low permeability to oxygen.
   d. The skin is exposed to an external environment relatively high in oxygen and low in carbon dioxide.
   e. The skin has more cells than other organs or structures in the body.

4. Given the structure of the stomach lining and the material presented in this chapter, which of the following would you most likely predict about the absorption of material across the stomach wall into blood vessels (page 101)?
   a. Most absorption would occur in the stomach.
   b. Absorption of proteins would occur in the stomach but not absorption of carbohydrates.
   c. Hydrochloric acid is concentrated in the stomach and would be absorbed.
   d. Absorption of carbohydrates would occur in the stomach but not absorption of proteins.
   e. Much less absorption would occur in the stomach than in the small intestine.

5. In the experiment shown in figure 9.3, what result would you expect if lactase were present (page 102)?
   a. Blood sugar would not rise when either lactose or sucrose is ingested.
   b. When lactose is ingested, blood sugar would decrease.
   c. Blood sugar would decrease when sucrose is ingested.
   d. When lactose is ingested, blood sugar would rise.
   e. The same results would be observed.

6. In the experiment described in figure 9.3, what was the purpose of consuming sucrose and measuring blood glucose (page 102)?
   a. To detect the presence of lactase in the small intestine.
   b. To determine time taken for ingested sucrose to reach the small intestine.

c. To see if a rise in blood sugar could be detected when glucose was produced in the intestine.

d. To detect the presence of sucrase in the small intestine.

e. To determine the amount of sugar that could pass through the intestine.

7. What would you expect to observe if both lactose and sucrose were consumed at the same time in a lactose intolerant individual (page 102)?

a. Blood glucose would rise.

b. After a decrease, blood glucose would rise.

c. Blood glucose would remain constant.

d. Blood glucose would fall.

e. Blood glucose could not be measured under these conditions.

8. A student proposed that humans could consume enzymes to increase the digestion of food or to digest material that could not normally be broken down. Which of the following most strongly contradicts this proposal (page 102)?

a. Most enzymes in the body are found in the small intestine.

b. Enzymes are composed of protein, a polymer of amino acids.

c. Acid and enzymes in the stomach would digest the ingested enzymes.

d. Enzymes catalyze only very specific reactions.

e. If enzymes are ingested, they should catalyze reactions inside the digestive system.

9. A student was skeptical that bacteria could live in the large intestine. He argued that as waste material is eliminated, the bacteria would also be eliminated, leaving no microorganisms in the body. Which of the following was this student most likely assuming (page 103)?

a. Bacteria can live in the small and large intestines.

b. More than one kind of microorganism is present in the digestive system.

c. As waste material is eliminated, bacteria would also be eliminated, leaving no microorganisms.

d. Bacteria and other microorganisms can digest material in the large intestine.

e. Microorganisms could not reproduce faster than they were eliminated.

10. Tadpoles, which feed only on plant material, have a very long, coiled intestine. Adult frogs that develop from the tadpole and feed only on insects and other small animals have a relatively short intestine. Which of the following is the best interpretation of these observations?

a. Tadpoles could not eat animal material even if ample material were available.

b. Plant material is more difficult to digest and absorb than animal material.

c. Intestine length is more difficult to measure in the adult frog than in the tadpole.

d. Tadpoles consume smaller particles of food than do adult frogs.

e. Animals in earlier stages of development (e.g., tadpoles) consume only plant material.

11. Root hairs are analogous to which of the following structures in the human digestive system (page 103)?

a. Stomach

b. Pancreas

c. Small intestine

d. Large intestine

e. Villi

12. When people consume dyes or pigments, the dyes sometimes appear later in the urine. Which of the following is the best conclusion drawn from this observation (page 107)?

a. The dye is digested in the small intestine before entering the bloodstream.

b. The dye is poisonous to the human body.

c. The dye molecules are smaller than proteins.

d. There is more dye in Bowman's capsule than in the blood.

e. The dye molecule is smaller than urea.

13. Under normal conditions in humans, all the glucose entering the kidney is reabsorbed and none is lost in the urine (table 9.4). When blood glucose increases, however, more glucose enters the kidney and some is lost in urine, as shown in the graph below. Which of the following best explains the loss of glucose in the urine?

a. More glucose enters Bowman's capsule than is present in the blood.

b. Higher glucose in the kidney tubules saturates the carriers that reabsorb glucose.

c. A greater volume of urine is excreted when glucose is high.

d. Glucose becomes concentrated in the fluid in the kidney tubules as water is reabsorbed.

e. When blood glucose is high, more glucose enters the filtrate.

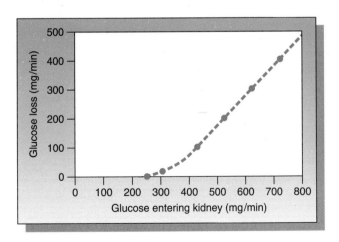

*Source: Data from D. C. Guyton,* Textbook of Medical Physiology, *3rd edition, W. B. Saunders Co., Philadelphia, PA.*

*Physiology of Plants and Animals*

14. Which of the following would most likely occur if dialyzing fluid were not changed frequently in an artificial kidney machine (page 109)?
    a. Blood would flow more slowly through the dialysis tube.
    b. Excess water would move from the dialysis fluid to the blood.
    c. Less urea would be removed from the blood.
    d. Blood would flow more rapidly through the dialysis tube.
    e. The salt concentration in the dialyzing fluid would decrease.

15. What would happen to a person's blood glucose if higher glucose were present in the dialysis fluid (page 109)?
    a. Blood glucose would increase.
    b. When the blood is returned to the body, less glucose would be present.
    c. Blood glucose would more likely remain stable.
    d. Blood glucose in the dialysis tubing would decrease.
    e. In the body, blood glucose would begin to decrease.

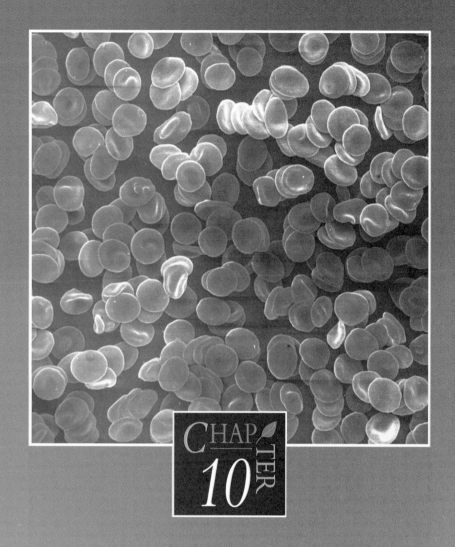

CHAPTER

10

# TRANSPORT AND CIRCULATION

e have learned how organisms use gas exchange organs, how plants manufacture organic molecules and obtain energy from sunlight, and how animals and plants obtain and eliminate food. We know that the materials organisms take in are used by individual cells, and that the wastes organisms give off are produced by cells. How do these materials get to and from the individual cells? For example, how does the oxygen inhaled by a wolf reach each cell of the wolf's body? How does water from the ground reach the leaves of a tree? The answer lies in the transport and circulatory systems.

When we examine a wide variety of organisms, we discover that they use different systems for transporting nutrients, gases, and water. Some very small organisms such as bacteria may be able to use only diffusion to obtain nutrients and to eliminate wastes directly to or from their environment. Other larger organisms have specialized organs or tissues that transport and circulate materials. These organisms have an extensive system of conducting tubes or vessels extending throughout their bodies. Examples are plants such as trees, grasses, and ferns, and animals such as earthworms, reptiles, amphibians, and mammals.

In many animals, these specialized organs make up the **circulatory system.** We will take a closer look at two types of circulatory systems, the open and the closed.

## OPEN CIRCULATORY SYSTEM

Many animals are too large for diffusion to provide sufficient nutrients directly through their skin. They must rely on a partial conducting system to transport materials to their vital organs. Insects, for example, have a single conducting vessel that runs along the length of their bodies (fig. 10.1). Muscular contractions of a portion of the vessel called the **heart** force fluid, or blood, out the front of the vessel and into the insect's body spaces. This fluid enters the opposite end of the tube through openings called **ostia.** Valves at the ostia allow fluid to move only *into* the conducting vessel.

If the ostia in insects did not have valves, how would fluid move when the heart pumps?

Once the insect's blood leaves the conducting tube it circulates freely around the organs in the body cavity. That is why this system is called an **open circulatory system.** As you might expect, fluid movement is somewhat sluggish in this type of system. It is sufficient, however, to transport food material and wastes to and from the insect's respiratory, digestive, and excretory systems.

If a dye were injected into an insect's body at a single location, what would be the dye's path of movement through the body?

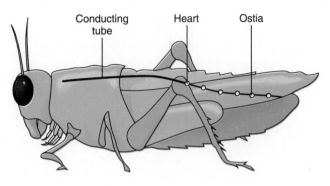

**FIGURE 10.1**    Insect circulation. Fluid is forced out the forward opening of the conducting tube by contractions of the heart. Fluid circulates freely around the body organs and reenters the conducting tube at the ostia.

## CLOSED CIRCULATORY SYSTEM

The sluggish movement and generally inconsistent flow of blood to all parts of the body in the open circulatory system means that delivery of material to all body cells is uncertain. Animals can achieve more consistent and reliable transport of materials if they have numerous conducting tubes extending throughout their bodies. Such an arrangement is called a **closed circulatory system.** It is typical of animals such as reptiles, amphibians, and mammals.

Several hundred years ago, scientists knew very little about the circulatory system in humans. They knew that a muscular **heart** pumps blood by contraction and that **arteries** and **veins** from the heart serve as a conducting system for blood movement (fig. 10.2). But they did not know how blood flows through this system. Some proposed that blood surges back and forth in the blood vessels, moving in two directions. Others argued that blood flows from the heart and is consumed when it reaches distant parts of the body. They believed new blood is constantly being formed from food material absorbed from the intestines.

In the early 1600s, English physician and anatomist William Harvey tested these hypotheses with a number of experiments. Harvey observed that the heart and many veins contain **valves** that control the movement of blood. When he pushed on a vein, he found that blood was pushed in only one direction (fig. 10.3).

In another experiment, Harvey attempted to force small rods through veins. He could easily push the rod through in one direction but was impeded by the valves when he pushed in the other direction. He argued that his results contradicted the hypothesis that blood surges back and forth in the blood vessels. Instead, blood must flow in only one direction, and the direction of blood flow is controlled by the valves.

In William Harvey's experiments, what were his observations? What were his interpretations?

*Physiology of Plants and Animals*

FIGURE 10.2    Some of the major components of the human cardiovascular system.

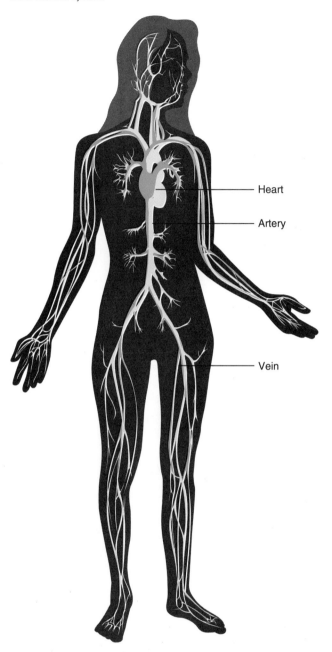

- Heart
- Artery
- Vein

FIGURE 10.3    Valves present in veins. (a) When a vein in the arm is pressed to force blood toward the hand, an expanded area is observed on the vein where blood is unable to pass. (b) Opening the vein at these expanded locations reveals valves that allow only one-way blood flow.

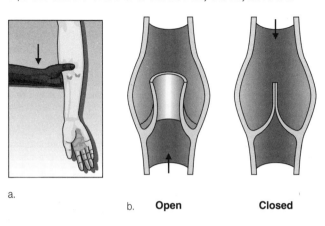

a.

b.   **Open**          **Closed**

FIGURE 10.4    Capillaries. Arteries and veins are connected by capillaries, blood vessels very small in diameter with thin walls. (a) A branched bed of capillaries connects arteries and veins. (b) An expanded view of a capillary shows the single-cell thickness of the wall. A red blood cell (shown for comparison) can just squeeze through the capillary.

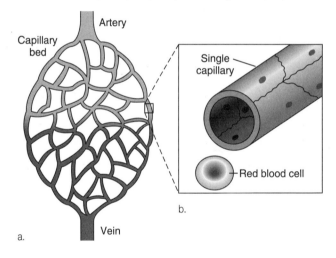

Artery

Capillary bed

Single capillary

Red blood cell

b.

a.    Vein

Harvey was also skeptical of the hypothesis that blood was constantly consumed by the body. He measured the blood pumped out of the heart in a single beat and estimated that over 80 pounds of blood are pumped to the body every half hour! This amount of material could not possibly be replenished by the food and water humans consume. His measurements were clearly inconsistent with this hypothesis.

How, then, does blood flow in the human body? Harvey proposed that there is some kind of connection between the arteries and veins. Rather than being used up or moving in both directions, blood flows in only one direction from the heart, through the arteries, through the unobserved connecting vessels, and back through veins to the heart. Harvey looked for these connecting vessels but could not find them. Many years later, when improved microscopes were developed, biologists discovered very small blood vessels called **capillaries** connecting the arteries and veins (fig. 10.4).

Should Harvey's proposed capillaries have been disregarded because they could not be seen?

**FIGURE 10.5** Human circulation. Blood moves through the four-chambered heart and circulates to the lungs (pulmonary circulation) and body (systemic circulation) as indicated by the arrows.

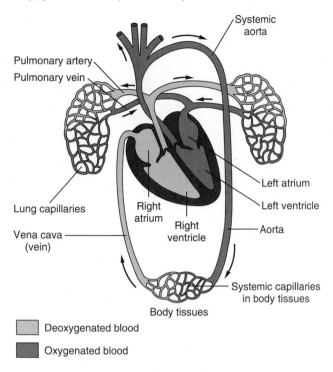

Deoxygenated blood

Oxygenated blood

**FIGURE 10.6** Blood vessel structure. The walls of arteries and veins are thick and include connective tissue and muscle. Capillaries have walls that are only one cell thick.

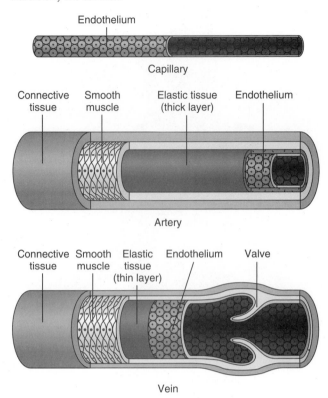

Extensive observations of the human circulatory system over the past century have revealed the structure diagrammed in figure 10.5. Blood returning to the heart from the body first enters the **right atrium** and moves into the **right ventricle.** This blood is low in oxygen and high in carbon dioxide. When the right ventricle contracts, the blood is forced into the **pulmonary artery** and carried to the lungs. The blood passes through capillaries around the alveoli where oxygen diffuses into the blood and carbon dioxide diffuses into the alveoli. Oxygen-rich blood returns to the heart via the **pulmonary vein** and enters the **left atrium.** After moving into the **left ventricle,** blood is forced by contraction into the **aorta,** the large artery leaving the heart, and on to the rest of the body. Repeated branching of the arteries leads to smaller arteries and then to capillaries. After passing through the capillaries, blood enters the veins and returns to the heart.

The capillaries are the location where oxygen, carbon dioxide, and other materials are exchanged with cells. To gain an idea of the extent of the capillary system, consider that no cell in the human body is farther than 130 micrometers (μm) from a capillary. Material can diffuse rapidly over this short distance. It is estimated that in muscle tissue, for example, every square centimeter of cross section exposes 60,000 capillaries!

Arteries, veins, and capillaries vary considerably in structure, as you can see in figure 10.6. The walls of capillaries are only one cell thick, allowing the shortest diffusion

distance between blood and body cells. The aorta leading from the heart is about 2.5 centimeters in diameter, whereas capillaries are only 10 μm in diameter. There are so many capillaries, however, that if their total cross-sectional area was added together it would total over 4,000 square centimeters (fig. 10.7).

In our discussion of open circulatory systems we noted that blood flow in such systems is rather sluggish. Blood flow in closed circulatory systems is much more efficient because blood is being channeled through the arteries and veins. In the closed system, blood reaches organs and tissues at a much more rapid rate, providing a greater supply of oxygen and nutrients. The velocity of blood flow decreases significantly in the capillaries, however, because the total cross-sectional area increases so dramatically.

Lower blood flow velocity in the capillaries increases the amount of exchange between the blood and body cells. Why do you think this is so?

An important aspect of the circulatory system is **blood pressure.** When the heart contracts, it forces blood through the vessels and back again. Figure 10.8 shows measurements

*Physiology of Plants and Animals*

FIGURE 10.7 Blood flow velocity. Flow velocity is very low in capillaries and much higher in arteries and veins (solid line). Differences in the cross-sectional area (dashed line) can account for these changes in velocity.

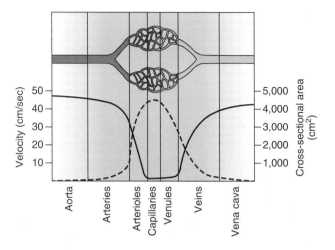

FIGURE 10.8 Blood pressure. In the arteries, blood pressure fluctuates between systolic and diastolic pressure. Pressure drops drastically in the capillaries due to resistance to flow and remains low in the veins.

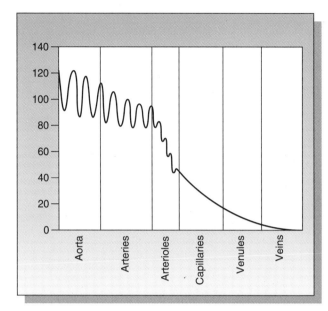

of blood pressure at different points along the circulatory system. Notice that the pressure drops as blood moves from the arteries through the capillaries to the veins. Why do you think this is so?

As the arteries branch into smaller and smaller vessels, the diameter of these vessels decreases dramatically. The resistance to blood flow through a small vessel is much greater than that through a large vessel. This resistance impedes the flow of blood and builds up pressure in the arteries. At the same time, the cross-sectional area of the vessels is increasing and resulting in a decrease in pressure.

You may notice in figure 10.8 that the blood pressure measurements in arteries fluctuate dramatically, as depicted by the wavy line in the first half of the graph. This fluctuation is caused by the pumping action of the heart. When the heart contracts, the blood is forced more rapidly through the arteries and the pressure rises. This is called **systolic pressure.** When the heart relaxes, the force decreases. This is called **diastolic pressure.** Physicians can measure both systolic and diastolic pressure to determine the health of an individual's circulatory system (Thinking in Depth 10.1).

# THE CIRCULATORY SYSTEM IN OTHER ANIMALS

Most animals rely on a closed circulatory system to transport nutrients to cells in their bodies, but not all circulatory systems rely on the same type of heart. The human heart has four chambers: the left atrium and ventricle and the right atrium and ventricle. Other animals have two- or three-chambered hearts. The circulatory systems of fish, amphibians, and birds and mammals are compared in figure 10.9.

Fish have only two heart chambers. They must pump blood through two capillary systems before it is returned to the heart. Since blood pressure decreases most when blood moves through smaller vessels, blood flow is slow in this system.

In amphibians such as frogs and salamanders, the heart has three chambers. Although blood is pumped separately through the lungs and capillaries, blood from the lungs and deoxygenated blood from the body are mixed before being pumped to the cells. As a result, blood circulating to the cells has less oxygen than blood returning from the lungs.

Why do humans and other mammals have a four-chambered heart? We can see the advantage after examining the two- and three-chambered hearts of fish and amphibians. In the four-chambered heart, blood is pumped through only one capillary system after it is oxygenated by the lungs, so more oxygen can be delivered to cells at a faster rate.

# BLOOD

What is blood? We know that it transports gases and nutrients to cells in the body, but how does it carry these materials? Why, for example, don't organisms simply use water to transport nutrients?

Simple observation and analysis of blood has shown that it consists of single cells suspended in a fluid. Red blood cells are the most common type. The fluid also contains a variety of ions and organic molecules in solution.

## THINKING IN DEPTH 10.1

# BLOOD PRESSURE

**FIGURE 1** Blood pressure measurements. (*a*) Under normal conditions, blood pressure keeps arteries expanded. (*b*) Pressure from the sphygmomanometer closes the artery. (*c*) Blood is just forced through the artery, at pressures between systolic and diastolic. (*d*) The artery remains open, at pressures less than diastolic pressure.

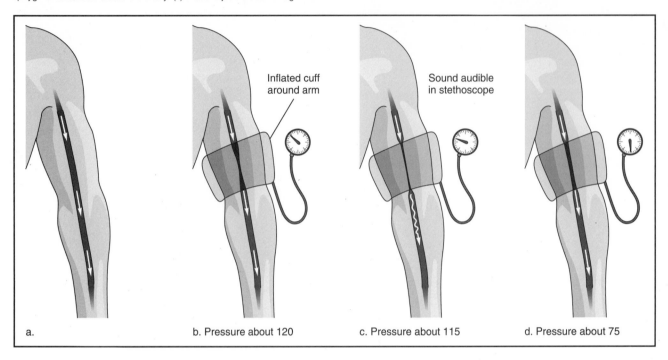

a.               Inflated cuff around arm     Sound audible in stethoscope

b. Pressure about 120    c. Pressure about 115    d. Pressure about 75

The steady pumping of the heart muscle moves blood through arteries, capillaries, and veins. As blood is forced through these remarkably small vessels, it creates pressure. A certain amount of pressure is necessary to keep the blood circulating, but if blood pressure rises too high, it may cause strokes or other injuries.

A blood pressure check is one of the mainstays of a physical examination. The test measures both systolic and diastolic pressure to gain an overall picture of the health of a person's circulatory system.

Measurements are taken with a **sphygmomanometer,** an inflatable cuff that fits around the upper arm (fig. 1). The cuff is first inflated to compress the artery so that blood cannot be forced through even at maximum blood pressure. As pressure in the cuff is gradually released, the systolic pressure forces some blood through the artery. These short spurts of blood can be detected by listening to the sound of blood moving in the artery.

As pressure on the cuff is further released, diastolic pressure keeps the artery open. At this point the sound of blood spurting through the constricted artery can no longer be detected.

The pressure in the artery is usually measured with a gauge attached to the cuff and is recorded as systolic pressure over diastolic pressure. Normal blood pressure is about 120/80. Pressures of 140/90 are considered high, and higher pressures can be dangerous. Pressures this high can burst small blood vessels, causing a stroke, or damage the glomerulus in the kidney. If high blood pressure is detected early, appropriate treatment can prevent such problems.

If blood pressure were measured at the wrist instead of the upper arm, what would you expect to observe?

---

We mentioned in chapter 7 that the oxygen content of water is relatively low. If the circulatory system carried oxygen in water, far too little oxygen would reach the cells. Blood, on the other hand, contains **hemoglobin,** which is localized in red blood cells. Hemoglobin is a red, iron-containing protein that readily combines with oxygen. As figure 10.10 shows, oxygen in the lungs diffuses into capillaries around the alveoli. This oxygen combines with hemoglobin (Hb) in the red blood cells to form oxyhemoglobin ($HbO_2$). When this blood is pumped to the body cells where

FIGURE 10.9 Comparative circulatory systems. Differences are observed in the number of heart chambers and in the pathway of blood through the system.

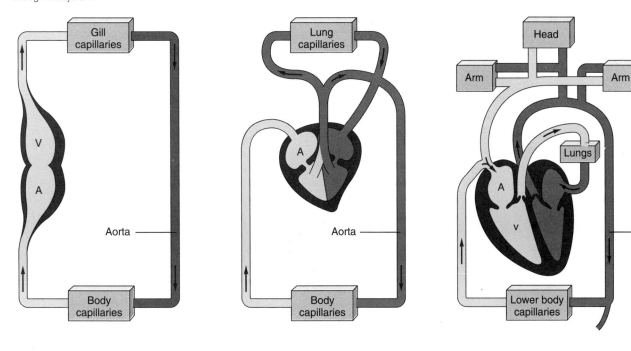

a. Fish  b. Amphibians  c. Birds and mammals

the oxygen concentration is much lower, the oxygen is released from the hemoglobin. Recall from chapter 4 that chemical reactions are in dynamic equilibrium. The speed of the reaction in each direction depends on the concentration of the reactants. Thus when oxygen concentration is low because respiring cells have utilized most free oxygen, the release of oxygen is favored. This released oxygen is then free to diffuse into cells where the oxygen content is low.

Hemoglobin transports oxygen as an absorbed substance that is not free to diffuse rather than as a gas dissolved in the liquid part of blood. Without hemoglobin, we would need about seventy-five times the amount of blood we now have for sufficient oxygen transport!

A student argued that if blood contained 10 percent more red blood cells, it could transport 10 percent more oxygen. What was this student assuming?

Hemoglobin also plays a role in controlling the pH of blood. Cells are highly sensitive to pH and will not survive major pH changes. Without the controlling effects of hemoglobin and other proteins, the pH in blood would change much more than it does. How does hemoglobin protect cells from changes in pH?

**FIGURE 10.10**  Oxygen transport. (a) Oxygen diffuses from the alveoli in the lung into capillaries and combines with hemoglobin in the red blood cells. (b) In the body tissues where oxygen is low, oxygen is released and diffuses to the cells.

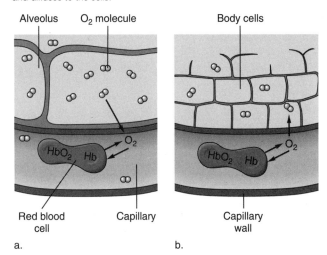

You know that carbon dioxide is released to the blood by respiration in the cells. When carbon dioxide ($CO_2$) dissolves in water in the bloodstream, carbonic acid ($H_2CO_3$) is formed. Carbonic acid dissociates to form hydrogen ions ($H^+$) and bicarbonate ions ($HCO_3^-$). The following equation shows both of these chemical reactions.

$$CO_2 + H_2O \rightleftharpoons H_2CO_3 \rightleftharpoons H^+ + HCO_3^-$$

**FIGURE 10.11** Conducting tissue in plants. (*a*) A cross section of vascular tissue in corn shows xylem and phloem cells. (*b*) A variety of hollow xylem cells from different plants are stacked end to end to form extended tubular structures. (*c*) Like xylem cells, phloem cells are stacked end to end with perforated plates at the junctions of cells. This structure allows a continuous pathway through the phloem.

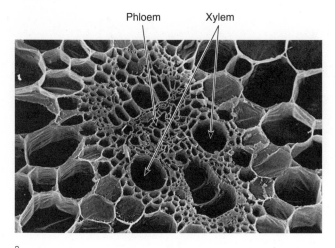

Phloem    Xylem

a.

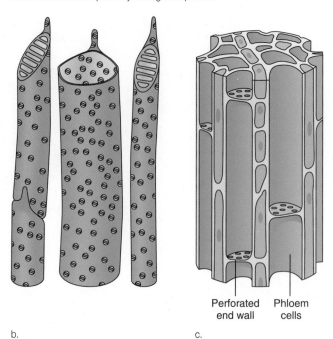

b.                Perforated    Phloem     c.
end wall    cells

The increase in hydrogen ion concentration means a decrease in pH. This is where hemoglobin steps in. Hemoglobin is normally present in blood as an ion, Hb⁻. Hb⁻ combines with H⁺ to form acid hemoglobin (HHb). Acid hemoglobin is a much weaker acid than carbonic acid. In other words, it will ionize much less than carbonic acid and release less H⁺. When carbon dioxide leaves the bloodstream and enters the lungs, the equation above reverses, or shifts to the left, and hydrogen ions in the blood decrease. The ability of blood to maintain a relatively constant pH balance is essential to life.

# TRANSPORT IN PLANTS

Like animals, plants must transport gases, nutrients, and water from cell to cell. But plants, as you may have guessed, do not have circulatory systems in the same sense that animals do. Plants do, however, transport materials through tubular conducting pathways. Not all plants have these pathways, but those that do, including all ferns and seed-bearing plants, are called **vascular plants.**

The conducting pathways in plants are not like the arteries and veins in animals in that they are not made up of many cells. Instead, these pathways are located *inside* elongated, cylindrical cells that are joined end to end (fig. 10.11). These cells form long chains that extend throughout the plant.

As figure 10.11*a* shows, the conducting cells in plants are of two types. **Xylem** cells transport water and some inorganic minerals. **Phloem** cells transport organic nutrients such as sugar and nitrogen compounds. Each of these tissues transports materials in different ways, so we will address them separately in the next sections.

## Xylem Transport

The movement of water in plants has long intrigued scientists. They have been particularly puzzled by how water moves to the top of very tall plants. Initially, biologists suggested that water might be pushed up by roots through the xylem, or pulled up from the top of the plant by some unknown mechanism. When they investigated the first option, researchers determined that many plants do not exert **root pressure.** Even when water was forced from the cut stump of some plants, tall plants did not exhibit nearly enough force to push water to the top.

Biologists have hypothesized that evaporation of water from leaves, a process called **transpiration,** "pulls" water through the xylem. As water molecules evaporate, other water molecules down the column move up to take their places. Water molecules tend to have a strong attraction to one another called **cohesion.** This cohesion allows the pull at the leaf to exert a force throughout the xylem without pulling apart the water column itself.

To test this hypothesis, scientists measured transpiration from the leaves and water uptake by the roots. They predicted that if pull on water in xylem cells was caused by evaporation, then water loss by transpiration would occur before water uptake by roots occurred. Remember that transpiration increases when stomates are open, and stomates are open during the day. The results of the experiment are shown in figure 10.12.

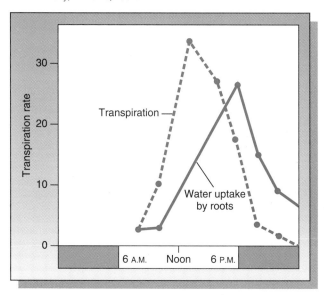

As you can see, transpiration increases rapidly in the early morning. Water uptake begins to increase a few hours later and continues to increase along with transpiration. A few hours after transpiration begins to decline, water uptake by the roots also declines. These results support the hypothesis that transpiration is causing the movement of water in the xylem.

In the experiment shown in figure 10.12, what results would you predict if the plant's stomates were kept closed during the day?

If root pressure were responsible for transporting water to plant leaves, what results would you predict in figure 10.12?

## *Phloem Transport*

In contrast to xylem transport, movement in the phloem occurs in both directions. Most of the time, sugars synthesized in the green leaves move downward in the phloem, but at other times transport occurs in the opposite direction. In the spring, for example, sap rises in trees. This sap is the upward movement of organic nutrients in the phloem. Since phloem transport can vary in direction and is not apparently related to water movement or transpiration, a different mechanism must be involved in transport in phloem.

FIGURE 10.13    The pressure-flow hypothesis of phloem transport. (a) A model system showing how increased sugar concentration at 1 causes water flow into the bulb and bulk flow through the system to 2. (b) The proposed operation of pressure-flow in the phloem tissue.

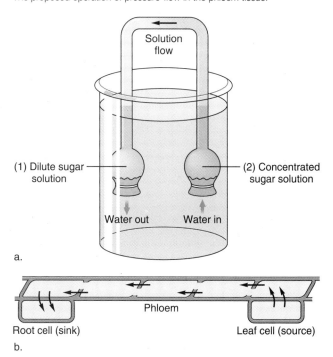

Biologists have hypothesized that phloem transport is caused by the flow of water into cells where sugar content is high. The model in figure 10.13 shows water moving by osmosis into a **source** cell that has a high sugar content. The entry of water increases the pressure in the source cell, forcing the material in the cell to flow toward parts of the system where the pressure and sugar content is lower. These places are called **sinks.** The effect is to transport sugar to other parts of the system. Figure 10.13*a* shows an artificial model of the system occurring in plants.

In plants, sugar content is high in the photosynthesizing cells, such as green leaf cells, because sugars are being produced there. The phloem transports the sugars to places in the plant where sugar concentration is low. This **pressure-flow** hypothesis accounts for the flow of nutrients to different parts of the plant and the speed of phloem transport. It is the most widely accepted explanation of phloem transport.

What kinds of experimental measurements would provide the strongest test of the pressure-flow hypothesis?

How would the pressure-flow hypothesis explain the rise of sap in trees before leaves are formed in the spring?

## Summary

*The circulatory system in animals is a marvel of complex function, delivering essential materials to every cell in the body and removing waste material. Without this system, multicellular organisms could not survive. In plants, the phloem and xylem transport systems move nutrients and water to cells throughout the plant, even to the tops of the tallest trees. These remarkable systems enable plants to produce the organic matter that other organisms rely on for food.*

1. Diffusion can account for adequate exchange of material between an organism's body and the external environment if the transport distance is sufficiently small.

2. In larger organisms, a **circulatory system** is necessary to transport nutrients, gases, and waste materials. Two major types of circulatory systems, open and closed, are found in animals.

3. An **open circulatory system** moves material and fluid in an organism's body but does not contain it in a conducting system.

4. A **closed circulatory system** completely encloses the circulating blood. Exchange between the body cells and blood occurs via diffusion across the blood vessel walls.

5. Circulatory systems are comprised of a muscular **heart, arteries** conducting blood away from the heart, and **veins** returning blood to the heart. **Capillaries,** blood vessels that are very small in diameter, connect the arteries and veins and are the site of exchange of oxygen and carbon dioxide with body cells.

6. The force exerted on the blood by the heart creates **blood pressure** that decreases with distance from the heart. The greatest decrease in blood pressure occurs at the smaller vessels.

7. **Hemoglobin** increases the oxygen-carrying capacity of the blood by absorbing oxygen at the lungs and releasing oxygen at other sites in the body. The pH of blood is regulated by the absorption and release of hydrogen ions.

8. In vascular plants, water and minerals are transported in **xylem** and nutrients are transported in **phloem.**

9. Water moves in the xylem in one direction, from the roots to the leaves. **Transpiration** from the leaves appears to create a pull on the water column in the xylem, pulling it upward.

10. Material moves in phloem in either direction, from a source, or high sugar concentration to a sink, or low sugar concentration.

## References

Schottelius, B. A., and D. D. Schottelius. 1973. *Textbook of Physiology* (St. Louis: C. V. Mosby).

Vander, A. J., J. G. Sherman, and D. S. Luciano. 1980. *Human Physiology: The Mechanisms of Body Function* (New York: McGraw-Hill).

## Exercises and Questions

1. Which of the following body forms would most favor transport of material by diffusion?
   a. a spherical body shape
   b. a wide, thin, ribbonlike form
   c. a cylindrical body with large diameter
   d. a square, dark-colored body form
   e. a short, fat body with numerous legs

2. If the ostia in an insect's circulatory system did not have valves, which of the following would describe the fluid flow (page 114)?
   a. Most fluid would be pumped toward the front of the insect.
   b. Much less fluid would be pumped by the circulatory system.
   c. Fluid would not reach the head of the insect.
   d. Most fluid would be pumped toward the rear of the insect.
   e. Equal amounts of fluid would flow in both directions from the heart.

3. If a dye were injected into an insect at the point shown in the diagram, which of the following would the dye next enter (page 114)?

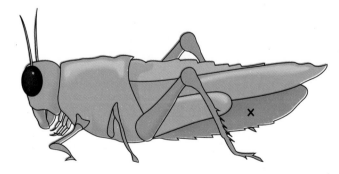

   a. ostia
   b. legs
   c. heart
   d. front part of the abdomen
   e. head

4. Which of the following best explains why you would not expect the body fluid of insects to contain hemoglobin?
   a. Movement of oxygen into the spiracles is controlled by valves.
   b. Insects have an open circulatory system.
   c. Carbon dioxide is produced when oxygen is utilized.
   d. Oxygen reaches cells by trachae.
   e. Insect hearts have no valves.

5. Which of the following is an interpretation made by William Harvey (page 114)?
   a. Valves are found in the hearts of many animals.
   b. Valves prevent blood from being forced in one direction through a vein, but not the other.
   c. New blood is constantly being formed from food.
   d. Valves in the veins permit blood to flow in one direction only.
   e. Many blood vessels contain valves.

6. Even though Harvey could not see his proposed capillaries, their existence was widely accepted. Which of the following is the best reason for this acceptance (page 115)?
   a. Capillaries connect arteries and veins and are the site where most oxygen is delivered to cells.
   b. Harvey was a careful observer and well known to scientists of his period.
   c. Movement of blood from arteries to veins was necessary to explain Harvey's observation of flow direction.
   d. Movement of blood in the veins and arteries is complex and was known even before Harvey's experiment.
   e. Harvey was the first scientist to propose the existence of capillaries.

7. Which of the following is the best reason lower flow velocity in the capillaries increases the amount of oxygen exchange between blood and body cells (page 116)?
   a. Slower flow provides a longer time for oxygen to be released and diffuse to cells.
   b. Little oxygen is lost by the blood in the arteries where the blood flow is high.
   c. Blood flow velocity is high in both the arteries and veins.
   d. The oxygen concentration is much lower near capillaries around body cells than near lung capillaries.
   e. Oxygen is combined with hemoglobin when carried in the blood.

8. If blood pressure were measured at the wrist instead of the upper arm, which of the following would you expect to observe (page 118)?
   a. The difference between systolic and diastolic pressure would be greater.
   b. Blood pressure would be lower.
   c. Systolic pressure would be higher.
   d. Blood pressure would be the same as in the upper arm.

9. Which of the following would characterize a two-chambered heart but not a three- or four-chambered heart?
   a. The heart pumps blood through a closed circulatory system.
   b. Valves in the heart control the movement of blood.
   c. Only deoxygenated blood passes through the heart.
   d. The heart pumps blood through capillaries before blood is returned to the heart.
   e. The heart pumps blood through an open circulatory system.

10. A student argued that if blood contained 10 percent more red blood cells, then it could transport 10 percent more oxygen. What was this student assuming (page 119)?
    a. Blood cells circulate faster if the heart beats faster.
    b. All blood cells carry the same amount of hemoglobin.
    c. Red blood cells are not the only cells in the blood.
    d. If the number of blood cells increases, the amount of oxygen that can be transported will increase.
    e. Oxygen is picked up by the blood in the lungs.

11. What observations would you predict in figure 10.12 if the stomates were prevented from opening (page 121)?
    a. Uptake would precede the increase in transpiration.
    b. Transpiration would increase but uptake would not.
    c. Transpiration and uptake would increase at the same time.
    d. Uptake would increase but transpiration would not.
    e. Transpiration and uptake would show little, if any, increase.

12. If root pressure pushed water to the leaves of a plant, what results would you predict in figure 10.12 (page 121)?
    a. Both transpiration and uptake would be increased.
    b. Transpiration would increase but uptake would not.
    c. Uptake would be delayed for a longer period after transpiration increases.
    d. Uptake would precede the increase in transpiration.
    e. Both transpiration and uptake would decrease.

13. Which of the following experimental measurements would provide the strongest test of the pressure-flow hypothesis (page 121)?
    a. the direction of fluid flow in the phloem
    b. the locations in the plant reached by phloem cells
    c. the sugar concentration in the phloem at the source and sink
    d. the sugar concentration in photosynthesizing cells
    e. the size of the proposed source and sink

14. According to the pressure-flow hypothesis, which of the following would be necessary for sap to rise in trees before leaves are formed in the spring (page 121)?
    a. high sugar use by the roots
    b. low sugar concentrations throughout the phloem
    c. low sugar concentration in the root phloem
    d. a high sugar concentration in the rising sap
    e. low sugar concentration in the twig phloem

CHAPTER
11

# MOVEMENT AND COMMUNICATION

FIGURE 11.1 Kymograph. An excised muscle is positioned in a holding device so that a lever arm is moved when the contracting muscle exerts a force on the arm. The moving lever arm traces a record of the force of contractions on the moving kymograph drum.

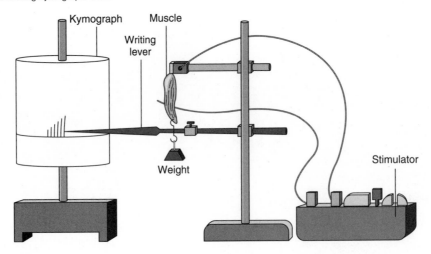

When you raise your arm, sit in a chair, or swallow food, muscles in your body contract or stretch to make these movements possible. Nerves send messages to the muscles from the brain and control muscle action. We tend to take our muscles and nerves for granted, but without them we would have no ability to move.

How do our muscles do these things? What causes them to contract and stretch? How does our brain communicate to our muscles the movements they should make? To answer these questions, we will examine the whole muscle, individual muscle cells, and the molecular processes that occur in muscle and nerve cells.

## WHOLE MUSCLE CONTRACTION

We can study an entire muscle by removing it from the body and attaching its ends to a **kymograph,** a machine that measures muscle contraction (fig. 11.1). When we stimulate the muscle electrically, it contracts, and the kymograph records the movement on a piece of tracing paper.

What effect does the electrical stimulus have on the muscle? Figure 11.2 shows the results of an experiment designed to find out what happens to the muscle when the electrical stimulus is steadily increased. At low voltages, the muscle does not contract at all. When we increase the voltage, the muscle contracts. As we continue increasing the voltage, the force exerted by the muscle increases until it reaches a maximum level. We call this maximum muscle contraction **tetanus.** We can increase the electrical stimulus beyond this point, but the muscle will not contract any further.

The muscle's reaction to the increasing stimulus is called a **graded response.** This response is important because it allows the force of contraction to vary. If our muscles could not contract with varying degrees of force, all our actions would always be the same. We would have much less control over our muscles than we do.

FIGURE 11.2 Muscle contraction with increasing stimulus. Kymograph tracings show that at low-stimulus strengths (0.2 and 0.4 volts), muscle does not respond. As stimulus strength increases, contraction strength increases until tetany is reached.

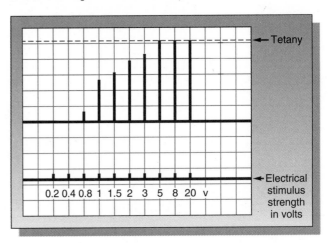

In another experiment designed to examine muscle contraction, electrical stimuli were applied to the muscle at the same intensity but at widely spaced intervals. The stimuli caused the muscle to contract and relax repeatedly, forming a **twitch response** (fig. 11.3a). When the intervals between stimuli were reduced, the twitch responses accumulated and increased the strength of the contraction. This **summation** of twitch responses eventually resulted in a maximum contraction force, or tetany (fig. 11.3b).

## SINGLE CELL CONTRACTION

Muscles are made up of single cells, so we naturally might expect muscle cells to react to stimuli the same way whole muscles do. In fact, we observe a very different response when we stimulate a single muscle cell with electricity. As figure 11.4

*Physiology of Plants and Animals*

**FIGURE 11.3** Muscle contraction with constant stimulus strength. A single stimulus will cause a twitch response in the muscle. If the time between stimuli is decreased so that the twitch does not have time to recover, summation will occur until tetany is reached. *Source: Data from S. Cooper and J. C. Eccles,* J. Physiol., *69:377, 1930.*

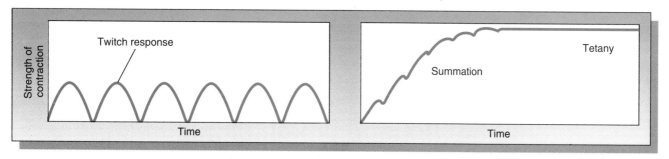

**FIGURE 11.4** Contraction of single muscle fiber. At low stimulus strengths (0.4 volts or less), the fiber does not contract. Once the fiber contracts, however, the strength of the contraction is always the same as shown by the twitch responses at 0.45 volts and higher.

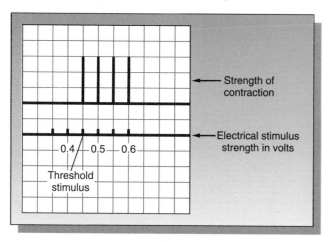

**FIGURE 11.5** Cross section of muscle fiber. The interior of muscle cells appears as a regular pattern of dark dots.

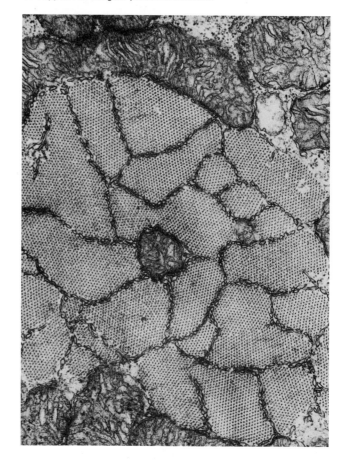

shows, the muscle cell does not react at all until a certain level, called the **threshold,** is reached. Once this level is reached, the cell exerts the same force in response to all stimuli. We call such a reaction an **all-or-none** response.

When you compare the contractions of a whole muscle (fig. 11.2) and a single muscle cell (fig. 11.4), what can you conclude about the contraction of single cells in a whole muscle?

How is it possible for muscles to have a graded response to stimuli when the cells that make up those muscles have an all-or-none response? Perhaps a closer look at muscle cells will provide an answer.

Muscle cells are long and narrow. If we cut a cross section through a muscle cell and examine it under a microscope, we observe a regular pattern of dark dots (fig. 11.5). If we cut the same cell down its length, or longitudinally, we observe a series of light and dark lines (fig. 11.6). A logical interpretation of these observations is that the dots observed in the cross section are the end points of lines or filaments observed in the longitudinal section.

If our interpretation is correct, we can diagram the muscle cell as shown in figure 11.7. The basic unit of the fiber repeats, as illustrated in figure 11.7b. Each one of these units is bounded on both ends by Z-lines. Inside these lines are I-bands. Between the I-bands is the A-band, which has an H-zone in the middle (fig. 11.7c).

**FIGURE 11.6** Longitudinal section of muscle fibers. The interior of muscle appears as a series of light and dark lines.

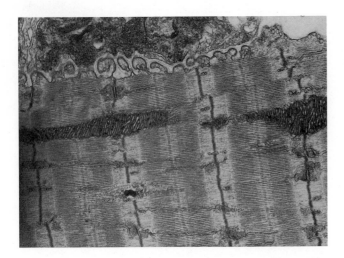

Biologists observing this pattern proposed that the bands were created by filaments that overlapped in some places but not in others. In other words, the darker A-band is a place where two filaments overlap, while the lighter I-bands are areas where only one filament shows.

To test this hypothesis, biologists made additional cross sections of the muscle fiber right through the bands. They predicted that a cross section of the A-band would reveal an overlap of filaments, while a cross section of the I-bands would show no overlap. Their cross sections are shown in figure 11.8.

As researchers expected, the A-band revealed overlapping filaments. They also observed that the A-band contained two kinds of filaments. The thicker filament appears to extend across the A-band, while the thinner filament extends only through the I-band. Their observations provide strong support for the hypothesis that muscle fibers are formed from overlapping filaments.

What results would you predict if the muscle fiber were cross-sectioned through the H-zone?

What is the purpose of muscle fibers having overlapping filaments? Scientists have hypothesized that these fibers are responsible for muscle contraction. When the muscle contracts or stretches, the thick and thin filaments slide past one another. This proposal is called the **sliding-filament hypothesis.**

If this hypothesis is true, we can make several predictions about how the bands will look when the muscle contracts or stretches. For example, if the muscle is stretched, the

**FIGURE 11.7** Interpretation of muscle fiber structure. (*a*) A section of muscle fiber. (*b*) and (*c*) The appearance of muscle fiber in a longitudinal section. Compare these to figure 11.6. (*d*) The proposed structure of filaments producing structure observed in (*c*).

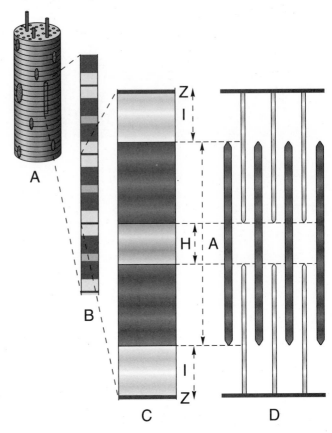

I-bands should increase in width, the A-band should remain the same, and the H-zone should increase in width. We can test these predictions by comparing muscle fibers under the microscope when the muscle is relaxed and when it is stretched. Figure 11.9 shows that our predicted changes do occur when the muscle is stretched, supporting our hypothesis.

What changes in the bands would you predict if the muscle contracted?

A student predicted that the muscle could not contract so far that the A-band would touch the Z-lines. What was this student assuming?

## THE CHEMICAL MAKEUP OF MUSCLE CELLS

We now know what happens inside muscles when they contract or stretch, but *how* do muscles contract? To begin answering this question, we will look at the chemical makeup of muscle cells. Analysis of muscle tissue reveals that it is primarily made up of protein, one of the biological

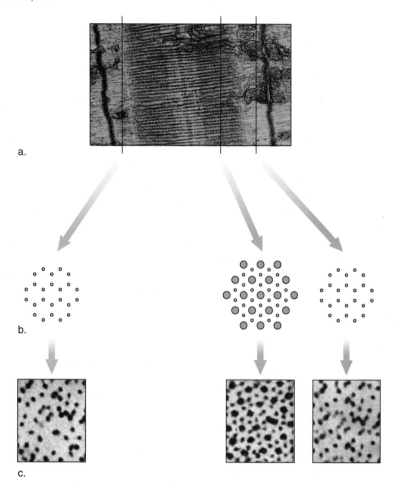

FIGURE 11.8    Cross sections of muscle fiber. (a) Longitudinal section of muscle. (b) Predicted cross-section view when sectioned through A-band and I-band. (c) Actual cross sections of A-band and I-band viewed with electron microscope.

a.

b.

c.

molecules we examined in chapter 3. About 90 percent of the protein can be separated into three different kinds: myosin, actin, and tropomyosin. Given the different appearances of the thick and thin filaments in muscle cells, it is logical to propose that thick filaments are made of one kind of protein and thin filaments of another.

We can test this possibility by first treating muscle cells with a salt solution that dissolves myosin. When we examine these cells, we find that the A-band is no longer visible, indicating that the thick filaments have been dissolved and

thus are composed of myosin. We can test muscle cells again with a solution that dissolves actin. When we examine these cells, we find that most of the thin filaments are removed. The third protein, tropomyosin, also appears to be associated with the thin filaments.

What hypothesis are we testing in these experiments on the proteins in muscle cells?

**FIGURE 11.9**  Structure of stretched muscle. (*a*) Muscle structure when relaxed compared to (*b*) structure when stretched. Notice increase in distance between Z-lines in (*b*) and increase in I-bands. There is no change in A-band.

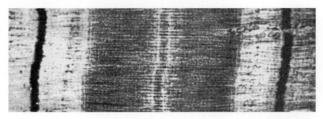

a.

b.

Recall the discussion of cellular respiration in chapter 6. In that chapter, we learned that a molecule called ATP (adenosine triphosphate) carries energy to cellular reactions. Knowing this, we can hypothesize that ATP provides the energy that makes muscle fibers contract. To test our hypothesis, we remove the membrane of muscle cells with a lipid-dissolving chemical. When we expose these membraneless cells to ATP, they contract. We conclude that ATP does indeed provide the energy to cause muscle contraction.

What would be a good control for this experiment? If there is no control, how should we interpret the results?

## MUSCLE TYPES

In our discussion thus far we have been referring to a particular type of muscle tissue called **skeletal muscle,** which affects movement of the limbs. Two other types of muscle tissue make up the bodies of vertebrates (fig. 11.10). **Cardiac**

**FIGURE 11.10**  Muscle cell types. (*a*) Smooth muscle, (*b*) cardiac muscle, and (*c*) skeletal muscle.

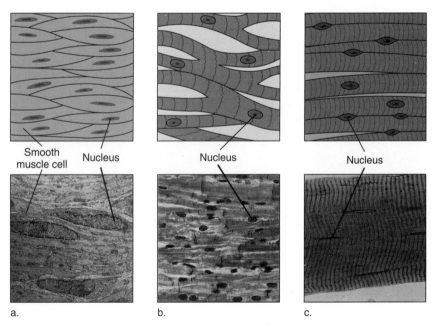

Smooth muscle cell    Nucleus

Nucleus

Nucleus

a.

b.

c.

FIGURE 11.11 Nerve structure. (a) Nerve cell typical of brain tissues with extensive branching of dendrites. (b) Motor neuron linking central nervous system to muscle tissue.

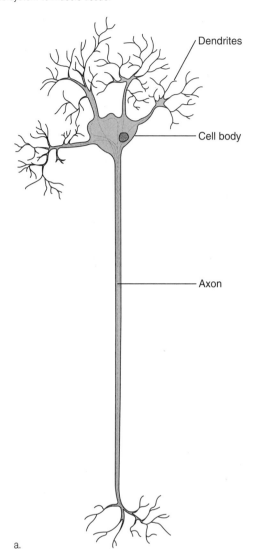

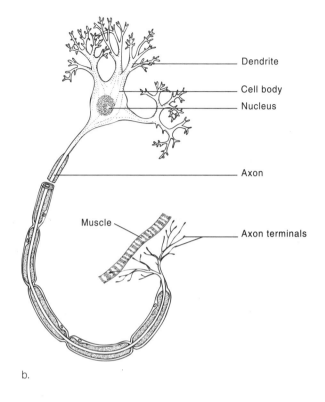

muscle is the tissue composing the heart. These branched and interlocked cells provide the pumping motion that circulates blood throughout the body (chapter 10). **Smooth muscle** is found in most of the internal organs, such as the intestine, bladder, and blood vessels. The cells in these muscles are long and spindle-shaped. The mechanism of contraction appears to be the same in all three muscle types.

Smooth muscles and cardiac muscle function automatically, without conscious decision on our part. That is, you do not have to tell your heart to beat or your stomach to digest food. All muscle, however, must be stimulated in some way. Recall the experiments at the beginning of this chapter in which electrical stimuli were applied to muscles. What stimulates muscles in our bodies? The answer to that question lies in the nervous system.

## THE NERVOUS SYSTEM

Nerves are cells in the body that transport messages from the brain to other cells. Without nerves, our bodies would not function. Two typical nerve cells are shown in figure 11.11. Their characteristically elongated shape is caused by the **axon** protruding from the body of the cell. The cell body contains the nucleus and most organelles, and the short

## MICROELECTRODES

Modern technology enables biologists to measure the electrical impulses in a single cell with a microeletrode (fig. 1). The microelectrode is a hollow glass tube pulled to a very fine, sharp point that can be inserted into a cell with minimal effect on the cell membrane. The tube is filled with an electrical conductor, such as a salt solution.

The microelectrode and a reference electrode are connected to a voltmeter and the cell is surrounded by a bathing solution. When both electrodes are in the bathing solution, the electrical potential reading on the meter is zero.

When the microelectrode is inserted into the cell (fig. 1b), the electrical potential reading immediately shifts to negative values. For a nerve cell, this resting potential is about -65 millivolts and will usually remain stable unless the cell is stimulated. If an action potential moves along the nerve or muscle fiber, the electrical potential will quickly change to positive and then return to the resting potential (see fig. 11.12).

Almost all plant and animal cells exhibit a negative resting potential. Only a very few cells, such as nerve and muscle cells and some large algae cells, have excitable membranes and thus the ability to generate an action potential.

**FIGURE 1**  Microelectrodes. (a) An experimental apparatus showing a microelectrode and a meter used to measure electrical potential across a membrane. (b) When inserted into a cell, a negative electrical potential is measured.

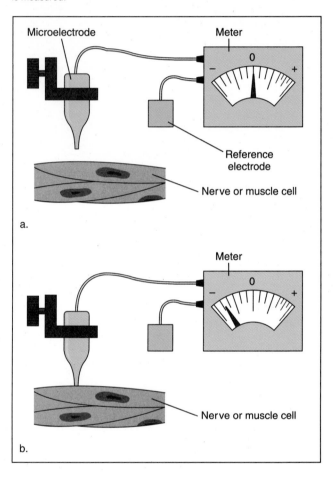

dendrites pick up signals from other nerve cells or sense organs. The branched end of the axon attaches to muscle fibers at the **neuro-muscular junction.** When the nerve cell is stimulated, the muscle fiber in contact with it contracts. But how does this stimulation occur?

We know that muscle cells are stimulated by electrical energy, thus it is logical to predict that the stimulation transmitted by nerve cells is electrical in nature. Investigations of nerve and muscle cells have shown that these cells maintain an electrical potential difference between the cell interior and exterior similar to the electrical potential difference between two poles of a battery. This electrical potential changes when the nerve and muscle are stimulated.

Small electrodes can be inserted into the cells to measure the potential (Thinking in Depth 11.1), just as the potential across the poles of a battery can be measured with a meter. When cells are unstimulated, the **resting potential** is approximately -65 millivolts (mv). When they are stimulated, the electrical potential quickly decreases, or **depolarizes,** to +40 mv and then recovers to the resting potential, all within a few milliseconds (fig. 11.12). This **action potential** passes along the nerve cell as a wave of depolarization and apparently stimulates the muscle to contract. A similar action potential is observed in muscle cells just before contraction.

What causes the action potential? How is it generated? After the action potential was discovered, scientists proposed conflicting hypotheses to explain how it works. One hypothesis

FIGURE 11.12 Action potential. At any one point on a nerve or muscle fiber, the action potential will appear as a "spike" of depolarization changing from the negative resting potential to positive values and back to the resting potential.

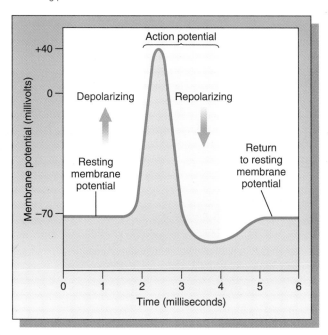

FIGURE 11.13 Axon preparation. (a) Cytoplasm can be squeezed out of an axon using a grooved roller. (b) The axon is filled with a salt solution similar to that found in axoplasm.

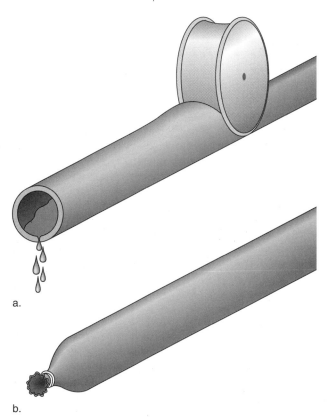

a.

b.

proposed that the observed electrical changes represent some special property of the nerve or muscle cell cytoplasm. A second hypothesis proposed that the electrical changes occur in the cell membrane and that the cytoplasm serves no function in generating an action potential.

Researchers designed an experiment to test these hypotheses. First, they squeezed the cytoplasm out of an axon, much like you would squeeze toothpaste out of a tube (fig. 11.13). This left a long, hollow cylinder bounded by membrane. Then they filled the cylinder with a salt solution of the same concentration as the cytoplasm in the axon. After they tied the ends of the cell to seal it, they inserted microelectrodes and measured the electrical potential. The prepared axon showed a normal resting potential. When it was stimulated, the action potential was identical to that in an untreated axon. The scientists concluded that the cell membrane generates and controls the action potential.

If the treated axon had not produced an action potential, would this have contradicted the second hypothesis?

In chapter 5 we mentioned that the concentration of ions can be very different inside a cell than outside. In particular, sodium ion ($Na^+$) concentration is low inside cells and potassium ion ($K^+$) concentration is high. Outside the cell, $Na^+$ is high and $K^+$ is low, as table 11.1 suggests. This ion imbalance is maintained by active transport of the $Na^+$ out of the cell (chapter 5).

**TABLE 11.1   Ion Concentrations in Nerve and Muscle Tissue (MM/1)**

| Ion | Frog Muscle | | Squid Axon | |
|-----|-------------|----------|-------------|----------|
| | Internal | External | Internal | External |
| $Na^+$ | 9.2 | 120.0 | 50 | 460 |
| $K^+$ | 140.0 | 2.5 | 400 | 10 |
| $Cl^-$ | 3.5 | 120.0 | 40 to 100 | 540 |

*Data from* Nerve, Muscle, & Synapse, *by Bernard Katz (McGraw-Hill, 1966).*

The details of how this process generates the resting potential are somewhat complicated and will not be discussed at length. It is important to remember, however, that the resting potential is maintained and is essential for the generation of the action potential.

The resting potential has a major effect on the cellular concentration of charged ions. The negative internal potential attracts positive ions and repels negative ones. Even though $K^+$ appears to be far from equilibrium due to the large difference between intracellular and extracellular concentrations, the negative resting potential attracts $K^+$ and maintains a high cellular concentration at equilibrium as shown in figure 11.14.

Internal $Na^+$, on the other hand, is maintained at a low concentration by active transport out of the cell. Low $Na^+$

FIGURE 11.14 Equilibrium conditions for K⁺ ions. Influx and efflux are equal. Influx is high due to negative internal cellular potential, and efflux is high due to high cellular K⁺ concentration.

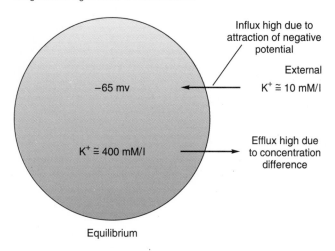

Equilibrium

FIGURE 11.15 Ion movements in action potential. Ion permeability changes and ion movements are shown during each phase of the action potential.

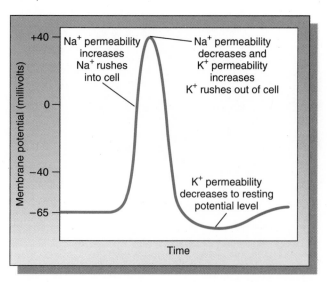

permeability of the cell membrane minimizes $Na^+$ influx even though both the electrical potential and concentration favor $Na^+$ influx. As a result, $Na^+$ is far from equilibrium even though influx and efflux are equal. This fact is critical for the generation of an action potential.

With this information on the state of $K^+$ and $Na^+$, we can investigate the generation of the action potential. The events during an action potential are illustrated in figure 11.15. The cell depolarizes because the $Na^+$ permeability of the cell membrane increases dramatically. As a result, external $Na^+$ rushes into the cell and neutralizes the internal negative potential. The membrane potential reaches +40 mv, the approximate equilibrium potential for $Na^+$ given the internal and external concentration. Just as quickly, the $Na^+$ permeability again decreases.

At this point, with a +40 mv potential, $K^+$ is far from equilibrium. It quickly moves out of the cell, restoring the -65 mv resting potential. These events—the rapid change in $Na^+$ permeability and the movements of $Na^+$ and $K^+$—will suffice in this text to explain the action potential changes.

What would happen to the action potential if there were no external Na⁺?

Membrane permeability to these ions is thought to be controlled by specific ion channels or pores in the membrane. When the channels are open, permeability is high; when they are closed, permeability is low. The ability to switch from a closed to an open state and back again would explain the rapid permeability changes of $Na^+$ in the action potential.

Interestingly, action potentials are relatively constant in size and shape. Reactions such as the amount of muscle con-

traction are not controlled by the *size* of an action potential. Instead, the *number* of action potentials apparently controls the intensity. In other words, the greater the number of action potentials, the greater the response.

## SYNAPSES

The nervous system is made up of many nerve cells that communicate with one another in order to pass messages between the brain and other cells in the body. Recall from figure 11.11 the branched projections, called **dendrites,** that extend from the body of a nerve cell. These dendrites make contact with numerous other nerve cells. The points at which axons make contact are called **synapses** (fig. 11.16).

When biologists first observed synapses with an electron microscope, they noticed a puzzling gap separating the synaptic knob from the post-synaptic membrane. They called it the **synaptic cleft.** How does electrical stimulation cross this gap? One hypothesis proposes that a chemical is released at the synaptic knob, diffuses across the cleft, and stimulates the post-synaptic membrane. Careful observation with an electron microscope reveals dense concentrations of **pre-synaptic vesicles** in the synaptic knob. These vesicles could contain a **transmitter substance** that is released into the cleft, diffuses to the post-synaptic membrane, and stimulates a permeability change. Such a change would alter the membrane potential of the nerve cell body.

Biologists have observed that when a nerve impulse reaches the synaptic knob, many of the internal vesicles fuse with the membrane at the cleft, indicating that the transmitter substance is released into the cleft. Researchers have also discovered enzymes at the post-synaptic membrane that destroy the transmitter substance after it interacts with the membrane. An important consequence of this type of

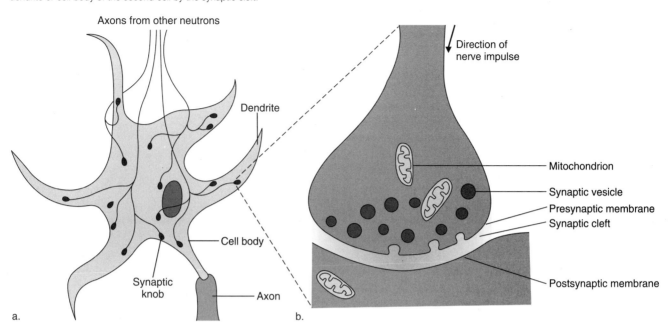

**FIGURE 11.16** Synapse. (a) Axons from several neurons may make contact with the cell body and dendrites of a single nerve cell. (b) The contact points are synapses where synaptic knobs are separated from the dendrite or cell body of the second cell by the synaptic cleft.

Axons from other neutrons

Dendrite

Cell body

Synaptic knob

Axon

a.

Direction of nerve impulse

Mitochondrion
Synaptic vesicle
Presynaptic membrane
Synaptic cleft

Postsynaptic membrane

b.

transmission, called **chemical transmission,** is that it occurs only one way. Stimuli cannot move from the post-synaptic membrane to the synaptic knob.

What would be the result if the transmitter substance were not destroyed by an enzyme?

The contact between nerve cells and muscle cells, called the **neuro-muscular junction,** functions the same way connections between nerve cells do. A transmitter substance is released from the nerve ends, diffuses to the muscle surface, and depolarizes the muscle cell membrane.

## SUMMATION AND DETECTION OF STIMULI

Microelectrode recordings at the post-synaptic cell body have shown that as stimuli are received at the dendrites, the cellular potential changes slightly with each stimulus. If enough stimuli are received in a short enough period of time, the cell will generate an action potential in the axon. Thus, transmission is not just the sending of one response for one stimulus. Instead, stimuli must combine to pass the message on to the next nerve cell. Moreover, some synapses can prevent or **inhibit** the excitation of the neurons. Whether the cells are inhibited or excited depends on the source of the stimuli. When the combination of incoming stimuli depolarizes the cell body to the threshold point, an action potential is generated and moves down the axon as shown in figure 11.15.

Incoming stimuli from several different sources can be combined or **integrated** and act together or in competition. The cell body, dendrites, and synapses are a way for the cell to "make decisions" about the transmission of stimuli.

Specialized nerve cells in the body are responsible for receiving information such as light, sound, heat, or touch from the external environment. These cells transmit this information to other nerve cells through a series of action potentials that are usually processed by synapses. In a fraction of a second, this information reaches the **spinal cord** and the **brain,** both of which are dense concentrations of nerve cells where far more complex decisions and responses are initiated.

In general, information processing in animals takes place through a vast number of nerve cell interconnections and synapses. While the organization of these cells is highly complicated, the basic events of transmission are the same for all cells in the system.

# Summary

*Communication and movement in animals depend on the highly specialized muscle and nerve cells found throughout the body. Together, these systems enable animals to move and think. Muscle and nerve cells function by a complex combination of molecular reactions. As you think about the biological processes discussed in this part of the book, keep in mind the following key points.*

1. The response of a whole muscle to a stimulus is called a **graded response.** The response of a single muscle cell is an all-or-none response.
2. The **threshold** is the minimum stimulus necessary to elicit a contraction response.
3. The contractile apparatus of muscle cells is comprised of thick and thin protein filaments. When the muscle cell contracts, the filaments slide past one another and shorten the cell.
4. ATP appears to provide the energy for contraction of the muscle cell.
5. Muscle tissue consists of three types.
   a. **Skeletal** muscle affects bodily movement.
   b. **Cardiac** muscle forms the heart.
   c. **Smooth** muscle is part of internal organs such as the intestines and blood vessels.
6. Nerve cells are generally comprised of a cell body, axon, and dendrites.
7. Nerve and muscle cells maintain a **resting potential** of about -65 mv across the cell membrane. When the cell is stimulated, the membrane **depolarizes** to about +40 mv and then quickly returns to the resting potential.
8. The rapid change in membrane potential moves along the nerve or muscle membrane as a wave of depolarization and is called the **action potential.**
9. Intracellular $Na^+$ is maintained at a low concentration by the active transport of $Na^+$ out of the cell. Internal $K^+$, however, is at a high concentration.
10. An increase in $Na^+$ permeability of the membrane causes depolarization. When $Na^+$ permeability decreases, $K^+$ permeability increases, and the resting potential is restored.
11. Connections between nerve cells occur at **synapses.** Stimuli are transmitted across the **synaptic cleft** by the release of a **transmitter substance** from **pre-synaptic vesicles.**
12. A consequence of chemical transmission across the synapse is that transmission occurs in only one direction.
13. Synaptic connections at the nerve cell body may either **excite** or **inhibit** the generation of an action potential. **Summation** of these stimuli will lead to the generation of an action potential or the suppression of stimuli transmission.

## References

Huxley, H. E. 1959. "The Contraction of Muscle." *Scientific American* 199, no. 5: 22.

Katz, B. 1966. *Nerve, Muscle, and Synapse* (New York: McGraw-Hill).

## Exercises and Questions

1. The contraction of a whole muscle is a graded response because the greater the stimulus strength, the greater the contraction. A similar response in nerve cells is achieved by:
   a. a greater movement of $Na^+$ out of the axon when the stimulus is stronger.
   b. producing a greater number of action potentials with stronger stimulus.
   c. greater depolarization of the axon when the stimulus is stronger.
   d. growing longer nerve cells to transmit stronger stimuli.
   e. cutting the axon so that an action potential can be transmitted partway on the axon.

2. The data shown in the graph below were obtained from a whole muscle stimulated with electric shocks of increasing strength.

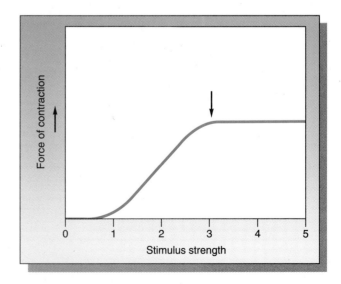

Three students (A, B, and C) observed these data and made the following statements.

   a. Different muscle cells in the muscle have different thresholds.

   b. All muscle cells in the muscle are contracting at stimulus strengths greater than the arrow (↓).

   c. The force of contraction increases gradually, reaching maximum at a stimulus of about 2.5.

Which of these statements is (are) the best interpretation(s) of the data? (See page 127.)

   a. A only

   b. B only

   c. C only

   d. A and B only

   e. A, B, and C

3. The tracings of muscle contractions in figure 11.2 (page 126) were made with a kymograph. The tracing shown below was also made with a kymograph.

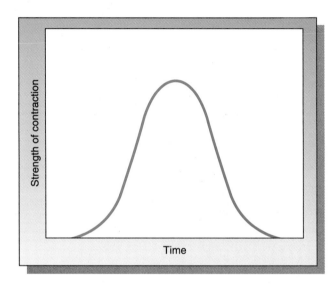

Why do the tracings appear different?

   a. In figure 11.2 the kymograph was turning more rapidly.

   b. In the figure above the kymograph was turning more rapidly.

   c. In the figure above the kymograph was not turning.

4. Why could it be argued that the diagram in figure 11.7 is an interpretation even though the diagram is intended to be an accurate description of muscle cell structure?

   a. The diagram shows skeletal muscle structure but not cardiac or smooth muscle.

   b. The diagram shows the uncontracted muscle, but not the contracted cell.

   c. The observations are limited to a part of the whole muscle.

   d. ATP is required to power the contraction of muscle.

   e. The sliding-filament model proposes that filaments slide past each other when contraction occurs.

5. Why would you expect the A-band to not change in width when the muscle contracts?

   a. The thick filaments do not change in length with contraction.

   b. Thin filaments are connected to the Z-line.

   c. The thin filaments usually do not touch when the cell contracts.

   d. ATP is required to power the contraction of muscle.

   e. The sliding-filament model proposes that filaments slide past each other when contraction occurs.

6. Based on figure 11.8, where would the section shown below be observed (page 128)?

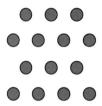

   a. cross section through Z-line

   b. longitudinal section through I-band

   c. cross section through the H-zone

   d. cross section through I-band

   e. longitudinal section through A-band

7. Which of the following would be a predicted change in muscle cell appearance when the cell contracts (page 128)?

   a. A-band decreases in width

   b. I-band increases in width

   c. A-band increases in width

   d. Z-line decreases in width

   e. H-zone decreases in width

8. A student argued that the muscle could not contract so far that the A-band would touch the Z-line. Which of the following was this student most likely assuming (page 128)?

   a. ATP is necessary to power muscle contraction.

   b. The muscle could not contract until the A-band touched the Z-line.

   c. The thin filaments could not slide past each other during contraction.

   d. The width of the I-band would decrease during contraction.

   e. Thick filaments are present in the A-band.

9. The experiment on page 129 describes the removal of specific muscle proteins from cells. Which of the following is most likely the hypothesis being tested with this experiment?
   a. The A-band contains thick filaments.
   b. Myosin and actin can be dissolved with salt solutions.
   c. ATP provides the energy for the contraction of muscle.
   d. Thick and thin filaments are composed of different proteins.
   e. Thin filaments are connected to the Z-line.

10. What would be an appropriate control in the experiment where the muscle cell membrane is removed and ATP is applied to the muscle (page 130)?
    a. Apply chemicals other than ATP to the muscle cell with the membrane removed.
    b. Fix the ends of the muscle fiber so that it could not contract when ATP was applied.
    c. Apply ATP to muscle cells with and without a cell membrane.
    d. Apply ATP to contracted muscle cells to see if ATP will cause relaxation.
    e. Observe the response of a muscle cell when nothing is applied.

11. In the experiment illustrated in figure 11.13, the axon cytoplasm was replaced with a salt solution. This axon was capable of generating an action potential. Suppose this axon did not generate an action potential. Would this result have disproved the hypothesis that the action potential originates at the cell membrane (page 133)?
    a. No, because a hypothesis cannot be disproven.
    b. Yes, because no action potential was generated without the cytoplasm present.
    c. No, because the treatment of the axon could have damaged the membrane.
    d. Yes, because the hypothesis must explain the function of the cytoplasm.
    e. No, because it was assumed that the membrane was the site where the action potential originates.

12. What would happen when the axon is stimulated if no external $Na^+$ was present (page 134)?
    a. Extra $K^+$ would rush into the axon.
    b. $Na^+$ would rush out of the axon.
    c. The resting potential would shift to zero.
    d. The axon would not depolarize.
    e. There would be no effect on the action potential.

13. Which of the following would most likely occur if the transmitter released from pre-synaptic vesicles was not destroyed by enzymes at the post-synaptic membrane (page 135)?
    a. Summation of stimuli could not occur.
    b. The post-synaptic membrane would continue to be stimulated.
    c. Stimuli could not be transmitted across the synaptic cleft.
    d. The stimulus across the synapse would be inhibiting.
    e. The synaptic cleft would be closed so that membranes would not be separated.

WRITING ACROSS THE DISCIPLINES

The exercises presented in this section provide an opportunity for you to explore relationships between biology and other academic disciplines such as law, literature, and psychology. There are no "right" answers to the questions posed in each exercise. Your response may be a short viewpoint or an in-depth research paper, depending on your teacher's instructions. References are provided with each exercise.

## Literature
### THE HEART

The human heart is a popular image in literature. Often, it is portrayed as having a dominant influence on the way an individual makes decisions or exercises judgment.

William Wordsworth wrote:

Thanks to the human heart by which we live,
Thanks to its tenderness, its joys, and fears,
To me the meanest flower that blows can give
Thoughts that do often lie too deep for tears.

A somewhat different view is expressed by Julie-Jeanne, Eleonore de Lespinasse:

The logic of the heart is absurd.

and by François de La Rochefoucauld:

The mind is always the dupe of the heart.

How would a biologist respond to these views of the heart?
Which of these quotes would a biologist most likely agree with?

## Medicine
### Organ Transplants

Success in transplanting human organs from one individual to another has grown impressively over the past forty years.[*] The kidney, lungs, liver, heart, and other organs can be transplanted with a reasonable expectation of favorable results. As success has grown, so also has the demand for organs to transplant. Unfortunately, this demand far outstrips the supply. For example, it is estimated that 6,000 people await a kidney transplant, but sufficient organs are not available from donors.

With such a shortage, who should be able to obtain organs? Should preference be given to the young who may have a longer and more productive life? Perhaps those who can most afford the high cost of a transplant should have first choice.

Cost is a major factor in transplants. A kidney transplant can cost $10,000, while a heart transplant can cost $150,000. Who should pay these expenses? Many insurance companies will not pay for transplant surgery, and the costs may be far beyond the capability of the recipient.

How should decisions be made about who will receive organ transplants? How should the costs of transplants be paid?

[*]Ricki Lewis, 1992, "Transplants—Past, Present, and Future," In *Readings in Biology to Accompany Life*. Ed. Ricki Lewis (Dubuque, Iowa: Wm. C. Brown), 170–74.

## History
### The Industrial Revolution

Why have countries in Europe and North America become conspicuously richer and more powerful than the rest of the world over the past 200 years?

Experts have proposed several factors as contributors to this success.[*] One is that the Western countries have made better organized efforts at scientific investigation and have applied the scientific method of hypothesis, experimentation, and replication more consistently.

A second factor is Western nations' ability to apply scientific discoveries to improve economic productivity. Even though Third World and Eastern European countries have access to scientific knowledge, they do not translate this knowledge into productivity.

Could governments encourage the translation of scientific knowledge into economic productivity?

[*]Nathan Rosenberg and L. E. Birdzell, Jr., 1990, "Science, Technology and the Western Miracle." *Scientific American* 263 (no.5): 42–54.

## History
### Food for Growth

Before the sixteenth century, the primary food in Europe was grains, such as wheat and oats. After the discovery of America, however, new food crops were introduced to the Old World that were to have a major impact. One of these crops was the potato, obtained from Native American farmers in the Andes.

The potato grew well in the cool, damp climates of Northern Europe. Moreover, an acre of land planted in potatoes produced almost twice the food energy of grain, required less work to grow, and matured twice as fast as grain. Since it is high in vitamins, potatoes and a little milk provided a complete and healthy diet.

Despite these advantages, the potato was slow to spread across Europe and was not widely adopted in Russia until the mid-1800s. When it was, the results were dramatic.

With the new calorie source and the new source of nutrition, the potato-fed armies of Frederick of Prussia and Catherine of Russia began pushing against their southern neighbors. During the Age of the Enlightenment these northern cultures wrestled free from the economic, cultural, and political domination of the south. Power shifted toward Germany and Britain and away from Spain and France, and finally all were eclipsed by Russia. Russia quickly became and remains the world's greatest producer of potatoes, and the Russians are among the world's greatest consumers of the potato. Their adoption of the potato as their staple food preceded their rise as a world power.[*]

How would population increase and better health lead to world power?

Is an aggressive, warlike diplomacy a necessary outcome of population increase?

[*]Jack Weatherford, 1988, *Indian Givers* (New York: Fawcett Columbine) 69–70.

PART
III

# REPRODUCTION AND INHERITANCE

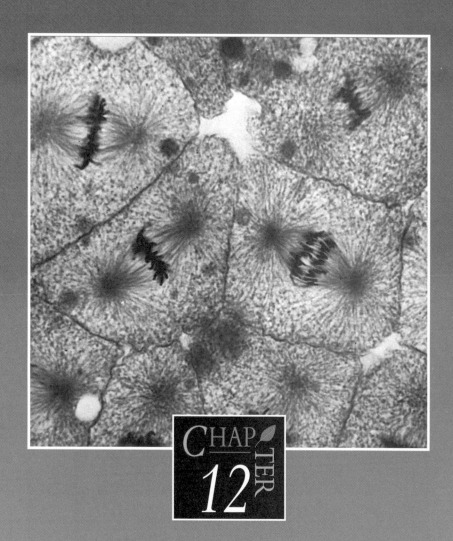

# CHAPTER 12

# CHROMOSOMES AND INHERITANCE

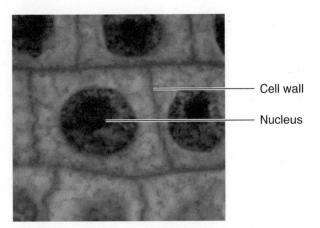

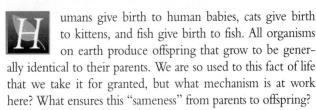

Cell wall

Nucleus

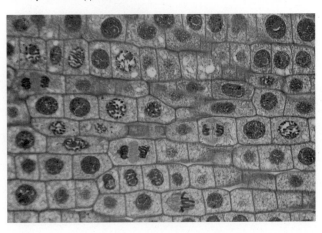

Humans give birth to human babies, cats give birth to kittens, and fish give birth to fish. All organisms on earth produce offspring that grow to be generally identical to their parents. We are so used to this fact of life that we take it for granted, but what mechanism is at work here? What ensures this "sameness" from parents to offspring?

Biologists know from years of investigation that organisms begin as one cell and grow to mature individuals by repeated cell divisions. Perhaps we can answer these questions by investigating dividing cells.

## CELL DIVISION

In most organisms, both plants and animals, cells in the body are continually dividing and forming new cells. If we observe these cells under a microscope, we see that most of them have the "typical" appearance described in chapter 2. Figure 12.1 shows a cell from an onion root tip that has this characteristic appearance. Note that the nucleus is well defined and appears to contain finely granular material. Biologists call this state of cell development **interphase.**

In other cells, we can observe distinct changes in the nucleus. Look at the cells in figure 12.2. Many of the nuclei in these cells appear to contain threadlike or rod-shaped structures. If we observe a single, typical growing and dividing cell for a length of time, we will see a sequence of changes in the nucleus that ultimately lead to the formation of two new cells. This sequence is called **mitosis,** and the steps are summarized in figure 12.3. Each stage in mitosis is shown as a photo and a diagram.

As we saw in figure 12.2, the nucleus material first begins to appear granular and condense into threadlike structures that gradually shorten. This first phase of mitosis

is called **prophase** (fig. 12.3a). In the next stage, called **early metaphase** (fig. 12.3b), the internal structures shorten even more. The resulting rodlike structures are called **chromosomes.** These are shown in more detail in figure 12.4. At this stage, the chromosomes have a characteristic X shape. The high magnification in figure 12.4 reveals the condensed structure of the chromosome, which appears to be formed from tightly coiled threads. The X shape is formed from two strands, called **chromatids,** connected at the **centromere.**

In the next stage of cell division, called **late metaphase** (fig. 12.3c), the chromosomes move to the center line of the cell. Fibers extend from each end of the cell and connect to the chromosomes. In **anaphase** (fig. 12.3d), the chromosomes appear to be pulled apart by the fibers so that the two chromatids from each chromosome move to opposite sides of the cell. In **telophase** (fig. 12.3e), after the chromosomes reach opposite sides of the cell, the fibers disperse, the chromosomes become indistinct, and two separate nuclei are formed.

The final stage of cell division occurs somewhat differently in plants and animals. In plant cells, once the two separate nuclei have formed, a new cell wall forms across the center of the cell. In animal cells, the membrane pinches at the midpoint. The final separation of the cell cytoplasm to form two new cells around the two nuclei is called **cytokinesis.** The two new cells enter interphase, the stage we observed in figure 12.1.

In mitosis, chromosomes are divided so that equal numbers of chromatids move to each new cell. For example, if the original cell contained 10 chromosomes, each of the new cells would contain 10 chromatids (now called chromosomes) immediately after division. Thus the new cells always contain the same number of chromosomes as the old cell.

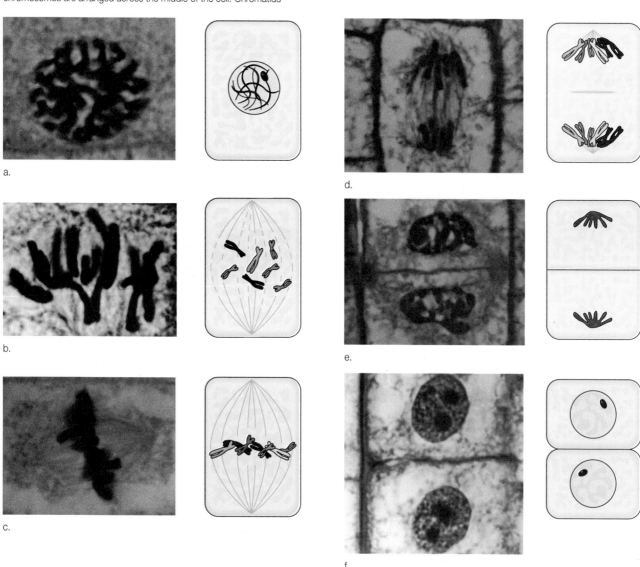

a.

b.

c.

d.

e.

f.

This constancy of chromosome number in cell division is observed in all organisms. Does this suggest that chromosomes carry the information that specifies what a cell or organism will become? Biologists have observed that all cells in an organism's body have the same chromosome number, except for a few exceptions that we will discuss later in this chapter. Perhaps this is another clue that chromosomes are responsible for how organisms develop.

## KARYOTYPES

Biologists make their studies of chromosomes easier by constructing **karyotypes,** photographic catalogues of the chromosomes in a cell (Thinking in Depth 12.1). When they compare karyotypes of many organisms they find that different species have different numbers of chromosomes, as well as variations in the sizes and shapes of their chromosomes. Humans, for example, have 46 chromosomes. Fruit flies have 8, and onions have 16. Other characteristics are apparent in the karyotype shown in Thinking in Depth 12.1. Each chromosome has a "mate" that is generally identical in size and shape. These "mates" or pairs are termed **homologous pairs.** One exception, however, is the **sex chromosomes.** In humans, for example, male karyotypes exhibit one homologous pair of different sizes, the X and Y chromosomes. Females have a pair of X chromosomes.

With the information we have, we can now return to our question—Do chromosomes carry directions specifying

## MAKING A KARYOTYPE

Among the many methods scientists have devised to investigate chromosomes and inheritance, the karyotype is one of the most useful. A karyotype is a photograph of all of a cell's chromosomes arranged in an orderly sequence. In theory, almost any cell could be used to make a karyotype, but in practice the white blood cells in animals and the root tip cells in plants are usually used.

The procedure we use to make a karyotype is illustrated in figure 1. The first step is to extract the cells and add colchicine, a poisonous alkaloid, to stop cell division at metaphase. Chromosomes at this phase are in their most condensed form, so we can detect variations in shape and size most easily.

In the cell, chromosomes appear to be arranged randomly. After they are photographed, however, we can cut out the photographs of each chromosome and arrange them by size and shape. When we do so, we find that each chromosome has a "mate" of similar shape and size.

Once we know the orderly sequence and arrangement of chromosomes in a normal individual, we can detect abnormalities in other individuals. Also, we can compare karyotypes of different species and discover differences and similarities in chromosome number, shape, and size.

FIGURE 1   Preparation of karyotype. Cells are extracted and cell division is inhibited at metaphase to reveal fully condensed chromosomes.

After photographing chromosomes, the individual chromosomes are cut out and arranged in sequence to show organization of homologous pairs.

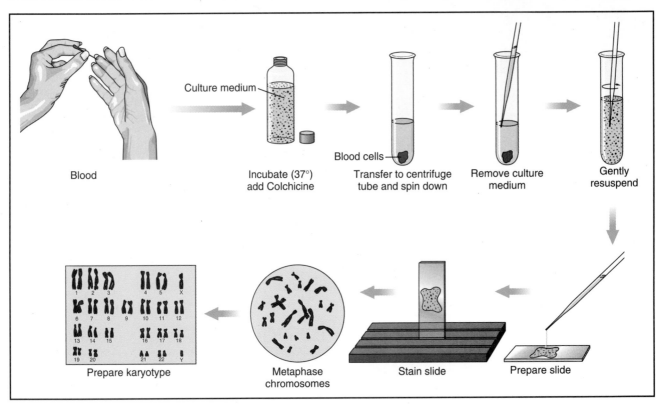

Blood

Incubate (37°)
add Colchicine

Transfer to centrifuge
tube and spin down

Remove culture
medium

Gently
resuspend

Prepare karyotype

Metaphase
chromosomes

Stain slide

Prepare slide

Culture medium

Blood cells

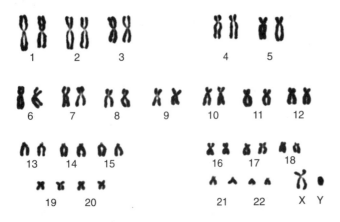

FIGURE 12.4 X-shaped chromosome. This structure of the chromosome is apparent in late prophase and metaphase. (*a*) Photomicrograph of chromosome in cell. (*b*) Diagram showing corresponding structures.

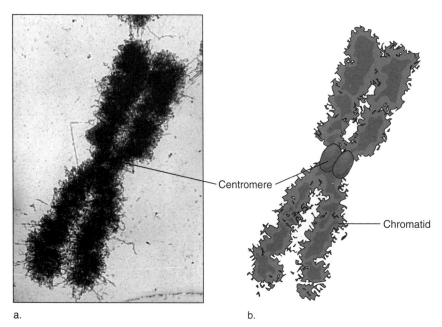

a.

b.

Suppose a cell with 14 chromosomes divides mitotically and one of the two new cells has 13 chromatids and the other 15 chromatids. At what stage in mitosis do you conclude a mistake occurred? If you observed this cell again and it had only 14 chromatids, what stage of mitosis would it be in?

what a cell or organism will become?—and make a prediction: If chromosomes direct what an organism will become, then changes in chromosomes should result in changes in the organism. Scientists have tested this prediction in a great variety of cases. Let's consider one example, shown in figure 12.5. The individual with this karyotype has three number 21 chromosomes. People who have this chromosome abnormality all exhibit the physical and physiological characteristics of Down's syndrome, including mental retardation and physical deformities.

In another example, shown in figure 12.6, the individual has three of each chromosome type. This is a serious chromosome abnormality called triploidy. Almost 1 percent of human conceptions are of this type, but more than 99 percent of these die before birth. Of those infants that do survive to birth, most die within a month. They display physical abnormalities such as an enlarged head, malformed mouth and eyes, and fusion of the fingers and toes.

In virtually every case where a chromosome or even part of a chromosome is added or deleted, plants and animals display physical or biochemical changes or both. The relation we predicted between observed chromosomal abnormality and changes in the organism is supported in every case that has been examined.

We made a prediction about the relationship between chromosome changes and changes in organisms. What hypothesis were we testing? What other interpretations could we make of the results that would contradict the conclusion that chromosomes carry information?

## FEMALE DOWN'S SYNDROME

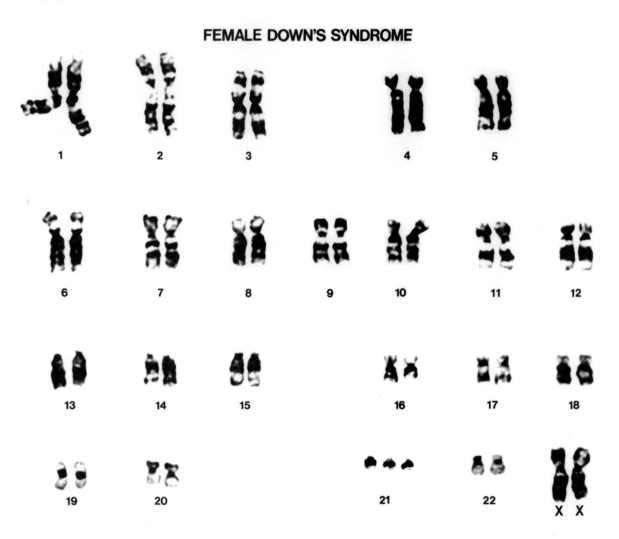

## TRANSMITTING INFORMATION

If chromosomes do carry information, how do they do it? Is something in the chromosomes themselves responsible? Analysis has indicated that about two-thirds of a chromosome is made up of protein and one-third of **deoxyribonucleic acid,** or **DNA.** Both protein and DNA are biological molecules (chapter 3). Does one of these molecules carry information, or is some other substance not found on the chromosome responsible?

To answer these questions, biologist Alfred Hershey and geneticist Martha Chase conducted an experiment using a virus, called a bacteriophage, that infects and destroys bacteria. The bacteriophage is a relatively simple organism composed only of protein and nucleic acid. As figure 12.7 shows, the nucleic acid is contained within a protein coat. These phages attach to bacteria cells and infect them (fig. 12.8). Apparently, nucleic acid, protein, or both are injected into the bacteria cell. Researchers have concluded that the injected material carries information to produce new phages, because after infection new phages are synthesized inside the bacteria. The bacteria cell breaks open and these phages are released to infect other bacteria.

In their experiment, Hershey and Chase grew phage in a nutrient medium containing radioactive phosphorus and radioactive sulfur. The radioactive phosphorus labeled the DNA in the mature phages, and the radioactive sulfur labeled the protein. These labels enabled the scientists to easily distinguish the protein and the DNA in the phages.

Why did radioactive phosphorus label only the DNA and radioactive sulfur label only the protein? (Hint: Review the structure of these biological molecules in chapter 3.)

**FIGURE 12.6** Triploidy. The karyotype of an individual with three of each chromosome type.

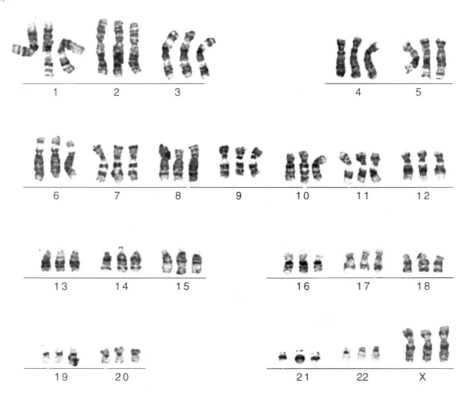

**FIGURE 12.7** Bacteriophage. (a) A scanning electron micrograph of phage. (b) Diagram with parts of the phage labeled.

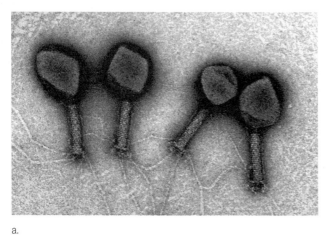

a.

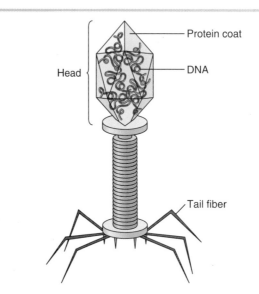

b.

Labeling the protein and DNA in the phage enabled Hershey and Chase to make a prediction: If only radioactive sulfur appeared inside an infected bacteria, protein must have entered the cell; if only radioactive phosphorus appeared inside, then DNA must have entered.

In their experiment, they mixed the radioactive phage they had grown with nonradioactive bacteria. The phage were given sufficient time to attach to the bacteria and inject material into them, as shown in figure 12.8b. Next, the phage and bacteria culture was agitated in a blender to remove the phage coats from the bacteria. After they separated the phage coats, they measured the radioactivity in the separated materials. Differences in the radiation produced by sulfur and phosphorus made it possible for them to determine which biological molecules were inside the infected bacteria.

FIGURE 12.8 Bacteriophage infection. (a) Electron micrograph of a phage attached to a bacterial cell. Structures appearing to be new phage "heads" can be observed inside the cell. (b) Proposed sequence of events in phage infection. Phage structures can be separated from bacterial cell to examine material remaining in phage coats.

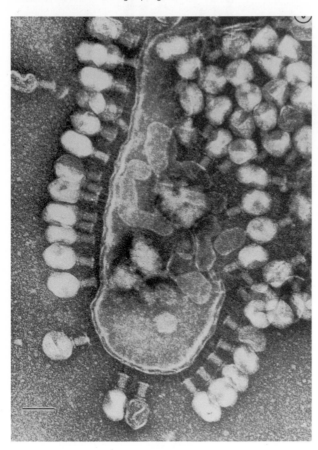

a.

Phage
Bacterium

Infection

Injection

Separation by
centrifugation

Agitation separates
phage and bacteria

b.

Hershey and Chase could have done two separate experiments, labeling only phosphorus in one and only sulfur in another. Dual labeling, however, allowed a full investigation with just one preparation of phage and bacteria.

When they had completed their experiment, the two scientists found radioactive sulfur in the separated phage coats and only radioactive phosphorus in the infected bacteria. They concluded that only DNA enters the bacteria and carries information to synthesize new phage.

If both radioactive phosphorus and radioactive sulfur were found in the separated phage coats, how would you interpret this result?

Suppose only radioactive phosphorus was used to label the phage. Why would the results be difficult to interpret?

Hershey and Chase's experiment gave a clear indication that DNA is the biological molecule that carries information to produce new bacteriophage. Many investigators, however, were hesitant to expand this knowledge to all plants and animals. Phage are rather unusual organisms. Just because DNA carries the directions for new phage does not necessarily mean it performs this function in all organisms.

The experiment does show, however, that DNA *can* carry information. In chapter 14, we will discuss *how* DNA carries information and how that information is copied from generation to generation.

## THE CELL CYCLE

Cell division, or mitosis, is a continual process. A cell divides, creating two cells, and those two cells divide again. Recall that when a cell divides during mitosis, it gives one chromatid to each new cell. If those cells later divide, how do they come to have a complete X-shaped chromosome so that they can give one chromatid to each new cell? Evidently, the DNA is copied before mitosis takes place a second time.

This process is illustrated in figure 12.9. The events that occur from the formation of a cell until that cell divides is called the **cell cycle.** For cells that repeatedly divide, the cell cycle takes about 20 hours (though some cells may divide much more slowly). Only one or two of these hours are taken up by mitosis, however. The rest of the time the cell spends in interphase, during which it produces new DNA, as well as new proteins and organelles. Thus when the cell enters prophase, the chromatids have become chromosomes, with the characteristic X shape. The two connected chromatids are presumably exact copies of one another.

A biologist measured the amount of DNA per cell in cells that were in prophase. If she measured the DNA per cell at other phases of mitosis, where would the amount of DNA be different?

If you examined a group of cells that were dividing mitotically, such as in an onion root tip, what stage would most of the cells be in?

Mitotic cell division distributes chromosomes so that each of the two new cells receives identical chromosomal material. Moreover, this process maintains a constant number of chromosomes per cell.

## MEIOSIS

Is mitosis the only kind of cell division? Are chromosome numbers always constant? For an answer, we can look at **sexual reproduction,** the process by which organisms propagate. When a new organism is created in sexual reproduction, two cells, one from each parent, combine to form the single, first cell of the offspring, called the **zygote.** For example, in humans a sperm from the male combines with an egg from the female. If the egg and sperm cells each contain 46 chromosomes, as other human cells do, then the offspring would have 92 chromosomes! What happens to prevent an increase in chromosome number from generation to generation?

The logical answer is that the human egg and sperm cells must have fewer than 46 chromosomes. How are these cells produced so that the chromosome number is reduced from the normal number in human cells?

The cell division that produces sex cells is called **meiosis.** Meiosis has proven somewhat more difficult to observe than mitosis. Nevertheless, studies have revealed that the process consists of two cell divisions rather than one.

In the first division, meiosis I, shown in figure 12.10, the chromosome number is reduced. The events are similar to mitosis except that in metaphase I homologous pairs of chromosomes line up along the middle of the cell. This pairing of homologues in early metaphase I is termed **synapsis.** In anaphase I the pairs separate, one member of each pair moving to each cell. In figure 12.10, the cell contains a total of 8 chromosomes in four homologous pairs. The two new cells each contain one-half the original number, or 4 chromosomes. Unlike in mitosis, these chromosomes are still in the X shape and the chromatids have not separated. In humans, the cells produced by meiosis I contain 23 X-shaped chromosomes (half of 46).

In the second stage of meiotic cell division, meiosis II, the movements of the chromosomes are basically identical to mitosis (fig. 12.11) but with half the original chromo-

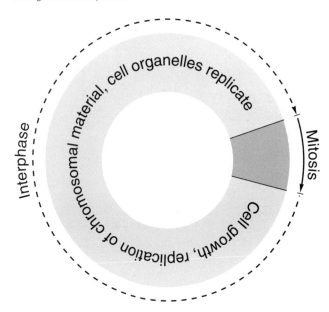

**FIGURE 12.9** Cell cycle. Mitosis occupies only a small portion of the cell cycle. During the remainder of the cycle, the cell is in interphase and the chromosomal material is replicated, the cell increases in size, and the cell organelles are replicated.

some number. The chromosomes move to the center of the cells in metaphase, and in anaphase they separate into chromatids. In our example, 4 chromatids move to each cell. In humans, 23 chromatids move to each cell. The final result of meiosis is four new cells, each containing chromatids from half the chromosomes in the original cell.

If you measured the DNA content per cell in each phase of the two divisions of meiosis, what would you find?

When an egg and sperm combine to produce a zygote, each parent cell contributes half of the zygote's chromosomes. In this way, the full number of chromosomes is restored. In humans, each parent contributes 23 chromosomes so that the resulting offspring has 46 chromosomes. Once the zygote is formed, that cell divides mitotically, increasing the number of cells and ultimately resulting in a mature organism. Throughout the organism's growth, the number of chromosomes per cell remains constant.

## HAPLOID AND DIPLOID NUMBERS

The reduced number of chromosomes in the egg and sperm cells is called the **haploid** number. The higher number of chromosomes in the zygote and in other body cells is called the **diploid** number. Biologists commonly denote the haploid number as n and the diploid number as 2n. For example, in humans n = 23 and 2n = 46.

FIGURE 12.10
Meiosis I. In metaphase I, homologous chromosomes pair along the cell midline and separate in anaphase I. At interphase II, nuclei have re-formed but contain only one member of each homologous pair. Both nuclei now enter meiosis II.

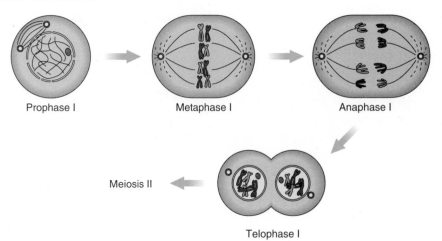

Prophase I      Metaphase I      Anaphase I

Meiosis II

Telophase I

FIGURE 12.11   Meiosis II. Events are essentially identical to mitosis. Chromosomes are separated into chromatids. The end product of the two cell divisions of meiosis are four haploid cells.

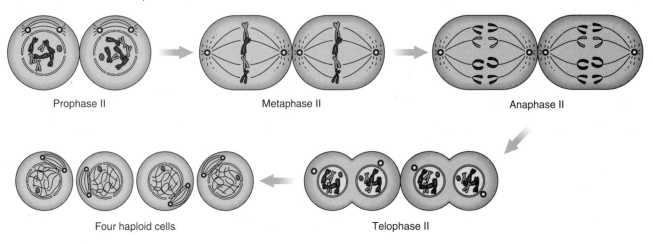

Prophase II      Metaphase II      Anaphase II

Four haploid cells.      Telophase II

A student observed a plant cell containing 17 chromosomes. Was it a haploid or diploid cell?

In humans and other animals, almost all body cells, which are called **somatic** cells, are diploid. These cells are produced by mitosis. Only a very small part of the body relies on meiotic cell division to produce haploid egg and sperm cells. A similar situation exists in flowering plants where meiosis is limited to a small part of the flower.

## SUMMARY

*In this chapter we have seen that cell division and the movement and distribution of chromosomes are critical events in the lives of organisms. In the next few chapters, we will examine reproduction and the molecular events in chromosome duplication in more detail, investigating such questions as how chromosomes are copied and how DNA carries information. Keep in mind the following key points from this chapter as you read further in the text.*

1.  The **cell cycle** describes the growth, development, and division of a single cell to produce two new cells.
2.  In actively dividing cells, about 90 percent of the cell's life is in **interphase.** The cell grows in size and the nucleus has a finely granular appearance. **DNA** is synthesized (copied) during this phase.
3.  **Mitosis** distributes **chromosomes** equally to the two daughter cells. Several distinct stages are apparent during this cell division process:
    a.  In **prophase,** nuclear material condenses so that chromosomes become apparent as X-shaped structures.
    b.  In **metaphase,** chromosomes line up across the midline of the cell.
    c.  The chromatids of chromosomes separate in **anaphase.**
    d.  In **telophase,** nuclei reform in the two new cells and chromosomal material disperses.
4.  **Cytokinesis,** division of the cytoplasm, occurs in plants when a new wall forms across the middle of the cell. In animals, the cell "pinches" in across the middle to separate into two cells.

5.  A **karyotype** can be constructed to display a cell's chromosomes, which are present as **homologous pairs.** The number, size, and shape of chromosomes varies in different organisms.
6.  **Meiosis** is a type of cell division that reduces the number of chromosomes in **diploid** cells by half to produce **haploid** cells. Meiosis occurs in two separate cell divisions:
    a.  In **meiosis I,** the phases are similar to mitosis except that homologous chromosomes form pairs (this is called **synapsis**) in metaphase I and these homologues are separated in anaphase I.
    b.  In **meiosis II,** events are identical to mitosis; chromatids separate at anaphase II.
7.  Hershey and Chase's experiments with bacteriophage indicated that DNA carries information to produce new phage. The generalization that DNA serves this function in all organisms requires extensive information on copying DNA and the structure of the DNA molecule.

## EXERCISES AND QUESTIONS

1.  Suppose a cell with 14 chromosomes divides mitotically and one of the two new cells has 13 chromatids and the other 15 chromatids. At which stage in mitosis do you conclude a mistake occurred (page 146)?
    a.  interphase
    b.  prophase
    c.  metaphase
    d.  anaphase
    e.  telophase
2.  If the cell described in question 1 was dividing normally and was observed to have 14 chromatids, what stage of mitosis would it most likely be in (page 146)?
    a.  interphase
    b.  prophase
    c.  metaphase
    d.  anaphase
    e.  telophase
3.  Which of the following is most likely the hypothesis being tested regarding the prediction that changes in chromosomes are accompanied by changes in an organism (page 148)?
    a.  DNA is the material in chromosomes that carries information.

    b.  Chromosomes carry information to form material and structures in cells.
    c.  Mitosis separates exact copies of chromosomes.
    d.  Chromosomes are composed of both DNA and protein.
    e.  DNA is a major chemical constituent of chromosomes.
4.  Which of the following would be an acceptable interpretation of the observation that chromosomal abnormalities are accompanied by physical abnormalities in an organism (page 148)?
    a.  When cells divide, chromosomes split and chromatids move to the two new cells.
    b.  Abnormalities in chromosomes are almost always accompanied by physical or chemical abnormalities.
    c.  Some unknown change occurs in the cell that causes both chromosomal and physical abnormalities.
    d.  Physical abnormalities usually shorten the lifespan of organisms.
    e.  Chromosomes should be observed before making an interpretation about physical abnormalities.

5. When phages were labeled with radioactive phosphorus, why was the phosphorus found only in DNA and not in protein (page 148)?
   a. Both DNA and protein are polymers composed of monomers.
   b. No amino acids contain phosphorus, but nucleotides do contain phosphorus.
   c. Phosphorus is found in several biological molecules and is present in all organisms.
   d. Protein makes up the coat of the phage and DNA is in the phage interior.
   e. Evidence has shown that only DNA and radioactive phosphorus enter infected bacteria.

6. Suppose both radioactive phosphorus and radioactive sulfur were found in the phage coats that had been separated from infected bacteria. Which of the following is an acceptable interpretation of this result (page 150)?
   a. Not all DNA from the phage was injected into the infected bacteria.
   b. New phage produced by the infected bacteria will contain radioactive phosphorus and sulfur.
   c. Protein is the material injected into infected bacteria.
   d. Protein and DNA are the only biological molecules that make up phage.
   e. DNA contains both phosphorus and sulfur.

7. A biologist examined mitotically dividing cells and found that the amount of DNA per cell in prophase was equal to $x$. What would be the DNA per cell in late telophase (page 150)?
   a. $x$
   b. $2x$
   c. $x/2$
   d. $x/4$
   e. $4x$

8. If you examined a group of cells that were dividing mitotically, such as in an onion root tip, what stage would most of the cells be in (page 150)?
   a. interphase
   b. prophase
   c. metaphase
   d. anaphase
   e. telophase

9. Suppose you measured the DNA per cell during each stage of meiosis (page 151). Which of the following would be equal to the DNA in prophase I/the DNA in prophase II?
   a. 2
   b. 1
   c. 1/2
   d. 1/4
   e. 1/3

10. A student observed a plant cell containing 17 chromosomes. Was it a diploid or haploid cell (page 152)?
    a. Diploid, since the chromosome number is greater than 10.
    b. Haploid, since the chromosome number is less than 20.
    c. Diploid, since the chromosomes are present in pairs.
    d. Haploid, since the chromosome number is odd.
    e. Diploid, since the chromosome number is less than 20.

11. A biologist successfully fertilized an egg with two sperm cells. Later study of the zygote formed in this way showed that it contained 27 chromosomes. How many chromosomes should normally be found in diploid cells of this organism?
    a. 9
    b. 18
    c. 27
    d. 36
    e. 54

12. A biologist marked a cell that she knew was about to undergo meiosis. A short while later she observed the four cells produced by the original marked cell. Their chromosome numbers were 17, 17, 18, and 16. She knew these numbers indicated that something abnormal had occurred during meiosis. Which of the following most likely occurred?
    a. One chromosome did not split apart into chromatids in the first meiotic cell division.
    b. Synapsis did not occur properly in the second meiotic cell division, but other events occurred normally.
    c. One homologous pair of chromosomes did not separate in the first meiotic cell division but moved to one cell and later separated.
    d. In the second meiotic division, one chromosome did not split apart but moved to one cell and later split.
    e. None of the above could account for these chromosome numbers.

13. The biologist in question 12 did not observe the chromosome number in the original marked cell. Which of the following was most likely the chromosome number of that cell?
    a. 33
    b. 34
    c. 35
    d. 36
    e. 68

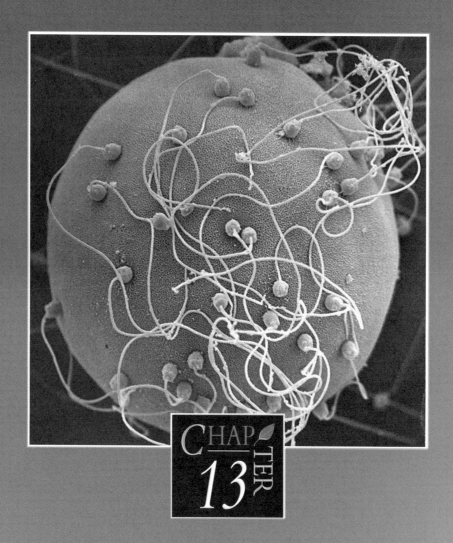

# CHAPTER 13

# REPRODUCTION

## FIGURE 13.1

Binary fission. (a) Diagram of cell division in *Paramecium*. Cell organelles and structures are reproduced and equally distributed into two new cells. (b) Photomicrograph of *Paramecium* division.

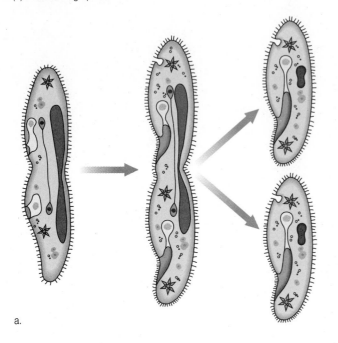

a.

b.

ost organisms on earth are able to create new organisms. Humans give birth to babies that grow to adults; trees produce saplings that grow into mature trees. This ability to generate new organisms from pre-existing organisms is called **reproduction.** Without it, life on earth not only would cease, it would never have begun. In this chapter we will explore the ways in which a variety of organisms accomplish the important task of reproduction.

In the last chapter we investigated how cells divide to form new cells. Cell division is the way organisms grow, but it also plays a role in how organisms reproduce. In the simplest form of reproduction, single-celled organisms such as bacteria and protozoans form new individuals by mitotic cell division (fig. 13.1). This type of division, called **binary fission,** simply splits the organisms into two new cells.

Can multicellular organisms reproduce in such a simple way? Some can. In primitive animals such as sponges and some worms, the parent can grow a fragment by mitotic cell division that breaks off and becomes a new individual. In hydra, reproduction can take place by budding (fig. 13.2), where an offspring "buds" or branches off the parent and later separates to form a separate organism. In a similar way, plants commonly grow from the cuttings or parts of a parent plant.

Reproduction by binary fission or from a fragment of the parent is called **asexual reproduction.** This type of reproduction requires only mitosis. A fragment grows and develops into a new organism by mitotic cell division of the cell or cells in the fragment. Because only mitosis occurs, the offspring are exactly like the parent.

## FIGURE 13.2

Budding in hydra. A new organism is formed as a "bud" from the older, mature parent. The new individual will detach and become a separate organism.

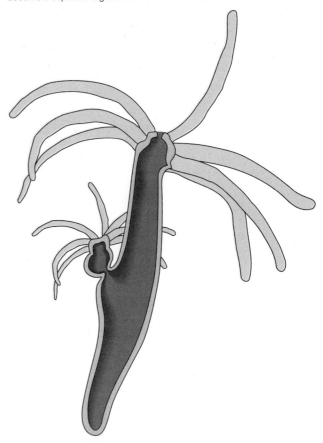

**156**

*Reproduction and Inheritance*

## SEXUAL REPRODUCTION AND VARIATION

There were 5.5 billion people on earth in 1992, and except for twins, no two of them looked alike. The variety in human physical characteristics is astounding, and even more amazing when we realize that all human beings have the same number of chromosomes. How does such variation in our species, or in any species for that matter, occur?

We saw in chapter 12 that during meiosis, homologous chromosomes are separated at anaphase I. Homologues contain information for the same characteristics (such as eye color), but these characteristics may be of different expressions. For example, in a homologous pair that contains information for eye color, one homologue may carry information for brown eye color and the other for blue eye color. When these two homologues separate, the two resulting cells will have different instructions for eye color.

Since several pairs of homologues usually pair and separate, many combinations can result. This process is illustrated in figure 1. Depending on the number of chromosomes, the number of variations in the gametes can be quite large. If there are 6 chromosomes in a diploid cell, eight combinations are possible. In humans, with 46 chromosomes, well over 8 million combinations are possible! (We will discuss variation in offspring in more detail in chapter 15.)

FIGURE 1    Variation in gametes. The form of information on different homologues may differ. For example blue eye color vs. brown eye color may be carried on different homologues. Under these conditions gametes produced from the same parent cell will carry different information.

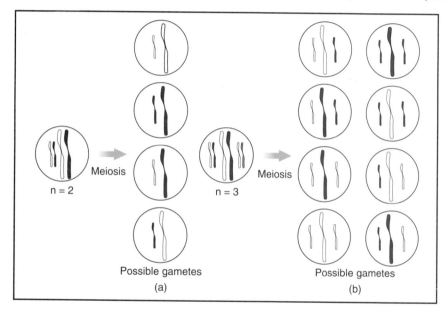

## SEXUAL REPRODUCTION

In most multicellular organisms, reproduction occurs by a very different method, called **sexual reproduction,** in which two cells combine to form a new individual instead of one cell dividing. Most often, the two cells come from two different parents, both of whom pass their traits on to the offspring. The resulting individual is unique, unlike either parent, though possessing characteristics of both.

Why does sexual reproduction occur? Compared with asexual reproduction, it seems inefficient and complicated. Two cells must meet and combine under just the right conditions, and those conditions are not always available. Why should any organism reproduce sexually when both plants and animals are able to reproduce asexually? Are there advantages?

The most important benefit of sexual reproduction is the variation in offspring that results. We will learn more about variation in chapter 17. For now it is enough to realize that sexual reproduction results in a variety of individuals in every species. Variation arises because of the separation of homologues in meiosis and the chromosomes coming together at fertilization (Thinking in Depth 13.1).

How does sexual reproduction occur? Recall our discussion of haploid and diploid cells in chapter 12. In sexual reproduction, two haploid (n) cells called **gametes** fuse to produce a diploid (2n) cell called a **zygote.** The zygote grows into a new organism. The point at which the two haploid cells fuse is called **fertilization.** This basic process occurs in both plants and animals, but in other respects reproduction in plants and animals can be strikingly different, so we will discuss them separately in the sections that follow.

# REPRODUCTION IN ANIMALS

The gametes, or sex cells, that combine to form a new organism are generally produced by two separate individuals of different sex: males produce sperm and females produce eggs. In some cases, such as earthworms, tapeworms, and barnacles, one individual produces both eggs and sperm. This is called self-fertilization. Such organisms are called **hermaphroditic.**

Would you expect more variability in offspring from asexual, hermaphroditic, or sexual reproduction?

Gametes in animals are almost always the direct result of meiosis. (In many plants and other types of organisms this is not the case.) Moreover, gamete production is limited to a small part of the organism's body. In other words, meiotic cell division is found only in those localized parts of the body that produce haploid gametes. All other cells in the body are diploid. In humans, ovaries are the site of gamete production in females and testes are the site in males (fig. 13.3).

A physiologist treated newborn hamsters with a substance that prevented any meiotic cell divisions in the animals for the rest of their lives. How will this affect the hamsters' growth and development?

## Fertilization

Bringing together an egg and a sperm is essential in sexual reproduction. As we will see, animals employ several reproductive strategies to increase the chances that egg and sperm achieve fertilization.

In animals, eggs and sperm are brought together in two basic ways. In **external fertilization** gametes are released into the external environment. In **internal fertilization** sperm are deposited in the body of the female.

External fertilization occurs almost entirely in animals living in aquatic environments, which includes most fish, aquatic invertebrates, and many amphibians. Contact between the egg and sperm is maximized when they are released at the same time and place by males and females. Large numbers of gametes are released by each sex to ensure a high probability of fertilization.

Most land animals use internal fertilization. They produce far fewer gametes than aquatic organisms do, but they deposit sperm directly into what is essentially an aquatic environment in the female's body in close proximity to egg cells. Internal fertilization avoids the release of gametes into harsh external environments where their survival is threatened. If

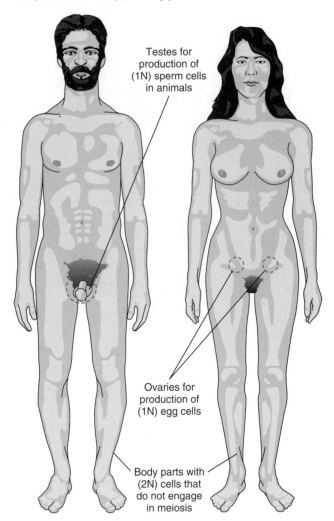

**FIGURE 13.3**    Meiotic cell division. In animals, meiosis is limited to small portions of the body producing gametes.

Testes for production of (1N) sperm cells in animals

Ovaries for production of (1N) egg cells

Body parts with (2N) cells that do not engage in meiosis

gametes of land animals were released into the environment, water loss from these cells would be a serious problem.

The release of gametes, particularly the release of eggs by females, occurs under complex internal controls. In mammals, several hormones interact with the reproductive organs to affect egg release and the survival of the fertilized egg (Thinking in Depth 13.2). Hormones are chemical substances released by the female's body that control processes within the body.

Once sperm and egg come in contact, a potentially serious problem arises: What happens if more than one sperm enters the egg? If more than one sperm enters, the zygote will contain more than the diploid number of chromosomes. As we saw in chapter 12, changes in the diploid chromosome number have serious consequences. How is this prevented?

One hypothesis is that more than one sperm does enter the egg, but only one set of male chromosomes survives to combine with the egg chromosomes. A second hypothesis is that a barrier forms around the egg after one sperm enters, blocking the entry of other sperm.

Let's look more closely at the second hypothesis. We can make a prediction that if fertilization creates a physical

*Reproduction and Inheritance*

## HORMONAL CONTROL OF REPRODUCTIVE CYCLES

In the reproductive system of the human female, shown in figure 1, eggs are produced and released by the ovaries. Egg-containing structures called follicles grow in the ovaries. When the follicles are mature, they burst and release eggs into the uterine tube. The burst follicle develops into a yellowish mass of cells on the surface of the ovary called the **corpus luteum.**

The maturing and releasing of eggs is apparently controlled by chemicals called **hormones** produced by the pituitary gland, which is attached to the brain. The growth of follicles is stimulated by the follicle stimulating hormone, or FSH. FSH and LH, a hormone produced by the pituitary gland, stimulate the growing follicle to produce **estrogen,** a female sex hormone (figure 2). The increasing estrogen appears to initiate a surge of LH that in turn triggers the release of an egg from the most mature follicle. The corpus luteum continues to produce estrogen, as well as a second female sex hormone called **progesterone** that causes several effects: the uterine wall thickens in preparation to receive a fertilized egg, follicle growth is inhibited, and the production of FSH and LH is inhibited.

If no egg is fertilized, the corpus luteum degenerates, progesterone decreases, and the FSH and LH initiate another cycle. If fertilization does take place, the growth of the embryo in the uterus secretes another hormone (HCG) that appears to preserve the corpus luteum and thus the higher progesterone levels. This in turn inhibits egg production.

These events are part of the complex, integrated responses of the human female reproductive system. They illustrate the impressive control that our bodies maintain over physiological processes.

Birth control pills contain substances similar to progesterone and estrogen. How do you think these substances prevent pregnancy?

**FIGURE 1**  Human female reproductive system. Eggs are produced by the ovaries. After release, eggs pass through the oviduct to the uterus.

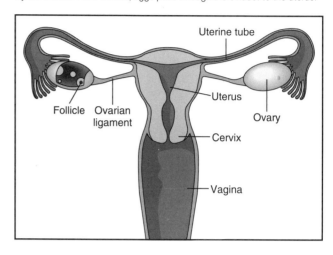

**FIGURE 2**  Hormonal control of egg release. Hormone levels in the blood are shown for hormones from the pituitary gland and the ovaries, as well as changes in the follicle and egg release.

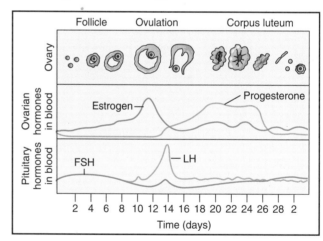

---

barrier, we could see the barrier under a microscope. Figure 13.4 illustrates what we see in sea urchin eggs under a microscope just before and after sperm entry. Before fertilization, the outer portion of the egg is comprised of the **vitelline layer** and the plasma membrane. Just inside the plasma membrane is the **cortical layer** containing thousands of cortical granules. When the sperm penetrates the plasma membrane, the cortical granules at the site of penetration appear to spill their contents into the space between the vitelline layer and the plasma membrane, forming a thickened layer called the **fertilization membrane** that sperm

cannot penetrate. The cortical granual reaction spreads over the surface of the egg from the site of sperm penetration and completely covers the egg in about 20 seconds.

Suppose two sperm cells attempted to enter an unfertilized egg from exactly opposite sides at exactly the same time. Which of these sperm cells is more likely to enter the egg?

FIGURE 13.4 Fertilization. (a) Before fertilization, the cortical granules can be observed just under the plasma membrane and vitelline layer. (b) After sperm entry, the cortical granules fuse with the plasma membrane and the vitelline layer is raised to form the fertilization membrane.

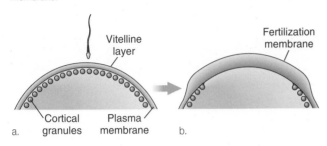

The visual evidence of the formation of an impermeable fertilization membrane strongly supports our second hypothesis. During the 20 seconds required for the membrane to form, however, several sperm could enter the egg. Could there be another mechanism preventing sperm entry?

In another experiment testing the barrier hypothesis, researchers found that a membrane potential of −60 millivolts (mv) is maintained in unfertilized eggs (see chapter 11 to review the techniques used to measure membrane potentials). When a sperm entered the egg, the membrane potential changed immediately to about +10 mv (fig. 13.5).

Researchers predicted that membrane potentials exercise some control over sperm entry and that if the potential of an unfertilized egg could be maintained at positive values, the egg would not be fertilized. To test this possibility, they passed an electrical current through a microelectrode to produce a membrane potential of +5 mv or greater in the egg. As they predicted, no sperm cells could fertilize the egg under these conditions. Apparently, the change in membrane potential provides a "fast block" to sperm entry, supporting our barrier hypothesis. These experiments, and our observations, suggest there are two barrier mechanisms to prevent multiple sperm entry.

In a further test of the barrier hypothesis, the membrane potential of a fertilized egg was maintained at −60 mv. If the membrane potential controls sperm entry, what would you expect to observe under these conditions?

## Development

After an egg is fertilized, mitotic cell divisions, foldings, and cell movements occur in a regular and apparently programmed sequence as the zygote develops into an **embryo.** An embryo is the early stage of an organism's development, during which organs are formed. These developmental events are highly complex and could be the subject of an entire book in themselves. In figure 13.6 we show a typical example of just the basic developmental stages of an embryo.

FIGURE 13.5 Electrical changes at fertilization. Before fertilization, the membrane potential of the egg is measured as about −60 mv. At fertilization, the membrane potential rapidly changes to about +10 mv.

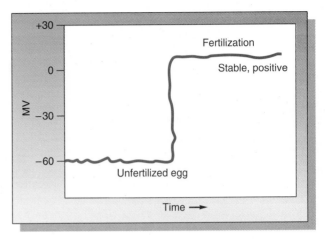

# REPRODUCTION IN PLANTS AND ALGAE

Plants and algae do not have the ability to move around as animals do. Because of this difference, and many others, the reproductive processes of plants differ from those of animals. Some elements are the same, however, as we will see in this section.

Plants rely on meiosis for sexual reproduction, but in plants meiosis produces **spores** rather than gametes. Spores are cells that divide by mitosis to produce a multicellular, haploid organism. For example, the pollen grains produced by flowering plants are the result of spores dividing. The haploid organism created from the spores produces gametes by mitosis. This reproductive cycle is summarized in figure 13.7.

This pattern, in which a diploid organism alternates with a haploid organism, is called the **alternation of generations.** The gamete-producing organism is called the **gametophyte** and the spore-producing diploid organism is called the **sporophyte.**

Within this general scheme of reproduction, different kinds of plants and algae exhibit an impressive variety. For example, the freshwater green alga *Ulothrix* is a filamentous algae often observed as a hairlike mass growing on rocks in streams. The filament thread is a chain of haploid cells (fig. 13.8). Mitosis in one of these filament cells can produce either zoospores that can grow directly into another filament or gametes that can be released and subsequently fuse into a diploid zygote. This zygote will divide meiotically to produce four cells. Each of these cells can grow into a filament. Notice that only one cell, the zygote, is diploid throughout the algae's entire life cycle.

After observing the life cycle of *Ulothrix,* a student stated that the gametes could be produced directly by meiosis. What is the best argument against the student's statement?

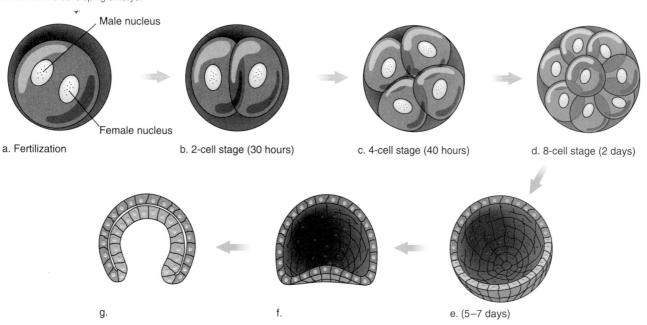

**FIGURE 13.6** Development: The fertilized egg (*a*) divides or cleaves repeatedly (*b, c,* and *d*) to form the blastula (*e*). The cross section of the blastula shows a hollow ball. Folding and cell migration (*f* and *g*) change the form of the developing embryo.

Male nucleus

Female nucleus

a. Fertilization

b. 2-cell stage (30 hours)

c. 4-cell stage (40 hours)

d. 8-cell stage (2 days)

e. (5–7 days)

f.

g.

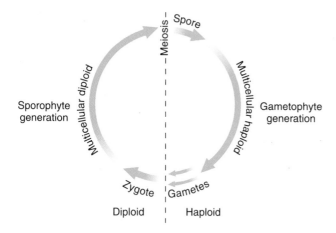

**FIGURE 13.7** Alternation of generations. In plants, meiosis produces haploid spores that grow into a haploid, multicellular organism that produces gametes by mitosis. Gametes fuse to form a zygote that usually grows into a multicellular diploid organism.

Meiosis — Spore

Multicellular diploid

Multicellular haploid

Sporophyte generation

Gametophyte generation

Zygote

Gametes

Diploid

Haploid

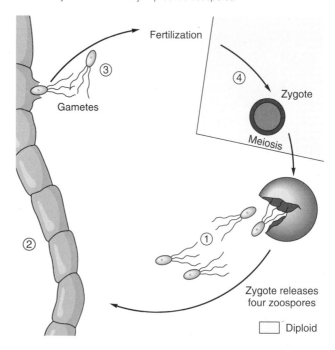

**FIGURE 13.8** Life cycle of *Ulothrix.* In sexual reproduction, a haploid zoospore (1) will grow into a haploid, multicellular filament. (2) Mitosis produces gametes (3) that fuse to form a diploid zygote (4). The zygote immediately divides meiotically to produce zoospores.

Fertilization

Gametes

③

④

Zygote

Meiosis

②

①

Zygote releases four zoospores

☐ Diploid

Another variation in the general scheme of plant reproduction is displayed by mosses. As figure 13.9 shows, their green, leafy, stemlike structures are haploid gametophytes. Separate plants are either male or female. Sperm produced by male plants are released into water after rain or heavy dew and move to female plants where eggs are fertilized. The resulting zygotes grow into diploid, stalklike sporophytes. Spores are formed by meiosis at the tip of the sporophyte and released to grow into another haploid plant.

Ferns produce spores on the underside of their green leaves. When the spores fall off the leaves onto moist earth, they divide mitotically and grow into a haploid gametophyte called the **prothallus** (fig. 13.10). Both eggs and sperm are produced on the prothallus. Sperm released into moisture around the prothallus swim to the eggs. The eggs are fertilized and grow into the large leafy structures called **sporophytes** that are the most recognizable part of the fern. Spores are again produced on the underside of the leaves, and the cycle begins anew.

FIGURE 13.9

Moss life cycle. (*a*) The green leafy haploid stage supports the stalk-like diploid stage. (*b*) Spores (1) grow into the green haploid stage (2) that produces gametes (3). Fertilization produces a zygote (4) on the haploid plant. This zygote plant. This zygote grows into the multicellular diploid (5).

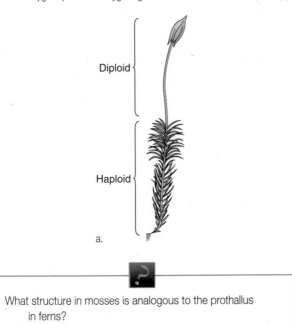

a.

 What structure in mosses is analogous to the prothallus in ferns?

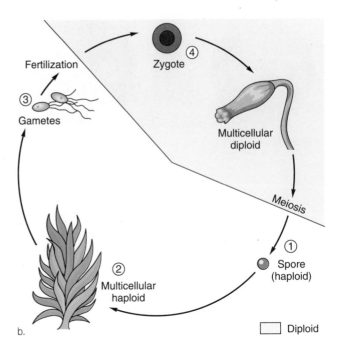

b.

In flowering plants, spores are formed by meiosis in the reproductive parts of the flower (fig. 13.11). The ovary produces female gametophytes, while the anther produces pollen grains, which are the male gametophytes. When a pollen grain is deposited on the stigma, a pollen tube grows down through the style into the ovary. Sperm nuclei move through the tube and enter the ovary, and one nucleus fertilizes the egg cell. The resulting zygote and other ovule tissue develop into a seed within the ovary of the flower. This seed later grows into the diploid, flower-producing plant.

In all these examples of sexual reproduction in plants, gametes, spores, or pollen are released into the external environment. It makes sense for plants to function this way because they are almost always stationary. Plants must release spores or gametes into the external environment and rely on wind or insects to carry them to other plants.

In the life cycle of a green algae, shown in figure 13.12, which structures would you expect to be haploid? Which would you expect to be diploid?

## CELL DIFFERENTIATION

As we mentioned earlier in this chapter, once an egg has been fertilized, the zygote divides to form an embryo, which continues to grow into a mature organism by repeated mitotic cell divisions that produce millions of cells. Since these cells are the result of mitosis, they should all appear the same.

FIGURE 13.10   Fern life cycle. A spore (1) divides mitotically to produce a multicellular haploid gametophyte, the prothallus (2). Both male and female gametes (3) are produced by the prothallus. The zygote (4) grows into the large leafy structure of the fern (the multicellular diploid (5). Spores are produced by meiotic division on the underside of the leaf.

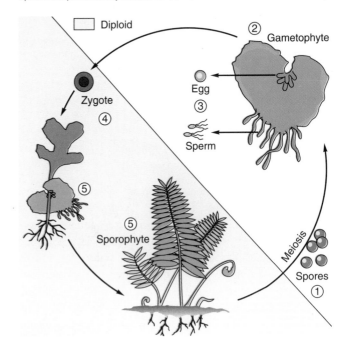

But this is not the case. As an organism grows, cells differentiate into several types. For example, in animals, some cells become nerve cells and some become muscle cells; the two are vastly different from one another. Similarly, in plants, some cells become phloem cells and others become xylem cells. Why should these cells develop so differently if all cells in an organism contain the same chromosomal information?

Biologists have proposed two hypotheses to explain this apparent contradiction.

> *Hypothesis 1*   As cells divide and become more specialized, information on the chromosomes is lost. The information remaining specifies the properties of the cell.
>
> *Hypothesis 2*   When cells become specialized, information on the chromosomes that specifies other cell types is inactive but still present.

What predictions can we make from these hypotheses? If hypothesis 1 is correct, then we can predict that it will not be possible to grow a complete organism from a specialized cell. If hypothesis 2 is correct, then we can predict that it will be possible to grow a complete organism from a single specialized cell.

Many biologists attempted to separate specialized cells and grow a complete organism from them. The first person to succeed was plant physiologist F. C. Steward in 1953. He was able to separate single cells from carrot phloem tissue and regenerate a complete carrot plant from one of those cells (fig. 13.13). Achieving regeneration required a tremendous amount of trial and error to discover the right combination of nutrients and other conditions for culturing the cells. Once Steward hit on this combination, he was able to grow new carrot plants, strongly supporting hypothesis 2.

Gurdon argued that the small amount of intestinal cell cytoplasm injected with the nucleus had no effect on the experiment. How could you test this claim?

After Steward's success, biologists wondered whether the same results could be obtained with animal cells. J. B. Gurdon, an English physiologist, attempted this experiment. Using unfertilized frog eggs, Gurdon transplanted a nucleus from a fully differentiated intestine cell into an egg, as shown in figure 13.14. Before he did so, however, he destroyed the haploid nucleus in the unfertilized egg with radiation. Next, he drew up an intestinal epithelial cell in a micropipette.

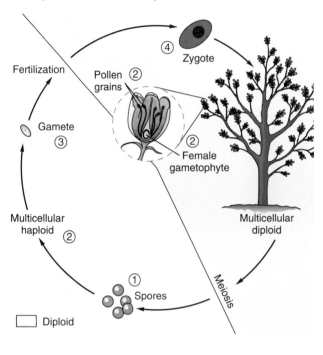

FIGURE 13.11   Flowering plant life cycle. Spores (1) produce a pollen grain or a female gametophyte (2). Gametes (3) are formed by mitosis. The male nucleus in the pollen tube is released into the ovary and fertilizes the egg cell. The resulting zygote (4), as well as associated tissues, grow into the seed that will grow into the mature plant.

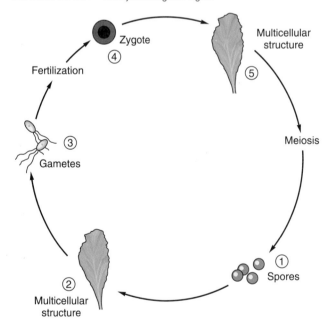

FIGURE 13.12   Life cycle of a green algae.

FIGURE 13.13   Steward's experiment. A single specialized carrot cell (1) is extracted and placed in nutrient medium (2). Under appropriate conditions, the cell will grow into a mature plant (3).

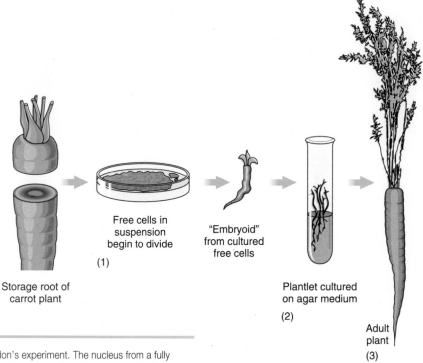

Storage root of carrot plant

Free cells in suspension begin to divide
(1)

"Embryoid" from cultured free cells

Plantlet cultured on agar medium
(2)

Adult plant
(3)

FIGURE 13.14   Gurdon's experiment. The nucleus from a fully differentiated intestine epithelial cell is implanted in an egg cell where the nucleus has been destroyed with ultraviolet radiation. Approximately 1 percent of these eggs will grow into mature frogs.

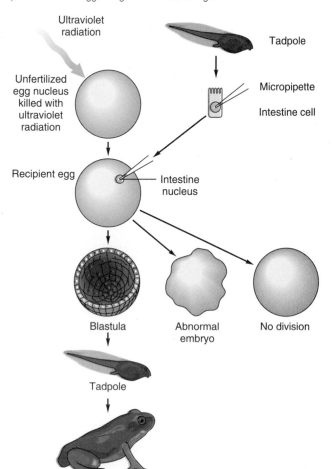

Ultraviolet radiation

Tadpole

Unfertilized egg nucleus killed with ultraviolet radiation

Micropipette

Intestine cell

Recipient egg

Intestine nucleus

Blastula

Abnormal embryo

No division

Tadpole

Mature frog

When the cell entered the pipette, the cell membrane ruptured, releasing the nucleus. Gurdon inserted the pipette into the unfertilized egg, injecting the nucleus with a small amount of cytoplasm.

About 1 percent of the eggs Gurdon injected developed into mature frogs that were identical to frogs formed from normal fertilization. In other words, the nucleus of the intestine cell contained all the information necessary to produce a complete organism. Gurdon's results support hypothesis 2 but are inconsistent with hypothesis 1.

Why did most of the eggs in Gurdon's experiment fail to develop into frogs? Gurdon examined them and discovered that they either did not divide or they developed into abnormal embryos. In many cases, technical problems caused failure. In some eggs, no nucleus had been injected, probably because it stuck to the micropipette. In other eggs, complete intestine cells were injected, apparently having failed to rupture when moving in and out of the pipette. These causes accounted for a large percentage of the failures. One percent may seem like a small amount, but it is significant because *none* of the eggs would have developed into adult frogs if nuclei from specialized cells did not contain information for the complete organism.

## SUMMARY

*Reproduction is a complex and diverse process in living organisms. We can, however, identify some fundamental, general features in both plants and animals. As you read further in this text, keep in mind the following key points about reproduction.*

1. **Binary fission** is simple cell division whereby single-celled organisms reproduce.

2. Many organisms can reproduce **asexually** by budding or when a fragment of the parent organism grows into a new individual.

3. Since only mitotic cell division occurs in asexual reproduction (including binary fission), all organisms from the same cell are identical.

4. In **sexual reproduction,** two haploid **gametes,** an egg and a sperm, fuse at fertilization to form a diploid **zygote** that grows into a new organism not identical to the parents.

5. **External fertilization** occurs in many aquatic animals, which release gametes into the external environment. Internal fertilization occurs in terrestrial animals, which deposit sperm in the body of the female.

6. The entry of sperm into an unfertilized egg appears to be controlled by physical barriers such as the **fertilization membrane** and electrical barriers.

7. Development from the zygote is characterized by highly regular cell divisions, foldings, and tissue movements to form the **embryo.**

8. Plant reproduction typically produces **spores** rather than gametes by meiosis. Spores grow into separate, haploid organisms called **gametophytes** that produce gametes by mitosis.

9. The alternation of a diploid **sporophyte** with a haploid gametophyte organism is called the **alternation of generations.**

10. The question of whether specialized cells retain all the information for a complete organism was tested by Steward and Gurdon. Steward was able to grow complete carrot plants from single carrot phloem cells, and Gurdon succeeded in raising mature frogs from frog eggs implanted with a specialized epithelial cell nucleus. Both Steward's and Gurdon's experiments support the hypothesis that differentiated cells with nuclei carry complete genetic information for the organism.

## REFERENCES

Epel, D. November 1977, "The Program of Fertilization," *Scientific American,* 128–38.

Gurdon, J. B. December 1968, "Transplanted Nuclei and Cell Differentiation," *Scientific American,* 24–35.

Marx, J. F. June 1978, "The Mating Game: What Happens When Sperm Meets Egg," *Science,* 1256–59.

## EXERCISES AND QUESTIONS

1. Which of the following characteristics of mitosis most likely indicates that it produces identical cells?
   a. Two cells are formed when the single parent cell divides.
   b. One chromatid from each chromosome moves to each new cell.
   c. Cytokinesis is completed after chromosome separation.
   d. Chromosomes condense and become shorter during prophase.
   e. During metaphase, chromosomes assemble along the midline of the dividing cell.

2. Which of the following best explains how sexual reproduction produces variability (page 157)?
   a. Homologous chromosomes exist in pairs carrying information for the same traits.
   b. Gametes from males and females carry the haploid number of chromosomes.
   c. Sexual reproduction is found in many plants and animals even though it is less efficient than asexual reproduction.
   d. Meiosis produces four cells rather than two cells as in mitosis.
   e. Homologues carry information for the same traits but not always in the same form.

3. Where would you expect more variability in offspring—from asexual, hermaphroditic, or sexual reproduction (page 157)?
   a. asexual
   b. hermaphroditic
   c. sexual

4. A physiologist treated newborn hamsters with a substance that prevented any meiotic cell divisions in the animals for the rest of their lives. How will this affect the hamsters' growth and development (page 158)?
   a. Their growth and mature size will not be affected.
   b. Their size increase will be about one-half that of untreated hamsters.
   c. The animals will exhibit no size increase after the chemical is applied.
   d. There will be a small effect on the animals occurring in all parts of the body.
   e. The animals will probably die shortly after the chemical is applied.

5. Birth control pills contain substances similar to progesterone and estrogen. How do these compounds prevent pregnancy (page 159)?
   a. They cause the corpus luteum to decrease in size.

b. Progesterone and estrogen interact and inhibit each other.
c. They initiate several surges of LH rather than just one.
d. They enhance the effects of FSH.
e. They inhibit follicle growth so no egg is produced.

6. Suppose two sperm cells attempted to enter an unfertilized sea urchin egg from exactly opposite sides at exactly the same time. Which of these sperm cells is more likely to enter the egg (page 159)?
a. Only the sperm attempting to enter from the left.
b. Only the sperm attempting to enter from the right.
c. Both will enter the egg.
d. Neither will enter the egg.

7. In a test of the barrier hypothesis, the membrane potential of a fertilized egg was held at −60 mv. If the membrane potential controls sperm entry, which of the following would you predict under these conditions (page 160)?
a. No sperm cells would enter the egg.
b. The membrane potential would become more negative after one sperm cell entered.
c. Only one sperm cell would enter the egg.
d. Several sperm cells would enter the egg.

8. A student mixes eggs and sperm from a certain fish (2n = 12). He finds that five of the resulting zygotes have the following numbers of chromosomes: 11, 16, 19, 23, and 20. He states that these chromosome numbers could not have resulted from multiple fertilization alone. Which of the following assumptions is he making?
a. Eggs and sperm all carry the haploid number of chromosomes.
b. Some chromosomes are accidentally lost when multiple fertilization occurs.
c. Egg cells possess mechanisms to prevent multiple fertilization.
d. Chromosomes come in pairs.
e. Homologous chromosomes are separated during mitosis.

9. After observing the life cycle of *Ulothrix,* a student stated that the gametes could be produced directly by meiosis. Which of the following is the best argument against the student's statement (page 160)?
a. *Ulothrix* is an algae that grows in an aquatic environment.
b. Meiosis is part of the life cycle in sexual reproduction.
c. Meiotic cell division reduced the chromosome number from diploid to haploid.
d. The zygote is diploid and is part of the life cycle.
e. A multicellular haploid organism is part of the life cycle.

10. What structure in mosses is analogous to the prothallus in ferns (page 161)?
a. Stalklike sporophyte
b. Green leafy structure
c. Spores
d. Gametes
e. Roots

11. A botanist was studying a colony of small, green, leafy plants growing in a damp area. He knew that these plants reproduce sexually. When he examined cells from the leafy portion of several plants, he found that all the cells contained 83 chromosomes. Which of the following is the most likely statement to make about these plants?
a. The diploid number for this organism is 83.
b. The haploid number for this organism is 164.
c. The haploid number for this organism is either 42 or 41.
d. The haploid number for this organism is 83.

12. In the life cycle of the green algae shown in figure 13.12, which structure(s) would be diploid (page 163)?
a. 1
b. 2
c. 3
d. 4
e. 5

The following information applies to questions 13 and 14:

In flowering plants, pollen grains are produced from spores. Shortly after reaching the female part of the flower, the pollen grain germinates and a pollen tube grows through the female structure to the egg cell. After germination, the single nucleus in the pollen grain divides and one of these nuclei divides again, resulting in three nuclei in the pollen tube. One of these nuclei will fuse with the egg cell to form a zygote.

13. In a plant that has a diploid chromosome number of 34, which of the following will be the chromosome numbers of the three nuclei in the pollen tube?
a. 34, 34, 34
b. 34, 34, 17
c. 34, 17, 17
d. 17, 17, 17

14. For the plant in question 13, what will be the chromosome number of the zygote?
a. 34
b. 6
c. 51
d. 17
e. It is not possible to deduce the chromosome number in this organism.

15. Gurdon argued that the small amount of epithelial cell cytoplasm injected with the nucleus had no effect on his experiment. Which of the following tests this claim (page 164)?
a. Inject the nucleus with an extra amount of cytoplasm and observe the effect.
b. Inject extra frog egg cytoplasm and see if the egg divides.
c. Inject only epithelial cell cytoplasm and observe the effects.
d. Insert the micropipette into the egg without injecting any material.
e. Use the micropipette to withdraw egg cytoplasm and observe the effects.

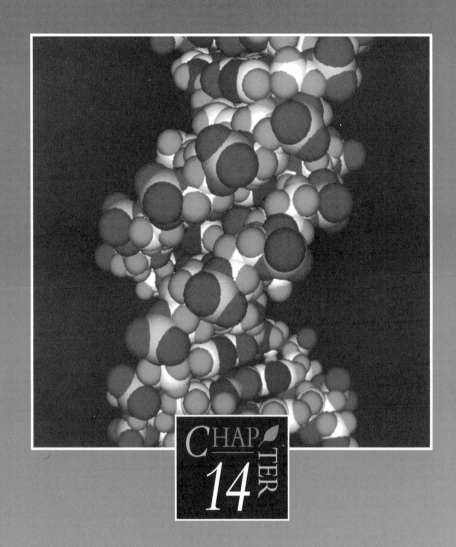

# CHAPTER 14

# MOLECULAR GENETICS

In chapter 12 we learned that a nucleic acid called deoxyribonucleic acid, or DNA, carries information from one cell to another when cells divide. The information carried by DNA determines the characteristics of organisms: the color of your hair, the size of a golden retriever, the shape of a daisy. But *how* does DNA carry information? What are the characteristics of this amazing molecule? And how is the DNA copied so that information is passed from one cell to the next?

In chapter 4 we had similar questions about enzymes. By looking at the enzymes' structure, we were able to determine their function. Perhaps the same will be true of DNA. In this chapter we will examine the structure of DNA for clues to how it works. Moreover, explaining how DNA transmits information will help us build a convincing argument that DNA is indeed the genetic material.

Early in the scientific investigations on inheritance of characteristics, the term "gene" was used to indicate the unit of information that specified a single characteristic. The word "genetics" is used to designate the inheritance of genes; the term "genetic material" refers to the molecules that carry information from cell to cell and organism to organism.

## DNA STRUCTURE

As we saw in chapter 3, nucleic acids are polymers made up of nucleotide monomers. There are four nucleotide monomers, each composed of three parts: a five-carbon sugar, a nitrogen-containing base, and a phosphate group (fig. 14.1). The four bases are called thymine, cytosine, adenine, and guanine. They vary slightly in structure, as you can see in figure 14.1. The monomers themselves are all linked together in the same way by their connecting phosphate groups and sugars.

The order of nucleotide monomers in DNA varies randomly. Any monomer may be linked to any other, or to a monomer of its own type. The monomers join to one another in chains that can be very long. Indeed, DNA can contain as many as 6 billion nucleotides!

Further chemical analysis of DNA from several varied sources reveal the data shown in table 14.1. Each DNA contains both **purines** (adenine or guanine) and **pyrimidines** (cytosine or thymine) in the molecule. What observations can you make of these data? One observation is that in every organism the percentage of adenine and thymine are almost identical, as are the percentages of guanine and cytosine. A second observation is that these percentages vary from species to species.

Why are the statements about table 14.1 observations (of results) rather than interpretations?

What predictions could you make about the DNA of organisms that are not listed in table 14.1?

**FIGURE 14.1** A portion of the DNA molecule. The molecule is a polymer composed of nucleotide monomers. The monomers are one of four nitrogenous bases: guanine, adenine, cytosine, or thymine.

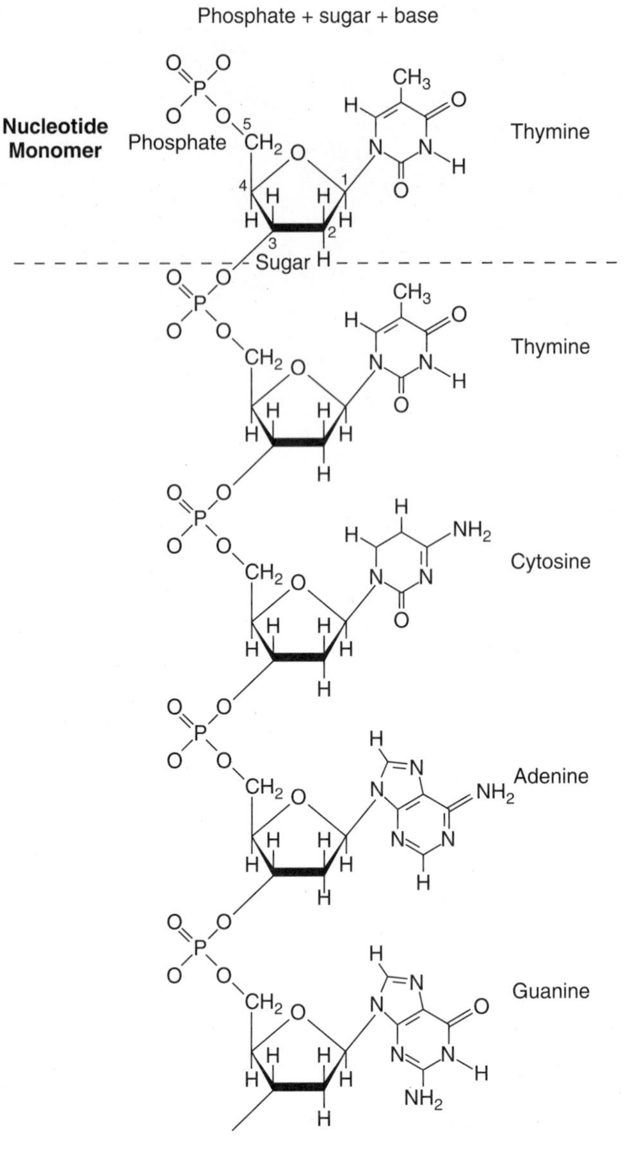

**TABLE 14.1** Percent of Purines and Pyrimidines from Purified DNA in a Variety of Organisms

| Source | Purines | | Pyrimidines | |
| --- | --- | --- | --- | --- |
| | Adenine | Guanine | Cytosine | Thymine |
| Human being | 30.4% | 19.6% | 19.9% | 30.1% |
| Ox | 29.0 | 21.2 | 21.2 | 28.7 |
| Salmon sperm | 29.7 | 20.8 | 20.4 | 29.1 |
| Wheat germ | 28.1 | 21.8 | 22.7 | 27.4 |
| *E. coli* | 24.7 | 26.0 | 25.7 | 23.6 |
| Sea urchin | 32.8 | 17.7 | 17.3 | 32.1 |

*Data from Helena Curtis,* Biology, *4th Ed Table 14–1, Page 285.*

*Reproduction and Inheritance*

FIGURE 14.2 X-ray diffraction patterns from crystallized DNA. The spots in the patterns represent repeating distances in the molecule.

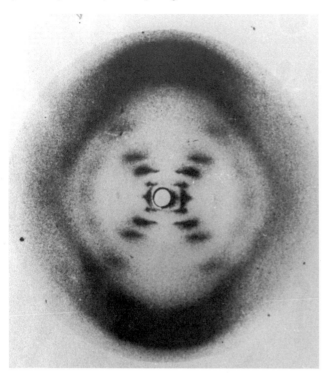

FIGURE 14.3 DNA helix. Watson and Crick proposed that the DNA molecule is composed of two polymer strands wound together in a helix structure. The length between turns of the helix would correspond to the 3.4 nm distance. The thickness of the molecule would be 2.0 nm.

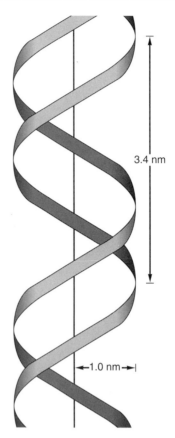

3.4 nm

←1.0 nm→|

The information in table 14.1 does not in itself explain how DNA functions. Perhaps the secondary structure of DNA will offer more clues.

More detailed studies of the molecular shape of DNA requires the use of a technique called **X-ray diffraction.** In this technique, a beam of X rays is focused on a crystalized material, in which the atoms and molecules are arranged in a highly regular order. X rays are reflected by this regular arrangement and the patterns of reflected rays are detected by photographic film. A DNA crystal can be produced by evaporating a concentrated solution of DNA in the same way that sugar crystals will form when water evaporates from a sugar solution.

When scientists performed X-ray diffraction studies on crystalized DNA, they obtained the patterns shown in figure 14.2. The spots on the X-ray diffraction pattern represent repeating patterns in the DNA crystal. By analyzing these patterns, British chemist Rosalind Franklin and British biophysicist Maurice Wilkins, using data produced by Franklin, were able to detect three primary repeating patterns in the DNA molecule: 0.34 nanometers (nm), 2 nm, and 3.4 nm (1 nanometer = $10^{-9}$ meters). Using these measurements, U.S. biologist James Watson and British molecular biologist Francis Crick attempted to construct a model of the molecule. Based on known interatomic distances, they proposed that the 0.34 nm measurement represents the distance between nucleotide monomers. The size of the nucleotide monomer molecules suggested that 2.0 nm would correspond to the width of the polymer molecule. What

structure would correspond to the 3.4 nm distance? Watson and Crick proposed that the DNA molecule is wound in a spiral or helix. The 3.4 nm would be the distance between turns of the helix.

Watson and Crick's proposal raised a problem, however. If the DNA molecule had a helix shape, then the **density** of DNA, or the weight per unit volume, would be only one-half the experimentally measured values. Watson and Crick's answer to this problem was that the DNA helix is actually two polymer chains wound together, as shown in figure 14.3. This structure accounts for the density.

Review Watson and Crick's reasoning up to this point. What observations do they use? What are their interpretations?

Watson and Crick's next task was to examine the helix structure more closely to see if the nucleotides fit together. They tried a great many possibilities and finally discovered that the purines and pyrimidines would fit well if a purine on one chain were paired with a pyrimidine on the opposite

**FIGURE 14.4** Purine and pyrimidine pairing. Specific base pairing of a larger purine with a smaller pyrimidine allows the two polymer strands to fit together with a constant distance between the strands.

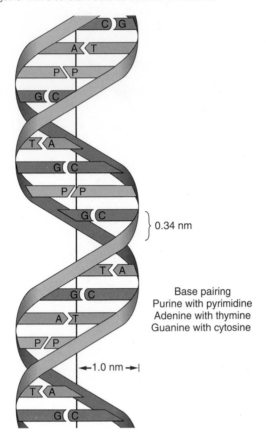

0.34 nm

Base pairing
Purine with pyrimidine
Adenine with thymine
Guanine with cytosine

1.0 nm

What would happen to the DNA helix structure if a purine were paired with another purine?

The hydrogen bonds between the bases are located in such a way that thymine always pairs with adenine and cytosine always pairs with guanine (fig. 14.5). Bases that pair this way are said to be **complementary.** This process, called **base pairing**, is clearly consistent with the data in table 14.1. If thymine and adenine always pair, we would expect to find equal amounts of thymine and adenine in DNA from a single species. We would expect the same for cytosine and guanine. The chemical analysis of purine and pyrimidine content provides strong support for the base pairing model of DNA molecular structure proposed by Watson and Crick.

## COPYING *DNA*

In chapter 12 we learned that chromatids separate during anaphase. In order to become X-shaped chromosomes during interphase, the chromatids must copy their DNA. This process of copying is called **replication.** The beauty of the Watson-Crick structure is that it indicates an obvious and simple mechanism for replication. If the two strands of

**FIGURE 14.5** Hydrogen bonding. The structure of the paired bases is such that the bases "fit" together at the appropriate distance to form hydrogen bonds (••••••). These paired bases are termed "complementary" bases.

········Indicates hydrogen bonds

**FIGURE 14.6**   DNA copying. A logical mechanism for DNA copying is for "new" individual nucleotides to pair with exposed bases when the two strands of DNA separate.

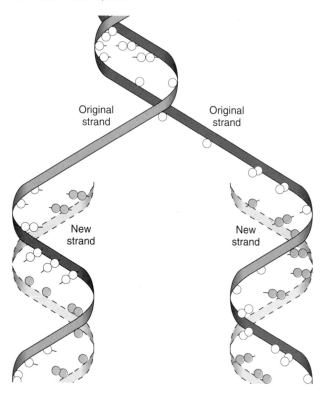

Original strand

Original strand

New strand

New strand

**FIGURE 14.7**   Meselson-Stahl experiments. (*a*) When parent cells are grown in N¹⁵ medium, all DNA is heavy form. When these cells are switched to N¹⁴ medium, the next generation of cells are predicted to contain one light and one heavy strand. (*b, c, d*) The position of the DNA after spinning in an ultracentrifuge indicates whether strands are light or heavy.

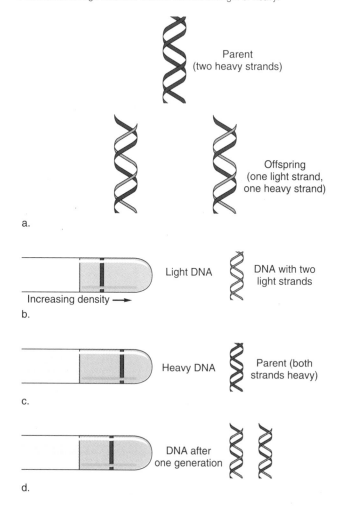

Parent
(two heavy strands)

Offspring
(one light strand,
one heavy strand)

a.

Increasing density ⟶

Light DNA

DNA with two
light strands

b.

Heavy DNA

Parent (both
strands heavy)

c.

DNA after
one generation

d.

nucleotides in the double helix are separated, individual nucleotides could "match" with the now exposed nucleotides on each strand, as shown in figure 14.6. An exposed adenine, for example, would pair only with a new thymine, and an exposed guanine would pair only with a new cytosine. In this way, two new double helixes would be formed that are exact copies of the single original helix.

The idea is certainly engaging, but is it true? Does DNA copy this way in the living cell?

Matthew Meselson and Frank Stahl, molecular geneticists working at the California Institute of Technology, formed the following hypothesis based on the Watson-Crick model:

*Hypothesis*   DNA replication occurs by separation of the two nucleotide polymers in the helical DNA molecule and the formation of a new matching strand on each of the separated old strands.

To test this hypothesis, Meselson and Stahl needed a way to distinguish the "old" strand from the proposed "new" strand. To do so, they grew bacteria on a nutrient medium containing a high concentration of nitrogen compounds with the heavy isotope of nitrogen ($N^{15}$). After a sufficient period of time, all the purines and pyrimidines in the bacterial DNA contained the heavy isotope rather than the light isotope, $N^{14}$. Next, they placed the bacteria on a medium with only the light isotope. Based on their hypothesis, Meselson and Stahl predicted that the DNA molecules in cells produced in the $N^{14}$ medium would have DNA molecules with one

heavy strand and one light strand, as illustrated in figure 14.7. The difference in these DNA molecules can be detected using an ultracentrifuge to separate the heavy and light molecules (see chapter 2 for a description of how molecules are separated with an ultracentrifuge).

If DNA is copied by some mechanism other than the one proposed in the hypothesis, perhaps some offspring would have all light strands and others would have all heavy strands, as shown in figure 14.8. When Meselson and Stahl carried out their experiment, however, their results supported the prediction illustrated in figure 14.7, providing clear evidence that DNA is indeed copied when the strands of the double helix separate and bases pair. Their results added support to the Watson-Crick model.

DNA is essentially the same in all members of a species. However, the sequence of nucleotide monomers in the DNA molecule can vary slightly from individual to individual. In humans, this slight variation can identify a person. The process by which people are identified by their DNA is called DNA fingerprinting (Thinking in Depth 14.1).

## DNA FINGERPRINTING

**FIGURE 1** DNA fingerprinting. Purified DNA (*a*) from a tissue sample is combined with restriction endonuclease enzymes to cleave the DNA into smaller fragments (*b*). When exposed to an electric field the DNA fragments separate into separate bands (*c*). The addition of radioactive probes (*d*) indicates the location of specific fragment.

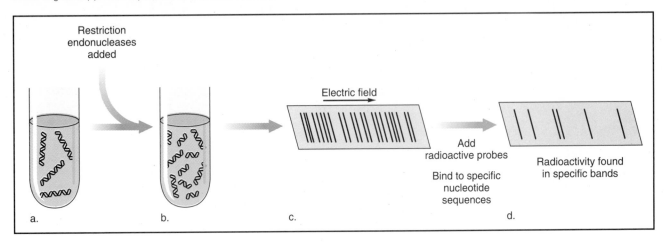

The sequence of nucleotide monomers in DNA is unique in every individual human being. Even though all humans are basically the same in terms of structure and physiology, genetic differences exist for different traits such as hair color and different physiological characteristics such as lactose intolerance (chapter 9). Moreover, there are about *3 billion* base pairs in a person's DNA. Some of these base pairs will be altered because of mutations, and these changes are not apparent unless the DNA is examined in detail. The process known as DNA fingerprinting provides such a detailed analysis; experts can detect the otherwise unknown differences that make each human being genetically unique.

The technique requires a very small sample of highly purified DNA. A single hair cell, for example, can provide enough DNA for analysis. Special enzymes, called restriction endonucleases, are added to the sample and cut the DNA molecules at a specific sequence of five to eight nucleotides, as shown in figure 1. In other words, at every location on the DNA molecule where the specific nucleotide sequence occurs, the enzyme will break the DNA polymer.

This treatment results in about 1 million DNA fragments of varying lengths. The fragments can be separated using an electric field. The size and the electric charge of the fragment will determine how far the fragment will move in the electric field. This procedure produces a series of bands; each band is composed of a different size fragment. The separated fragments are exposed to radioactively labeled and known nucleotide sequences called **probes** that are prepared in the laboratory. When the nucleotide sequence of the probe is complementary to a sequence of bases within a fragment, the two will bond by base pairing. By detecting the location of the radioactivity on the band pattern, the presence of matching probe and fragments can be established. The observed band pattern is characteristic of and unique to each individual, just as fingerprints are unique to each individual.

DNA fingerprinting has been used to identify individuals in criminal cases. For example, in rape cases, DNA from a sperm sample of the assailant can be compared to DNA from suspects. A match in DNA reveals the identity of the attacker.

With this one elegant proposal, Watson and Crick provided an explanation for one of the major questions in biology: How is the genetic material copied? Moreover, Watson and Crick's model led to numerous hypotheses about other genetic questions and opened doors to a great many discoveries, some of which are discussed in the sections that follow.

## DNA CONTROL OF CELL FUNCTION

How does DNA control the events that occur in cells? In earlier chapters we learned about the metabolic processes that take place in cells. The chemical reactions are controlled by enzymes, but what directs the enzymes? It is logical to hypothesize that genes, the individual units of heredity carried

**FIGURE 14.8** *Second prediction of the Meselson-Stahl experiment.* (a) A second possible copying mechanism might produce offspring with a heavy molecule in one daughter cell and a light molecule in the other cell. (b) This mixture of light and heavy molecules would appear as two separate bands in the ultracentrifuge.

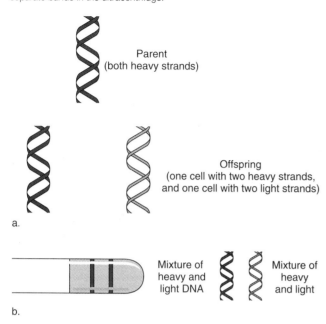

Parent
(both heavy strands)

Offspring
(one cell with two heavy strands,
and one cell with two light strands)

a.

Mixture of
heavy and
light DNA

Mixture of
heavy
and light

b.

on the DNA, correspond to the information needed to form one enzyme. In other words, the link between DNA and cellular control is the ability of genes (DNA) to direct synthesis of specific enzymes.

In the early 1940s, U.S. geneticists George Beadle and Edward Tatum tested this hypothesis by inducing changes in an organism's DNA and then observing changes in the organism's metabolic pathways. Actually, Beadle and Tatum performed their experiments before the role of DNA as the carrier of genetic information was even established. They did know, however, that genes carry genetic directions. The organism they used was a bread mold called *Neurospora crassa.* This organism can reproduce either sexually or asexually and is convenient to use in metabolic and genetic tests. The mold reproduces by forming haploid spores (chapter 13).

First, Beadle and Tatum treated *Neurospora* spores with radiation to induce changes in the genes as shown in figure 14.9a. Apparently, radiation can damage small portions of the DNA molecule so that a specific part of the DNA is nonfunctional. Such changes are called **mutations.** Beadle and Tatum predicted that if a gene were damaged, then some metabolic function would be lost. The metabolic function would be lost because a specific enzyme could not be produced.

Next, the two scientists placed the radiated spores on a complete growth medium with added vitamins and amino acids (fig. 14.9b). If the mold had lost the ability to synthesize a vitamin or amino acid when it was irradiated, it could still grow because the necessary materials were supplied in the growth medium. On the other hand, if the spores had

been damaged so badly by the radiation that they could not grow for other reasons, then they would not grow even in the complete medium.

Once Beadle and Tatum established that radiated spores could grow in the complete medium, they transferred the mold to a minimal medium that had no added vitamins or amino acids (fig. 14.9c). Some of the molds grew on the minimal medium, but others did not. Beadle and Tatum interpreted this result to mean that the molds that did not grow lacked the ability to synthesize some vitamin or amino acid.

What interpretation could you make about the molds that did grow in the minimal medium?

Next the scientists prepared several different media (fig. 14.9d) to discover whether vitamin synthesis or amino acid synthesis had been damaged. Their results, shown in figure 14.9d, indicated that the mold could not synthesize a vitamin.

If the mold had lost the ability to synthesize both a specific vitamin and a specific amino acid, what results would you predict?

What was the purpose of the two controls in Beadle and Tatum's experiment (fig. 14.9)? What did they show?

To discover precisely which vitamin was missing, Beadle and Tatum used the growth media shown in figure 14.9e. Their results indicated that only vitamin $B_6$ was deficient. They argued that the mold had lost the information in the gene that specified the enzyme that synthesizes vitamin $B_6$. In the next chapter we will look at further experiments that investigate the inheritance of a specific characteristic.

# DNA CONTROL OF ENZYME SYNTHESIS

Suppose Beadle and Tatum's interpretation is accurate. How is information on the gene (DNA) translated into an enzyme? This question suggests three more specific questions: How is information carried on the DNA? What does this information specify about enzymes? What events lead from DNA to a completed enzyme?

The answers to these questions have been discovered through a great number and variety of experiments. Consider the question on how information is carried in the DNA molecule. Early on in the study of molecular genetics, investigators proposed that the sequence of the four monomers in DNA could somehow specify the sequence of amino acids in a protein molecule, like a code in which one

FIGURE 14.9    Beadle-Tatum Experiment. (a) The mold *Neurospora* is irradiated to produce mutations. (b) Spores are transferred to complete medium and produce mold. (c) A portion of each mold colony is transferred to minimal medium. (d) A portion of a mutant colony is transferred to a variety of mediums. (e) A mutant colony is transferred to media where different vitamins are supplied.

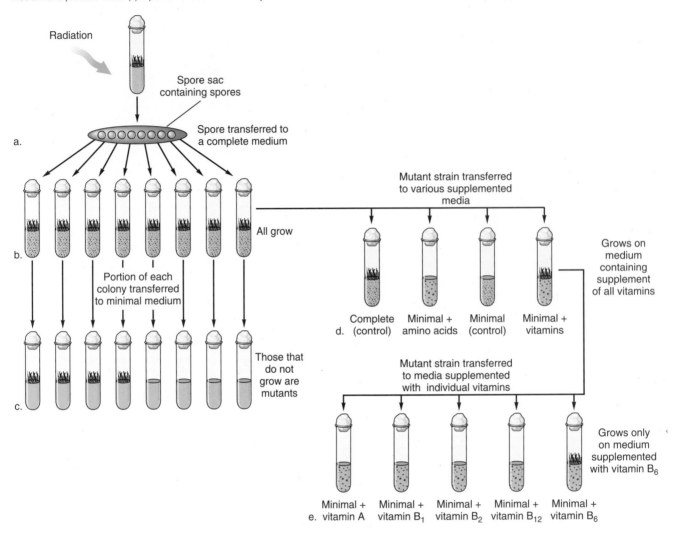

character stands for another. Biologists knew that cellular proteins were made up of about twenty different amino acids, so they knew one nucleotide could not specify one amino acid. Even sequences of two nucleotides would not be enough since only sixteen different pairs could be formed with four nucleotides. At least three nucleotides must code for each amino acid. Groups of three nucleotides would allow sixty-four different combinations—more than enough to specify the twenty amino acids. This language or scheme for carrying information is called the **triplet code.**

While this makes an interesting and plausible explanation, is this actually the way information is carried? To answer this question, biologists had to examine the events that result in protein synthesis more closely. (Recall that enzymes are proteins.)

The simplest process we can propose is that protein is synthesized in direct association with DNA. We observe, however, that protein synthesis occurs in the cytoplasm, not in the nucleus where DNA is located. Something must be transferring information from the nucleus to the sites of protein synthesis in the cytoplasm.

Several biologists observed a second kind of nucleic acid in cells, called **ribonucleic acid**, or **RNA.** They suggested that RNA played an essential role in protein synthesis.

*Reproduction and Inheritance*

**FIGURE 14.10**  RNA. As in DNA, the RNA molecule contains four different bases, but uracil replaces thymine, and the ribose sugar is present instead of deoxyribose.

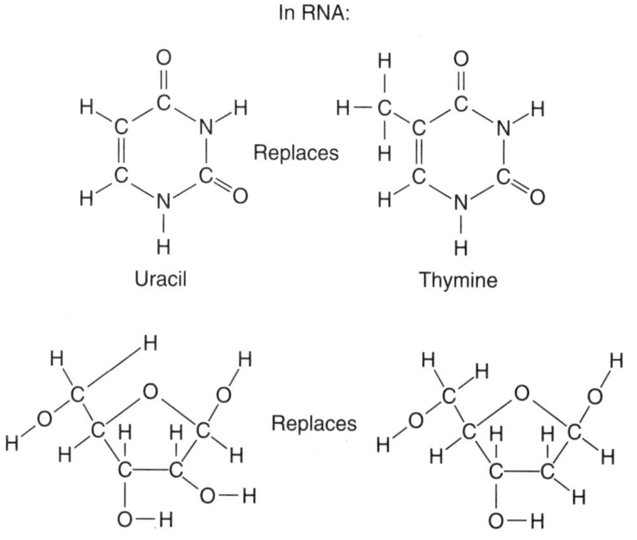

In RNA:

Uracil    Replaces    Thymine

Ribose    Replaces    Deoxyribose

RNA is similar to DNA except that the base uracil replaces thymine and the sugar component is slightly different (fig. 14.10). Given their similar construction, we can hypothesize that base pairing occurs between the two nucleic acids. Uracil in RNA can pair with adenine in DNA. In this way, information can be copied from DNA to RNA and the triplet code can specify amino acids by transferring information to RNA. This transfer of information from DNA to RNA is called **transcription.**

RNA can be distinguished from DNA by labeling the uracil with a radioactive tracer. Recall in chapter 4 that radioactive molecules could be used to trace metabolic pathways. Radioactive uracil can be used in the same way to trace the location of RNA in the cell. As we see in figure 14.11, within 5 minutes after labeled uracil is added to a cell, newly synthesized RNA is localized in the nucleus. After 60 minutes, however, we find RNA containing labeled uracil only in the cytoplasm.

Does this RNA receive the message from DNA and transfer it? Biologists have answered this question with an experiment involving a **cell-free system**, in which a culture of cells such as bacteria are ground with an abrasive and the cells are ruptured (fig. 14.12). If we add amino acids to this preparation, we find that protein can still be synthesized. Furthermore, if we destroy the DNA in the cell-free material with a DNA digesting enzyme (DNAase), we observe that protein is *still* synthesized. Biologists have interpreted these results to mean that the RNA in the cell-free preparation is directing protein synthesis. In a modification of this experiment, biologists added a crude extract of RNA (but no DNA) to the cell-free preparation. When they did so, protein synthesis increased.

**FIGURE 14.11**  RNA synthesis. (a) Cells are exposed to radioactive uracil and after five minutes the free uracil is washed away. Radioactivity is found only in the nucleus. Uracil has apparently been incorporated into RNA synthesized in the nucleus. (b) Exposure to radioactive uracil for five minutes, then placement in nonradioactive uracil, results in the appearance of radioactivity only in the cytoplasm after 60 minutes.

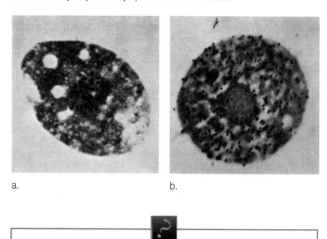

a.    b.

After a period of time, the preparation in figure 14.12b would no longer synthesize protein even if amino acids were added. How do you interpret this observation?

More detailed analysis of RNA revealed not one, but three different kinds of RNA in the cell. One kind, **messenger RNA (mRNA)** is formed in the nucleus in association with DNA and appears to carry the information from DNA to the cytoplasm. A second kind, **ribosomal RNA (rRNA)**, is a major component of the ribosomes in the cytoplasm (recall from chapter 2 that ribosomes are an organelle in the cytoplasm). A third kind of RNA, **transfer RNA (tRNA),** is a smaller molecule that links to specific free amino acids in the cytoplasm during protein synthesis.

What exactly are the roles of these RNAs in protein synthesis? Biologists have sought to answer this question through many experiments. In one experiment, U.S. biochemist Marshall Nirenberg predicted that if synthetic mRNA composed of only one base were added to a cell-free preparation, then a protein composed of only one kind of amino acid would be synthesized. That is because there would be only one code word on the mRNA. For example, if mRNA were composed only of a polymer containing uracil, the resulting protein would contain only the amino acid coded by -u-u-u-, if the code is a triplet. When Nirenberg conducted the experiment, the protein produced was made entirely of phenylalanine. Not only had Nirenberg found evidence that mRNA carries information, he had also discovered the code for the amino acid phenylalanine!

This "reading" of the code in mRNA to produce proteins is called **translation.** Biologists revealed the codes for all twenty amino acids in similar experiments using a variety of synthetic mRNAs.

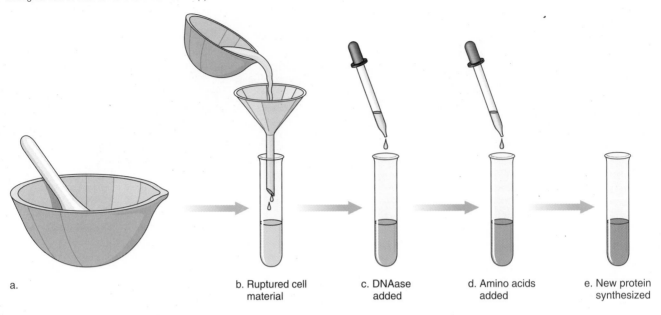

**FIGURE 14.12** Cell-free systems. (*a*) Cells are ground with an abrasive to rupture them and release their contents. (*b*) Material is put through a coarse filter to remove the abrasive. (*c*) DNAase is added to destroy DNA. (*d*) Amino acids are added. (*e*) Protein synthesis is observed utilizing added amino acids.

a.

b. Ruptured cell
   material

c. DNAase
   added

d. Amino acids
   added

e. New protein
   synthesized

As biologists discovered the code words for amino acids, they discovered that many amino acids are coded by more than one word, or triplet. Why is this not surprising?

What is the role of transfer RNA? Why should it be part of the protein synthesis process? Scientists argued that there must be some mechanism for amino acids to recognize the appropriate positions on the mRNA. Since nucleic acids can form base pairs, the tRNA composed of nucleotides would "recognize" a specific set of three bases along the mRNA and match with the mRNA at this position, thus placing the amino acid in the right spot. In other words, by attaching itself to a specific tRNA, an amino acid can be positioned in the correct location in the protein.

Suppose that we are correct about amino acids specifically bonding with tRNAs. Does this mean that tRNA provides the recognition mechanism? Perhaps the tRNA is attached for some other, unknown function and a completely different amino acid recognition mechanism is in operation.

Biologists examined this question in an elegant experiment using the amino acid cysteine (fig. 14.13). After cysteine was bonded to the cysteine-specific tRNA, the sulfur atom was removed from the amino acid. This converted cysteine to the amino acid alanine without removing the cysteine-specific tRNA. Would the alanine be placed in a protein where cysteine should be, or would it be placed where alanine should be? When researchers analyzed the resulting protein, the alanine was located in positions nor-

**FIGURE 14.13** The role of tRNA. After the amino acid cysteine is attached to the cysteine-specific tRNA (*a*), a catalyst removes the sulfur group and thus converts cysteine to alanine (*b*). It is observed that the alanine is now placed in the position in the protein normally occupied by cysteine.

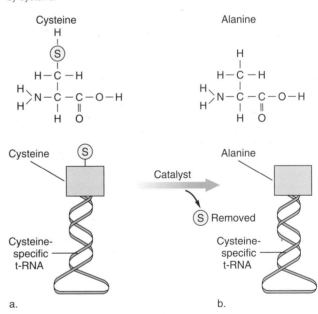

mally occupied by cysteine. This result clearly supported the proposal that specific tRNAs paired with triplet codes on the mRNA to specify the location of amino acids in a protein.

Ribosomal RNA is a major component of ribosomes. Chemical analysis and the use of radioactive amino acids have indicated that newly synthesized protein first appears in

**FIGURE 14.14** Ribosomes and protein synthesis. In the electronmicrograph, a single mRNA molecule extends from the lower right to the upper left of the figure. Numerous ribosomes are observed attached to the mRNA. Structures extending from the ribosomes are interpreted as growing protein molecules in the process of synthesis.

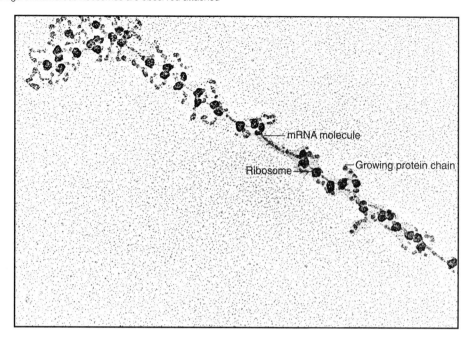

**FIGURE 14.15** Summary of protein synthesis. mRNA synthesized on DNA in the nucleus moves to the cytoplasm. mRNA combines with ribosomes and tRNA in the cytoplasm where protein is produced.

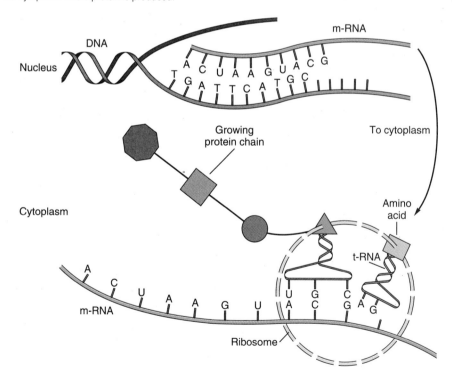

the cytoplasm in close association with ribosomes. Electron micrographs have further shown that ribosomes attach to mRNA and growing protein polymers. Biologists have interpreted these observations as an indication that the ribosomes function to bring together mRNA and tRNA and assemble the amino acids to form proteins (fig. 14.14).

From these and numerous other experiments, an overall scheme of protein synthesis has been developed; it is illustrated in figure 14.15. The sequence of nucleotides in DNA and mRNA uniquely specify the sequence of amino acids in proteins. Base pairing allows the transcription of information to mRNA and the translation of information by tRNA.

## SUMMARY

*Discovering the simple yet elegant process of protein synthesis is one of the great achievements of modern biology. Our understanding of this process has generated many hypotheses about genetics and cell metabolism that scientists continue to investigate, increasing our understanding of biology. As you continue to study these processes, keep in mind the following key points about molecular genetics.*

1. **Deoxyribonucleic acid (DNA)** is comprised of four kinds of **nucleotides.** Each kind of nucleotide contains a different base: adenine, cytosine, thymine, or guanine.

2. In the DNA of any given organism, the amounts of adenine equal thymine and the amounts of cytosine equal guanine.

3. **X-ray diffraction** experiments led to the model of the DNA molecule developed by James Watson and Francis Crick. The model shows DNA as two strands held together by hydrogen bonds between adenine and thymine and between cytosine and guanine.

4. DNA **replication** occurs by **base pairing,** in which the bases in a single strand of DNA are copied onto another strand.

5. Experiments by Matthew Meselson and Frank Stahl indicated that DNA is copied when the two strands in the original DNA molecule separate and the bases on the existing strands pair with new bases to create two new double-stranded DNA molecules.

6. DNA appears to control cell function by specifying the enzymes and other proteins present in the cell.

7. **Messenger RNA (mRNA)** is formed on DNA when RNA pairs with bases in the DNA molecule. This process, called **transcription**, transfers information to the mRNA.

8. Messenger RNA moves from the nucleus to the cytoplasm where it associates with **ribosomes.**

9. **Transfer RNA (tRNA)** combines with specific amino acids. Bases in tRNA pair with sets of three bases, called **triplet codes**, in the mRNA molecule to establish the order of amino acids.

10. The ribosome appears to bring together the mRNA and tRNA with attached amino acids and to facilitate the linking of amino acids to form the new protein.

## REFERENCES

Allfrey, Vincent, and Alfred Mirsky. September 1961. "How Cells Make Molecules." *Scientific American* 205, no. 3: 74–82.

Darnell, James, Jr. October 1985. "RNA." *Scientific American* 253, no. 4: 68–78.

Felsenfeld, Gary. October 1985. "DNA." *Scientific American* 253, no. 4: 58–67.

Nirenberg, Marshall. March 1963. "The Genetic Code: II." *Scientific American* 208, no. 3: 80–94.

## EXERCISES AND QUESTIONS

1. Which of the following best explains why the statements about purine and pyrimidine percentages in table 14.1 are observations rather than interpretations (page 168)?
   a. The results show clear differences between organisms.
   b. The statements show that adenine pairs with thymine and cytosine with guanine.
   c. Verbal statements would not be used to present interpretations.
   d. The data in table 14.1 show more details of DNA composition.
   e. The statements describe the data in the table.

2. Which of the following is an interpretation made by Watson and Crick when they constructed their model of DNA (page 169)?
   a. DNA is composed of four different monomers.
   b. The molecule is composed of two strands.
   c. X-ray diffraction shows distances in the DNA molecule.
   d. The DNA molecule carries genetic information.
   e. DNA is localized in the nucleus of the cell.

3. If each nucleotide in DNA paired only with itself (A with A, C with C, etc.), which of the following statements would most likely be true?
   a. A pairs with T and C pairs with G.
   b. DNA could not replicate.
   c. The two strands of each DNA molecule would be identical.
   d. DNA could not carry genetic information.
   e. More monomers would be required to synthesize DNA.

4. Which of the following would most likely happen to the DNA helix structure if purines paired only with other purines (page 170)?
   a. The molecule would be longer.
   b. The molecule would be lighter.
   c. The helix structure would be more obvious in the X-ray diffraction patterns.
   d. The two strands of the molecule would be further apart.
   e. The molecule would be shorter.

5. A student states, "If A paired with G and C paired with T, DNA could not be replicated." How would you contradict this statement?
   a. A pairs with T and C pairs with G.
   b. As long as each nucleotide pairs specifically with only one other nucleotide, replication will occur.
   c. The two strands must separate before replication can occur.
   d. If A paired with C and G paired with T, DNA could not be replicated.
   e. The student is assuming that DNA carries the information specifying a cell's characteristics.

6. In the Meselson-Stahl experiment, if a cell with DNA composed of one light and one heavy strand divided and replicated its DNA in a medium containing only the light isotope, $N^{14}$, what would be the composition of the double-stranded DNA molecules in the two resulting cells?
   a. One cell with all light DNA, the other with one heavy and one light strand.
   b. Both cells with one light and one heavy strand.
   c. One cell with all heavy DNA and one cell with all light DNA.
   d. Both cells with all heavy DNA.

7. What interpretation could you make about the molds that did grow on the minimal medium in figure 14.9c (page 173)?
   a. These molds had lost the ability to synthesize more than one vitamin.
   b. These molds had lost the ability to reproduce.
   c. The DNA of these molds was not affected by the radiation.
   d. Growth of these molds would have been even greater in a complete medium.
   e. These molds had lost the ability to synthesize more than one amino acid.

8. If the molds in figure 14.9d had lost the ability to synthesize both a specific amino acid and a specific vitamin, what results would you predict (page 173)?
   a. Mold would not grow on either a minimal medium with vitamins or a minimal medium with amino acids.
   b. Mold would grow on both controls.
   c. Mold would grow in all kinds of media.
   d. Mold would not grow on the complete medium.

9. Which of the following was most likely the purpose of the two controls in figure 14.9d (page 173)?
   a. To measure the amount of mold growth in complete medium.
   b. To determine whether the mold had lost the ability to synthesize both a vitamin and an amino acid.

   c. To see if the culture medium could support the growth of organisms other than the mold.
   d. To check if the ability to grow on complete medium and not grow in minimal medium had remained unchanged.

10. Consider the hypothesis that RNA is synthesized on DNA. To test this hypothesis, some investigators injected radioactive monomers into organisms A and B to form radioactive RNA. Then they purified the labeled RNA from each organism. They also purified the DNA from organism A, heated it to cause the strands to separate, and chemically attached the single strands to plastic beads. When the beads were mixed with RNA from organism A, radioactivity stuck to them. When the beads were mixed with radioactive RNA from organism B, radioactivity did not stick to them. Many scientists accept observations such as these as evidence that RNA is synthesized on DNA. Which of the following assumptions are they making?
    a. Radioactivity causes mutations in organism B.
    b. Heating DNA changes the sequence of its monomers.
    c. RNA molecules from organisms A and B have the same sequence of monomers.
    d. Only nucleic acids with complementary sequences of monomers can stick together by base pairing.
    e. Only nucleic acids with the same sequence of monomers can stick together by base pairing.

11. Suppose one strand of mRNA coded for a protein molecule 347 amino acid monomers in length. If the first two monomers were lost from the mRNA, how many of the amino acid monomers would be changed in protein molecules later synthesized using this information?
    a. 1
    b. 2
    c. 4
    d. 0
    e. More than 4.

12. Suppose you discover an organism in which all proteins are composed of only thirteen amino acids. Which of the following predictions would be least acceptable?
    a. Amino acids will be coded by sets of two nucleotides in DNA.
    b. There will be thirteen different tRNAs.
    c. mRNA will be transcribed from DNA.
    d. Amino acids will be coded by single nucleotides in DNA.
    e. There will be as many mRNAs as there are kinds of protein.

13. Before human red blood cells are released into the bloodstream, they lose their nuclei. Yet it can be shown that for a considerable time after they are in circulation they are capable of synthesizing additional hemoglobin. That is, if red blood cells without nuclei are incubated in a medium containing radioactive amino acids, some of the labeled amino acids are incorporated into hemoglobin. Based upon what you know of transcription and translation of genetic information, which of the following rationales is most justified?
    a. Hemoglobin must be capable of self-replication.
    b. A portion of the nucleus containing the hemoglobin genes remains behind when the rest of the nucleus is lost from the developing red blood cells.
    c. Some of the RNA that encodes the information for hemoglobin remains in the cytoplasm after the nucleus is lost.
    d. Hemoglobin is not translated directly from nucleic acid but is made by enzymes that were synthesized before the cell nuclei are lost.

14. Human hemoglobin differs significantly in amino acid sequence from frog hemoglobin. When human hemoglobin messenger RNA is carefully injected into frog eggs, traces of human hemoglobin can be found in the eggs after one hour. From this experiment it can be concluded that:
    a. Uninjected frog eggs must contain the genes for human hemoglobin.
    b. The same code words (groups of three nucleotides on mRNA) specify the same amino acids during translation in humans as in frogs.
    c. The rate of RNA synthesis (transcription) must be very high in uninjected frog eggs.
    d. Injecting mRNA possibly damages the RNA molecule and changes the sequence of monomers.

15. After a period of time the preparation in figure 14.12*b* would no longer synthesize protein even if amino acids were added. Which of the following is the best interpretation of this observation (page 175)?
    a. The mRNA has broken down.
    b. Uracil is no longer available in the preparation.
    c. DNA in the preparation is no longer functional.
    d. Grinding the cells destroyed the ability to synthesize protein.
    e. Adding amino acids is not necessary for protein synthesis.

16. As biologists discovered the code words for amino acids, they found that many amino acids are coded by more than one word, or triplet. Why was this not surprising (page 176)?
    a. tRNA matches with triplets on mRNA.
    b. Uracil is found only in RNA.
    c. Twenty amino acids are commonly found in cells.
    d. Triplets are composed of three bases.
    e. Sixty-four different triplet codes are possible.

17. Suppose you discover an organism containing only six amino acids. Which of the following inferences about this organism is most justified?
    a. This organism produces only six different forms of messenger RNA.
    b. Amino acids must be coded by sets of at least two nucleotides in DNA in this organism.
    c. The organisms cannot produce more than six different kinds of protein.
    d. The protein of this organism will contain monomers other than amino acids.
    e. This organism could produce protein molecules no more than six amino acids in length.

*Reproduction and Inheritance*

CHAPTER
15

# MENDELIAN GENETICS

You don't have to be a biologist to recognize that organisms of all species pass their characteristics to their offspring. Some of these characteristics are easy to see: If your parents are blond, chances are you have blond hair too. Other physiological and biochemical characteristics, like lactose intolerance, may be less obvious. Nevertheless, the inherited nature of these qualities is clear.

Are there patterns or "rules" for the transmission of these traits? In the nineteenth century, most scientists believed that parental traits were "blended" in the next generation, that is, the genetic contributions of parents were mixed to produce offspring with intermediate characteristics. According to this **blending hypothesis,** the mixed traits could not be separated in subsequent generations. The mechanism would work somewhat like mixing red and white paint; the resulting pink color could not be "unmixed."

A few scientists of this time proposed a different explanation, called the **particulate hypothesis.** They believed that characteristics are inherited as discrete units that remain separate. This type of inheritance would be like mixing red and white marbles. Once mixed, the marbles could be separated.

In the mid-nineteenth century, an Austrian monk named Gregor Mendel carried out a series of experiments that tested these hypotheses. Mendel studied inheritance in pea plants. His carefully controlled experiments took the better part of his life. In one experiment, he interbred, or crossed, red-flowered plants with white-flowered plants, as shown in figure 15.1. He collected the seeds of this mating and planted them. When these plants reached maturity, they all had red flowers. Next, Mendel crossed two of these red-flowered offspring. When those seeds reached maturity, most of the plants had red flowers, but a few had white flowers.

The results of Mendel's experiments favor the particulate theory of inheritance. Only if the traits retained their individual character in the offspring, or first generation, could they reappear in the second generation.

What results in Mendel's experiment would have supported the blending hypothesis?

What happened to the white trait in the first generation? There was no evidence of it at all in the red-flowered plants. To explain this observation, Mendel proposed that some traits are **dominant** and some are **recessive.** In the case of the pea plants, the red flower color is dominant and the white flower color is recessive. That means that when both traits are present, as they are in the first-generation plants, only the dominant trait is displayed.

Mendel studied the seven traits in pea plants listed in table 15.1. In all cases, only one trait was observed in the

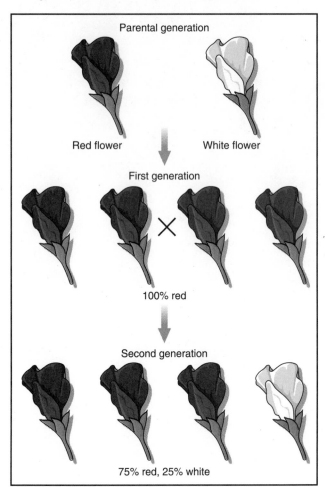

**FIGURE 15.1** Mendel's experiments with a single characteristic. Two parents of different flower color (red and white) are crossed and the resulting seed planted. These seeds grow into the first generation and all exhibit red flowers. When first generation plants are crossed and the resulting seed planted, about one quarter of the plants have white flowers.

first generation, but the other trait reappeared in the second generation. Even more striking, the numbers of plants in the second generation exhibiting the dominant trait were always about three times the number of plants exhibiting the recessive trait. In other words, the ratio of dominant to recessive was about 3 to 1.

What could explain this ratio? In the 1850s, no one had observed chromosomes under a microscope or proposed genes. Mendel realized, however, that something was controlling inheritance in his garden peas. He proposed that some "factor" controlled each trait, and that these factors separated when gametes were formed. Recall from chapter 14 that gametes are the haploid cells that combine to form a zygote, or offspring. In sexual reproduction, each parent contributes a gamete. Mendel believed that each gamete would have the factor for one trait. At fertilization, the factors would combine to produce the observed ratio of 3 to 1. Figure 15.2 illustrates Mendel's hypothesis.

**TABLE 15.1**   Mendel's Results of Crosses with Single Character

| Parental Characters | First Generation | Second Generation | Second Generation Ratio |
|---|---|---|---|
| 1. Round × wrinkled seeds | All round | 5,474 round: 1,850 wrinkled | 2.96:1 |
| 2. Yellow × green seeds | All yellow | 6,022 yellow: 2,001 green | 3.01:1 |
| 3. Red × white flowers | All red | 705 red: 224 white | 3.15:1 |
| 4. Inflated × constricted pods | All inflated | 882 inflated: 299 constricted | 2.95:1 |
| 5. Green × yellow pods | All green | 428 green: 152 yellow | 2.82:1 |
| 6. Axial × terminal flowers | All axial | 651 axial: 207 terminal | 3.14:1 |
| 7. Long × short stems | All long | 787 long: 277 short | 2.84:1 |

*From: W. T. Keeton,* Biological Science, *3rd Edition Norton Publishers. Table 14–1, page 584.*

**FIGURE 15.2**   Separation and combination of factors. (a) The white parent gametes with only the white factor and the red parents only gametes with the red factor. These factors recombine in the first generation. (b) First-generation plants form gametes with either the red or white factor. These gametes combine randomly to form some plants with red flowers and others with white.

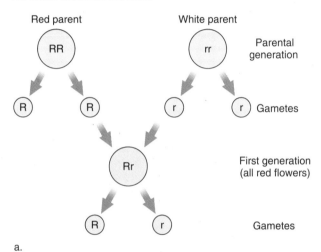

a.

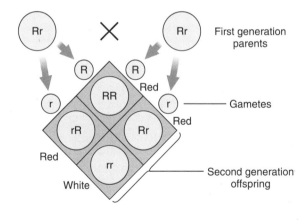

b.

**Genetic Notation**
As a convenient shorthand, a gene trait is indicated by a single letter. In figure15.2, the dominant red trait is shown as a capital R while the recessive white trait is shown as r.

R = dominant red trait
 r = recessive white trait

White-flowered plants produce gametes with only the white factor, and red-flowered plants produce gametes with only the red factor. In the first generation, all plants contain both red and white factors, but we see only the red because it is dominant. Biologists have developed a way of designating dominant and recessive traits. Notice in figure 15.2 that the dominant red trait is denoted with a capital letter: R. The recessive white trait is denoted with a lower-case letter: r. An offspring that has received a dominant trait from one parent and a recessive trait from the other parent is denoted with Rr.

When the first-generation plants reproduce, they produce equal numbers of dominant and recessive gametes. If they are crossed with each other, the random combination of gametes causes three-quarters of the offspring to have the dominant trait and one-quarter to have the recessive trait. In our example, illustrated in figure 15.2, three-quarters of the second generation have red flowers and one-quarter have white flowers. The numbers predicted in Mendel's hypothesis agree almost perfectly with the numbers observed in table 15.1.

> If one of the first-generation plants was crossed with a plant that had white flowers, what would the offspring look like?

Mendel's insightful explanation of inheritance was an impressive success, but when he published his results in 1866, they attracted almost no interest. Historians have suggested several reasons for this lack of response. One is that other scientists of Mendel's day did not understand plant reproduction and fertilization; thus they did not understand the significance of Mendel's experiments. Another possibility is that other scientists did not understand Mendel's use of mathematical ratios. In 1900, however, Mendel's research was "rediscovered" and recognized for its important contributions.

| TABLE 15.2 | Inherited Traits Exhibiting Mendelian Ratios |
|---|---|
| **Trait** | **Description** |
| Dwarfism in corn plants | Decreased height |
| Albinism in corn plants | White plant color instead of green |
| Huntington's disease in humans | Degeneration of nervous system |
| Widow's peak in humans | Hairline comes to point in middle of forehead |
| Albinism in humans | Lack of pigmentation in skin and eyes |

The patterns Mendel observed in pea plants have been observed in countless other sexually reproducing plants and animals. Table 15.2 lists several examples. Apparently, Mendel's explanation of inheritance is universally applicable.

## INHERITANCE AND MEIOTIC CELL DIVISION

Are the rules of inheritance discovered by Gregor Mendel related to the behavior of chromosomes we studied in chapter 12? Can Mendel's observations be explained in terms of meiosis, chromosome movements, gamete formation, and fertilization?

We hinted in the previous section that the answer to these questions is yes. If we reexamine the chromosomes in a diploid cell, we can logically propose that Mendel's "factors" correspond to genes on the chromosomes (fig. 15.3). Recall that genes for a single characteristic (such as flower color) are found on homologous chromosomes (chapter 12). The genes on each homologue, however, may be for different expressions of the trait. For example, the gene for white flower color could be located on one homologue while the gene for red flower color could be located on the other. These expressions of the same trait are called **alleles.** As Mendel proposed, each plant will contain two alleles for the same trait. The trait that we see, such as red flower color, is called the **phenotype.** The genes that are actually present on the chromosomes are called the **genotype.** In figure 15.2, for example, the first generation plant has a phenotype of red and a genotype of Rr. A plant that had a genotype of RR would also have a phenotype of red; a plant that had a genotype of rr would be white.

Why must the genotype always be shown as two alleles?

If the alleles of a plant are the same—that is, the genotype is RR or rr—the plant is referred to as **homozygous** for that trait. If the alleles are different—Rr—the plant is called **heterozygous** for that trait.

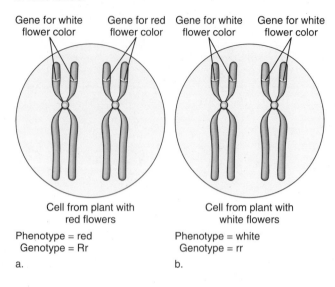

**FIGURE 15.3** Factors corresponding to genes. (*a*) A cell from a plant with red flowers shows one pair of homologous chromosomes that carry the gene for flower color. Different alleles may be carried on the different homologues. (*b*) A cell from a plant with white flowers illustrates the case for white flowers.

Gene for white flower color    Gene for red flower color    Gene for white flower color    Gene for white flower color

Cell from plant with red flowers
Phenotype = red
Genotype = Rr
a.

Cell from plant with white flowers
Phenotype = white
Genotype = rr
b.

Why does each homologue have two copies of each gene? In a plant with white flowers, what would be the alleles on the homologues?

Do genes for a characteristic separate when gametes are formed? Recall that in meiosis, homologues separate as shown in figure 15.4. Each gamete is left with only one allele of the gene. At fertilization, two gametes combine, bringing together the alleles, one from each parent, so that the diploid cells have two genes for a given trait. Since the combination of gametes is random, the chance of one of Mendel's second-generation pea plants getting two white genes is one in four. That explains why only one-fourth of the second-generation plants had white flowers.

Mendel's explanation of his results—that flower color is controlled by some unknown "factor"—corresponds to what we know today about meiosis and fertilization. We observe **independent assortment** of traits because of the separation of homologues in meiosis. We observe the combination of traits because each gamete contributes one allele at fertilization. Independent assortment and combination are key terms that describe the relation between Mendel's explanation and the events of meiosis and fertilization.

## FIGURE 15.4  Separation of alleles. Haploid cells resulting from meiosis will contain only one allele even though the parental cells contained both alleles.

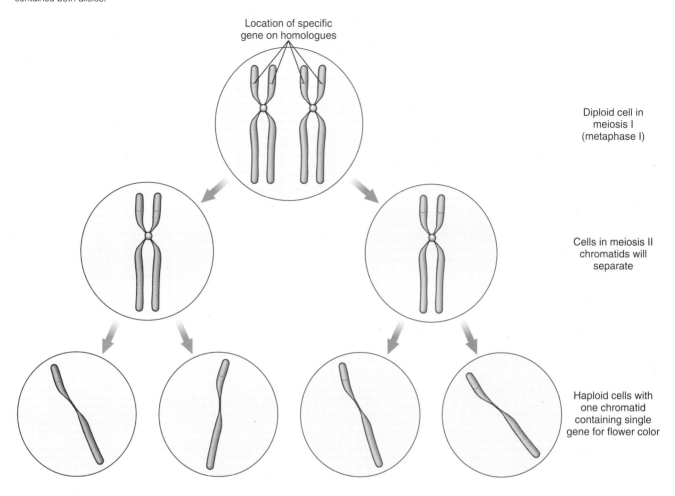

Location of specific gene on homologues

Diploid cell in meiosis I (metaphase I)

Cells in meiosis II chromatids will separate

Haploid cells with one chromatid containing single gene for flower color

## INHERITANCE AND PROTEIN SYNTHESIS

In the last chapter, we learned how genetic information carried by DNA in the chromosomes is translated into enzymes and other proteins. Do these molecular events have any bearing on inheritance? Mendel proposed that traits are either dominant or recessive. Can molecular mechanisms explain why this is so?

Consider the example of albinism. Every now and then an individual appears in a population who lacks pigmentation, or color, of the skin or eyes (fig. 15.5). Normal pigmentation is caused by the pigment melanin, which is produced from the amino acid phenylalanine. Albino individuals lack one of the enzymes that convert phenylalanine to melanin (fig. 15.6). Recall figure 4.10, which showed how lack of an enzyme can block a metabolic pathway. Apparently, the lack of the enzyme is caused by deficient information in the DNA of the individual.

Is albinism a recessive or dominant trait? We can answer this question by examining the chromosomes. The gene that codes for the enzyme that produces melanin should be found on both homologous chromosomes. The position of this enzyme in the melanin production pathway is illustrated in figure 15.7. Even if the gene is defective on one homologue, the enzyme could be produced from information on the other homologue. In other words, an individual could be heterozygous for melanin production,

A student argued that an organism heterozygous for albinism should produce one-half as much of the essential enzyme as a homozygous normal individual. What is the student assuming?

a.

b.

FIGURE 15.6 Deficiency of melanin production in albinism. Lack of the functional enzyme to catalyze conversion of tyrosine to melanin will result in a blockage of the pigment production pathway.

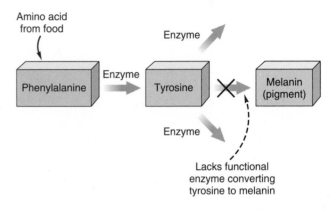

FIGURE 15.7 Melanin production. Even if an individual has a defective gene for the melanin-producing enzyme, a functional enzyme can be produced by the other homologue and normal coloration will be exhibited.

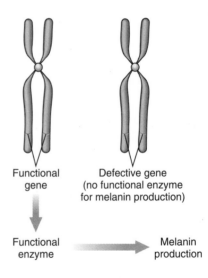

with one allele for production and the other for albinism (no melanin production). Melanin production would be expressed, but albinism would not. Thus, albinism should be a recessive trait.

If albinism is indeed a recessive trait, what predictions can we make about its inheritance? One prediction is that if an albino and a normal homozygous individual are crossed, they will have only offspring with the normal phenotype. These offspring will be heterozygous. If two of the offspring interbreed, they may have some albino offspring.

Biologists have tested these predictions many times and have found that albinism is indeed recessive. They have studied

this trait in both laboratory animals and humans, though of course humans have much longer life spans, making it hard to examine several generations. To track albinism in human families, biologists use a **pedigree chart,** which shows the relationships between individuals in a family and allows researchers to track a genetic trait (Thinking in Depth 15.1). A pedigree chart can reveal whether a trait is dominant or recessive, as well as the genotypes of many family members.

Assuming that albinism is under the control of a single gene pair, could two albino individuals ever produce any offspring with normal pigmentation?

## PEDIGREE CHARTS

The inheritance of genetic traits in human families is often investigated using pedigree charts. Such a chart can be constructed by examining family members from several generations and by reading family records and histories that may contain descriptions of phenotypes. Once family relationships and the appearance of the trait are established, the family can be diagrammed in a chart. Figure 1 is a pedigree chart showing the inheritance of albinism in a family.

Do the data in the pedigree chart indicate whether the trait is dominant or recessive? We can examine several relationships to find out. For example, female 4 in generation 2 is an albino and the offspring of two normal parents. The only way this daughter could be albino is if both parents are heterozygous for albinism. The daughter must have inherited a recessive albino allele from each parent. The same argument applies to males 1 and 2 in generation 3.

The offspring of male 3 and female 4 in generation 2 are consistent with the proposal that albinism is recessive. If albinism were dominant, all their offspring would be albinos. This is not the case, however. Some of their children are normal, therefore albinism must be recessive.

What can you conclude about the genotype of male 3 in generation 2? For any of this male's children to be albino, they must inherit an albino allele from their father and one from their mother, female 4 in generation 2. So the father must have at least one albino allele. Since his phenotype is normal, he must be heterozygous for the trait.

> ?
>
> What can you conclude about the genotypes of male 4 and female 6 in generation 3?

**FIGURE 1**   The following symbols are used in pedigree charts:

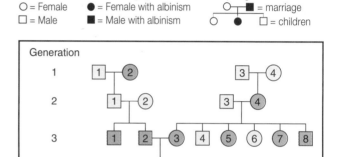

## MUTATIONS

Most of the time, the genetic information on chromosomes codes for the right protein. What happens when the DNA is changed and results in a different protein, possibly even a defective one? We can answer this question by studying **sickle-cell anemia,** a disease in which red blood cells change shape when oxygen concentration is low.

Normal cells are disc-shaped, but in individuals with sickle-cell anemia they can become crescent or sickle-shaped (fig. 15.8). The altered red blood cells are more fragile than normal cells and can be easily broken. The blood cannot replace cells as fast as they are lost, so the oxygen-carrying capacity of the blood is reduced. Moreover, the sickled cells can block small blood vessels, reducing the flow of blood to some parts of the body.

All humans have an oxygen-carrying protein in their blood called hemoglobin. If we examine individuals with sickle-cell anemia, we find that the hemoglobin in their cells is different from that in normal cells. Apparently, a change in the hemoglobin molecule causes the drastic changes in red blood cells and the destructive effect on the

body. A substitution in only one base in the DNA would be enough to cause this effect, since it would change the triplet code for the amino acid.

In chapter 14, we saw how George Beadle and Edward Tatum used radiation to induce genetic changes in the spores of a mold. The mold they used lost the ability to synthesize vitamin $B_6$ after it was irradiated. Beadle and Tatum hypothesized that in the irradiated spores the enzyme that produces vitamin $B_6$ in the metabolic pathway was no longer functioning. Furthermore, they proposed that the enzyme was not functioning because the genetic information necessary to make it function had been changed by the radiation. We call such a genetic change a **mutation.** The organism whose genetic makeup has been changed is called a **mutant.**

Suppose that the vitamin $B_6$ deficiency in Beadle and Tatum's mold was indeed caused by a mutation. What would you predict if the mutant mold were crossed with a normal mold? The results of the cross should be similar to the albino cross, that is, it should produce all heterozygotes with the normal phenotype (normal meaning they can synthesize vitamin $B_6$). If two of the heterozygote offspring are crossed, their offspring should be three-quarters normal and

FIGURE 15.8

**FIGURE 15.8**  Red blood cells from a person with sickle cell anemia. (*a*) Normal, disc-shaped cells. (*b*) Deformed, "sickled" cells.

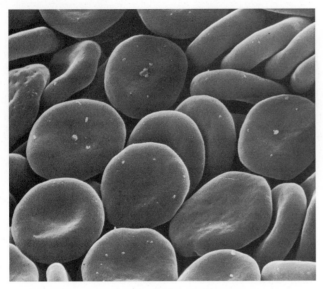

a.

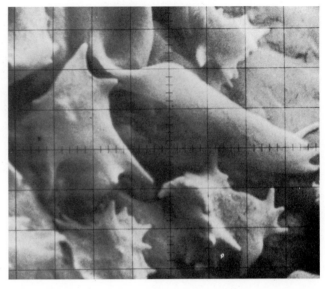

b.

one-quarter mutant. Beadle and Tatum executed these crosses and got the predicted results, thus gaining further support for their hypothesis that a change in the genetic material causes a change in a specific enzyme.

Using what you have learned about the relationship between protein synthesis, meiosis, and inheritance of traits, you should be able to make predictions about inheritance in cases that are more complex than the examples we have studied so far. For example, what rules govern the inheritance of two traits? What rules govern the sex of offspring, given that sex is determined by the X and Y chromosomes? These questions and others will be considered in the next sections.

## INHERITANCE OF TWO TRAITS

How would you predict the inheritance of two traits at the same time? A cross between two individuals that are heterozygous for two traits is called a **dihybrid cross.** Consider, for example, the inheritance of both flower color and seed pod color in garden peas. Earlier in the chapter we saw that red flower color is dominant over white flower color. In the same way, green seed pod color is dominant over yellow seed pod color.

Suppose we start with parents that are homozygous for each trait (fig. 15.9*a*). In other words, one parent is RRGG: red flowers and green seed pods. The other parent is rrgg: white flowers and yellow seed pods. When these plants are crossed, each gamete contains a gene for each trait. All the offspring in the first generation will be heterozygous for both traits: RrGg.

Now suppose we cross two of the first-generation plants (fig. 15.9*b*). Genes for the two traits will be present in all possible combinations in the second generation. Since we are dealing with two traits, sixteen combinations of genotypes are possible instead of four. As the grid in figure 15.9 shows, nine of the offspring are dominant for both traits (RRGG, RrGG, or RRGg). Three of the offspring are dominant for flower color (RRgg or Rrgg) and three of them are dominant for seed pod color (rrGG or rrGg). Only one offspring in the second generation is recessive for both traits (rrgg). Thus the ratio of dominant and recessive traits is 9:3:3:1. Scientists have performed numerous dihybrid crosses and have found this ratio to be correct. We can conduct crosses for three or more traits in the same way; the number of combinations will be even greater.

## LINKED GENES

For the most part, the rules of inheritance Gregor Mendel discovered hold true in nature. However, sometimes when traits are crossed, the offspring do not match Mendel's ratios. Why should such deviations occur?

Consider crosses of fruit flies. The usual traits observed in fruit flies are red eyes (R) and long wings (L). These are the dominant traits. Figure 15.10 indicates the predicted offspring in this case. How do these predicted results agree with actual results?

In table 15.3, a heterozygous fruit fly was crossed with a fly exhibiting recessive traits: purple eye color (r) and short wings (l). What do the data in this table indicate?

FIGURE 15.9 Dihybrid cross. (*a*) In the first generation, all plants exhibit both dominant characteristics. (*b*) In the second generation, numerous combinations of traits are possible in gametes and second-generation offspring.

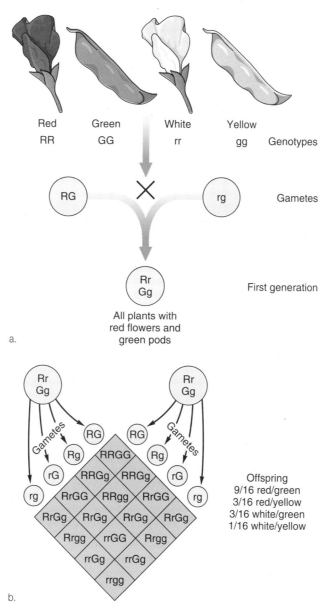

a.

b.

Offspring
9/16 red/green
3/16 red/yellow
3/16 white/green
1/16 white/yellow

| TABLE 15.3 | Offspring of a Cross between a Fly Heterozygous for Both Traits and a Homozygous Recessive Fly |
|---|---|

| Phenotype | Number |
|---|---|
| Red eyes, long wings | 1,339 |
| Red eyes, short wings | 151 |
| Purple eyes, long wings | 154 |
| Purple eyes, short wings | 1,195 |
| Total | 2,839 |

The results observed in table 15.3 are considerably different from the ratio we expect (fig. 15.10). Do these data contradict the relationship between genes, chromosomes, and gamete formation? Perhaps not. Biologists have proposed that the genes for eye color and wing length in fruit flies are on the *same chromosome*. Such genes are said to be **linked.** The genes would be arranged as shown in figure 15.11. In normal organisms, all the genes carried on the same chromosome would be linked and could amount to several hundred.

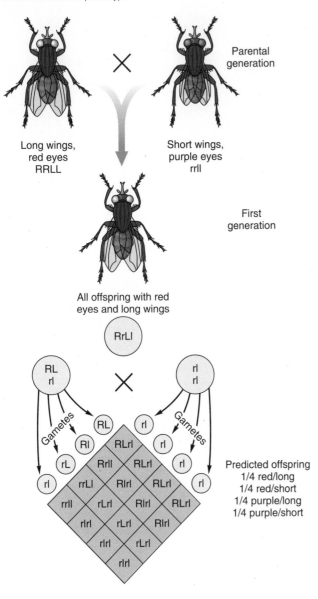

If linked genes were the whole story, we would expect no offspring with Rl or rL phenotypes, but these phenotypes did appear in table 15.3. To explain them, we need to consider a process called crossing over.

A biologist was studying an organism with only two pairs of homologous chromosomes. She was also studying the inheritance of three traits:

height — tall or short
color — blue or green
weight — heavy or light

What can you conclude about the inheritance of these traits?

FIGURE 15.11   Linked genes. (a) Genes for different traits located on the same chromosome are linked. (b) Inheritance of linked genes indicates that only a limited number of genotypes and phenotypes are possible.

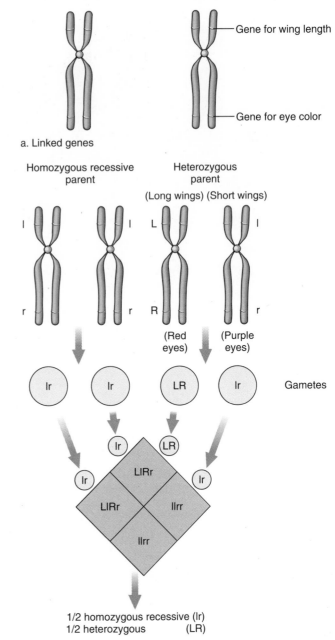

## CROSSING OVER

If a fruit fly with red eyes and long wings is crossed with a fly with purple eyes and short wings, all offspring should exhibit the dominant phenotype and be heterozygous for both traits. If one of these offspring is crossed with a homozygous recessive fly, half the offspring should have both dominant phenotypes and the other half should have both recessive phenotypes, as predicted in figure 15.11. These ratios do not agree, however, with the data in table 15.3.

**FIGURE 15.12** Crossing over. (a) At synapsis, homologues can become entangled. Parts of chromatids can be exchanged. (b) When the chromatids separate, genes on the portions of the exchanged chromatid fragments are now on different chromosomes. (c) This exchange permits different combinations in gametes.

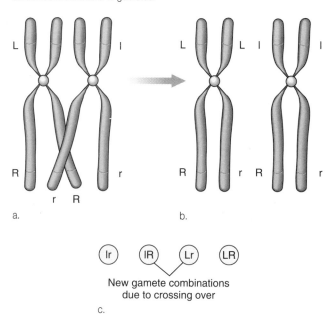

New gamete combinations
due to crossing over

Our prediction for linked genes anticipated no offspring with Rl or rL phenotypes, yet these phenotypes do appear.

How do individuals acquire these phenotypes? One possible explanation is that homologues can exchange genes in the process of meiosis. At synapsis, homologues are sometimes observed to become entangled. Apparently, pieces of the chromosomes can be exchanged by **crossing over,** allowing different combinations of alleles (fig. 15.12). These new combinations can also come together at fertilization to produce the anomolous offspring. While linkage and crossing over result in initially puzzling deviations from Mendel's principles, we can explain them in terms of chromosome behavior.

## SEX-LINKED TRAITS

As we saw in chapter 12, human males and females exhibit different karyotypes: females have two X chromosomes while males have an X and a Y chromosome. At synapsis, either the two X chromosomes pair in a female or the X and Y pair in a male. The gametes of the male will be different from those of the female at the sex chromosomes (fig. 15.13). When males and females mate to produce offspring, the combination of X and Y chromosomes should result in equal numbers of males and females.

How do collected data compare with this prediction? Table 15.4 shows the ratios of males and females at different ages in the population.

U.S. geneticist Thomas Hunt Morgan investigated the question of sex inheritance. Morgan was working with

**FIGURE 15.13** Sex linkage. (a) Males exhibit an x and y chromosome pair, females two x chromosomes. (b) Gametes produced by males and females. (c) Result of a cross.

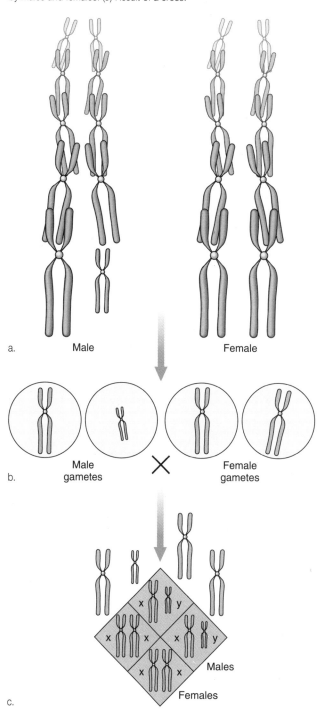

fruit flies when he discovered a single white-eyed male among his experimental organisms. He immediately thought the white eye color was caused by a mutation. To test this possibility, he crossed the white-eyed male with a red-eyed female. All the offspring had red eyes, as you

Some scientists have proposed that, because the Y chromosome is lighter weight, sperm having a Y chromosome can swim faster and reach the egg more quickly than sperm having an X chromosome. Thus, more Y chromosomes would enter eggs and more males would be conceived. What kinds of experimental evidence would support this hypothesis?

| TABLE 15.4 | Ratios of Males and Females in the U.S. Population | |
| --- | --- | --- |
| **Age** | **Sex Ratio** | |
| Conception (estimated) | 1.6:1 | 160 males for 100 females |
| Birth | 1.05:1 | 105 males for 100 females |
| 20 to 25 years | 1:1 | 100 males for 100 females |
| Older than 25 | less than 1:1 | less than 100 males for 100 females |

| TABLE 15.5 | Offspring of a Red-eyed Male and a Red-eyed, Heterozygous Female | | | |
| --- | --- | --- | --- | --- |
| | **Red-eyed** | | **White-eyed** | |
| | **Male** | **Female** | **Male** | **Female** |
| Number of offspring | 1,011 | 2,459 | 782 | 0 |

would expect if the white eyes were a recessive trait. Next, Morgan crossed two of these red-eyed offspring and obtained the results shown in table 15.5.

The results of Morgan's second cross were clearly unusual: The white-eyed trait appeared only in males! What did the results mean? Would any cross ever result in a white-eyed female?

Morgan proposed that the gene for eye color is present only on the X chromosome. He knew that females have two X chromosomes while males have only one. Since males have only one X chromosome, he reasoned, a recessive gene on the X chromosome would not be masked by the dominant gene (fig. 15.14). In females, on the other hand, a recessive gene, such as the one for white eyes, could be present but not show if the female is heterozygous.

Could a female fly ever exhibit white eyes? What kind of cross would produce white-eyed females?

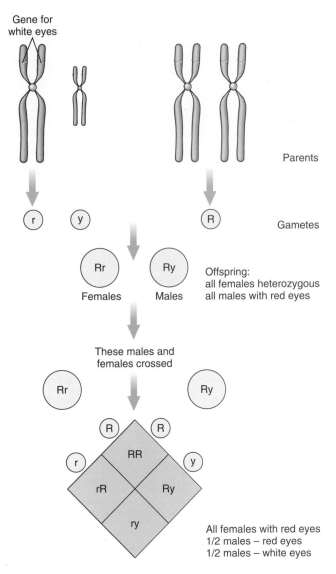

**FIGURE 15.14** Sex-linked trait. The inheritance of the white-eyed trait carried on the X chromosome is shown. This scheme can account for the observed phenotypes in table 15.5.

Gene for white eyes

Parents

Gametes

Offspring:
all females heterozygous
all males with red eyes

These males and females crossed

All females with red eyes
1/2 males – red eyes
1/2 males – white eyes

Genes that are connected to the X and Y chromosomes are called **sex-linked.** In further studies, Morgan revealed nearly 100 sex-linked traits in fruit flies. Many other organisms exhibit traits that are sex-linked. In humans, for example, sex-linked traits include color blindness and hemophilia, a condition in which excessive bleeding results from even small wounds. As we would predict for sex-linked traits, color blindness and hemophilia appear much more frequently in males than in females. They appear in females only if the trait is present on both X chromosomes.

## Summary

*In this chapter we have studied the heritability and expression of genetic traits, from easily observable phenotypes to the molecular level of enzyme action. With your knowledge of patterns of inheritance you can deduce genotypes and predict the frequency of traits in offspring. For the most part, these principles of inheritance hold true, though actually determining traits in offspring can be complicated by the number of genes being examined and environmental factors. As you learn more about the field of genetics, keep in mind the following key points.*

1. In the mid-nineteenth century, two hypotheses were accepted as possible explanations of inheritance. The blending hypothesis proposed that characters mixed when passed from parents to offspring and could not be separated. The particulate hypothesis proposed that characters retained their identity and could later separate.

2. Gregor Mendel's experiments supported the particulate hypothesis when they demonstrated the reappearance of traits in the second generation.

3. Mendel explained his results by proposing that (a) traits exist as **dominant** or **recessive,** (b) two "factors" for each trait are present in parents and separate at gamete formation, and (c) these factors combine randomly at fertilization. These principles enabled Mendel to account for the number of different offspring produced.

4. Twentieth-century understanding of chromosome assortment in meiosis and recombination in gamete formation and fertilization are consistent with Mendel's principles on inheritance of traits.

5. Mendel's "factors" turned out to be the genes on homologous chromosomes. Each gene may have two expressions or **alleles** for the trait. If the genes are identical, the organism is **homozygous** for the trait. If they are different (i.e., two alleles are present), the organism is **heterozygous.**

6. The expression of a trait is the **phenotype,** while the genetic composition is the **genotype.**

7. Information on the chromosomes is expressed as the ability to produce specific enzymes or other proteins.

8. A cross involving two traits is called a **dihybrid cross.** Dihybrid crosses result in more combinations of offspring.

9. If information for different traits is carried on the same chromosome, the traits are said to be **linked** and are inherited together. At synapsis, genetic material can be exchanged between homologues in a process called **crossing over**.

10. Genes on the X chromosome are called **sex-linked.** Traits that are sex-linked exhibit different patterns of inheritance.

## References

Cummings, Michael R. 1988. *Human Heredity: Principles and Issues* (St. Paul: West Publishing).

Singleton, W. Ralph. 1967. *Elementary Genetics* 2nd ed. (London: Van Nostrand).

## Exercises and Questions

Solving genetic problems can be exceptionally challenging for biology students. Predicting genotypes and phenotypes of offspring or working backward to infer information about parents requires using your knowledge of meiosis (the separation of genes) and fertilization (combination or reunion of genes). Moreover, to answer many questions on the production and quantity of enzymes you must use information on chromosome structure and molecular biology. Keep these points in mind as you work on the exercises for this chapter.

1. In the experiment illustrated in figure 15.1, what results would have supported the blending hypothesis (page 182)?
   a. Some white flowers on both first and second generation plants.
   b. Both red and white flowers on the same plants.
   c. Some plants with white flowers in the first generation but all red flowers in the second generation.
   d. All plants with pink flowers in the first generation but some plants with white flowers in the second generation.
   e. All first and second generation offspring with pink flowers.

2. In the cross illustrated in figure 15.2, what would the offspring look like if one of the first generation plants was crossed with a pea plant that had white flowers (page 183)?
   a. Half the offspring with red flowers and half with white flowers.
   b. Red flowers on 3/4 of the offspring and white flowers on 1/4.
   c. White flowers on 3/4 of the offspring and red flowers on 1/4.
   d. All offspring with white flowers.
   e. All offspring with red flowers.

3. Which of the following is the best reason why genotypes must be shown as two alleles (page 184)?
   a. There are two genes on the X-shaped homologue.

b. Homologues separate in meiosis and genes combine at fertilization.

c. With dominant and recessive genes, only one allele is expressed.

d. The two homologues sometimes carry genes for different forms of the trait.

e. The genotype will not always indicate which allele is present in a gamete.

4. Which of the following is the best reason why each homologue contains two copies of each gene (page 184)?

a. The X-shaped homologues are characteristic of diploid chromosomes.

b. The cells of organisms contain several pairs of homologous chromosomes.

c. A diploid cell contains two alleles that are sometimes different.

d. One chromatid is copied from the other.

5. A student argued that an organism heterozygous for albinism should produce one-half as much of the essential enzyme as a homozygous normal individual. What is this student assuming (page 185)?

a. A homozygous normal individual would have homozygous normal parents.

b. Each homologue will always code for the production of equal amounts of the enzyme.

c. An individual heterozygous for albinism produces one-half as much of the enzyme.

d. A person heterozygous for albinism will have normal pigmentation.

e. Among the offspring of a heterozygous individual, one-half will produce no melanin.

6. Examine the pedigree chart on page 187. Which of the following can you most likely conclude about the genotypes of male 4 and female 6 in generation 3?

a. Both are homozygous for albinism.

b. The female is heterozygous; the male is homozygous normal.

c. The male is heterozygous; the female is homozygous normal.

d. Both are heterozygous for albinism.

7. Assuming that albinism is under the control of a single gene pair, could two albino individuals ever produce any offspring with normal pigmentation (page 186)?

a. Yes, but most offspring would be albino.

b. No. Neither parent contains a gene for normal pigmentation.

c. Yes. Since albinism is recessive, the dominant, normal gene could be expressed.

d. No. Many infants will die before birth.

8. A biologist was studying an organism with only two pairs of homologous chromosomes. She was also studying the inheritance of three traits:

height — tall or short
color — blue or green
weight — heavy or light

Which of the following can you most confidently conclude about the inheritance of these traits (page 190)?

a. The tall, blue, and heavy traits are dominant.

b. These organisms could produce only two different kinds of gametes.

c. Genes for at least two of the traits are linked.

d. The tall, blue, and heavy traits are recessive.

9. When crossing over occurs, chromosomes are changed in specific ways, creating combinations that would not be observed before crossing over. Which of the following would you most likely find only as a result of crossing over?

a. Two homologous chromosomes have information for different traits.

b. Two homologous chromosomes have identical information.

c. One chromosome contains two different alleles of the same trait.

d. One chromosome has information for one allele of a trait.

10. A blood analysis of a small child indicated the presence of two different forms (R and S) of a particular sugar-digesting enzyme. The two forms differed in only a few amino acid positions, and both forms were active in catalyzing the same reaction. Forms R and S were present in approximately equal amounts in the blood sample used for analysis. Assuming that information for this enzyme is carried on a single homologous pair of chromosomes, which of the following statements is most justified?

a. At least one parent possessed genetic information for form S.

b. Neither parent can be homozygous for form R or form S.

c. It is possible that neither parent possessed genetic information for form S.

d. At least one parent is homozygous for form S.

11. In question 10, suppose it was later discovered that the child's father produced only form S of the enzyme. Which of the following best states the genetic information for the enzyme possessed by the child's mother?

a. She would also be homozygous for form S.

b. She would be homozygous for form R.

c. It is possible that the mother has no genetic information for either form of the enzyme.

d. She would be heterozygous, i.e., have genetic information for both forms of the enzyme.

12. Suppose the parents described in questions 10 and 11 had another child. Which of the following is the most likely prediction regarding the genetic information for the sugar-digesting enzyme possessed by this second child?

    a. The child would be homozygous for form R.
    b. The child would be homozygous for form S.
    c. If crossing over occurs, the child would have no information for either form.
    d. The child would be heterozygous, i.e., have genetic information for both forms.

13. Hemophilia is the "bleeder's" disease. Hemophiliacs have blood that does not coagulate well and often die at a young age. The gene for the disease is recessive and sex-linked. Which of the following is the best prediction about the offspring of a hemophiliac male and a nonhemophiliac female?

    a. All children would be hemophiliac.
    b. All females would be hemophiliac.
    c. All males would be hemophiliac.
    d. Hemophiliac children would occur in a 3:1 ratio, normal to hemophiliac.
    e. Some males and females could be hemophiliac.

14. Assume that the nonhemophiliac mother in question 13 is a carrier, i.e., she has the hemophilia gene on one of her chromosomes but the other is normal. Which of the following is the best prediction about the offspring between her and the hemophiliac male?

    a. All males would be hemophiliac.
    b. All females would be hemophiliac.
    c. Half of all the offspring (whether male or female) would be hemophiliac.
    d. Half the males would be hemophiliac but no females would be.
    e. Half the females would be hemophiliac but no males would be.

15. In one family there were three brothers and all were color-blind. Suppose the fourth child was a daughter. Which of the following inferences could you most confidently make about the color vision of this child?

    a. Even if the father is color-blind, the daughter will have normal vision.
    b. If the father has normal vision, the daughter will have normal vision.
    c. The daughter will be color-blind.
    d. If the mother is color-blind, the daughter will be color-blind.
    e. Answers b and d are equally likely choices.

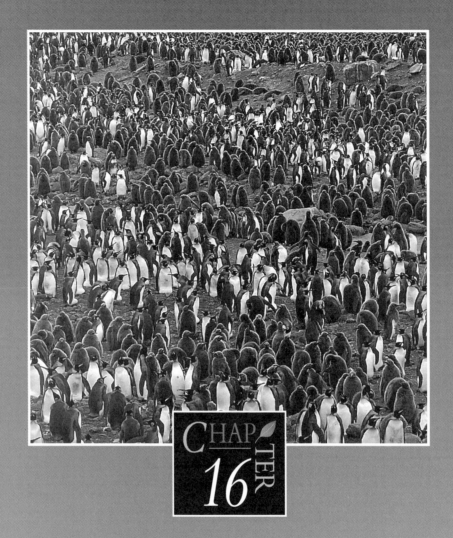

# POPULATION GENETICS

In chapter 15, we examined how characteristics are passed from parents to their offspring. We limited our examination to specific crosses between two parents. Now we can broaden our examination: What happens to genes in an entire interbreeding population? For example, would recessive alleles gradually be eliminated? We can find answers to these questions by looking at **population genetics,** the study of the genes in an entire population.

Let's consider albinism, the trait we studied in chapter 15. About one in 10,000 human births is an albino. To produce an albino, two recessive alleles must combine in the zygote during fertilization. We know that there are albino alleles in the population because we see the results in albino children. Most of the albino alleles, however, are carried in heterozygous individuals and are thus undetected.

How many albino alleles are present in any given population, and what are the chances that two will combine to produce an albino offspring? Is the number of albino alleles decreasing or increasing? We will address these questions in this chapter.

## GENE FREQUENCY

As we begin to study inheritance in populations, we will find it useful to think of all the genes for all the traits that an organism exhibits as the **gene pool.** For example, in Mendel's garden peas, all the traits—flower color, height, seed pod color, and all the rest—are the plants' gene pool. The genes for one of those traits, for example, flower color, are *part of* the gene pool. The proportion of one allele for a gene relative to other alleles for that gene is the **gene frequency.** In our example, the gene frequency measures the proportion of alleles for red flower color to the alleles for white flower color.

Consider the imaginary case of a population of butterflies on a small island. The butterflies are isolated from outside influences and are freely interbreeding.

Suppose the butterfly population carries a gene for wing spot color (fig. 16.1). The phenotype is either blue spots (B) or white spots (b), and blue is dominant. All the alleles for wing spot color are part of the gene pool.

In this population, three genotypes for wing spot color are possible: BB, Bb, and bb. Suppose that scientists capture a random sample of 100 butterflies and find that 84 have blue spots and 16 have white spots. Breeding experiments with the blue-spotted butterflies indicate that 36 have the genotype BB (homozygous) and 48 have the genotype Bb (heterozygous).

What is the gene frequency of the wing spot alleles in this population? Table 16.1 shows how scientists calculated the gene frequencies based on the information they gathered.

Notice that the number of B alleles for the BB genotype is 72, or 2 × 36. The number of b alleles for the bb genotype is 32, or 2 × 16. That is because homozygous individuals contribute two of the same alleles. Notice also that the total number of alleles is twice the total number of individuals (100 × 2 = 200).

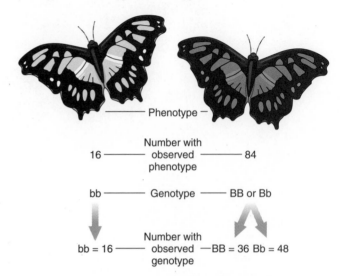

**FIGURE 16.1** Wing spot alleles in the gene pool. When a representative sample of 100 butterflies are collected from the island population, 16 are found with white spots and 84 with blue spots. Breeding experiments are necessary to detect which of the dominant phenotypes are heterozygous or homozygous.

**TABLE 16.1    Calculating Gene Frequency**

|  | Genotypes | | | Total |
|---|---|---|---|---|
|  | BB | Bb | bb |  |
| Number of individuals | 36 | 48 | 16 | 100 |
| Number of B alleles | 72 | 48 | 0 | 120* |
| Number of b alleles | 0 | 48 | 32 | 80* |

Gene frequency of B = 120/200 = 0.6 = $p$.
Gene frequency of b = 80/200 = 0.4 = $q$.

*Total number of alleles in the population (120 + 80) = 200

To calculate the gene frequency of an allele, we divide the number of a specific allele by the total number of alleles in the population. For the blue allele, that is 120/200 = 0.6. For the white, it is 80/200 = 0.4. Thus the gene frequencies for spot color in butterflies are 0.6 and 0.4. The frequency of the dominant allele is generally indicated by the letter $p$, and the frequency of the recessive allele is indicated by $q$.

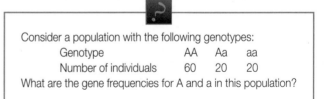

Consider a population with the following genotypes:
| Genotype | AA | Aa | aa |
|---|---|---|---|
| Number of individuals | 60 | 20 | 20 |
What are the gene frequencies for A and a in this population?

Measuring gene frequencies of actual populations in this way is difficult, mostly because it is difficult to determine whether individuals who exhibit the dominant phenotype are heterozygous or homozygous. Breeding experiments can be laborious and time-consuming, and with organisms that

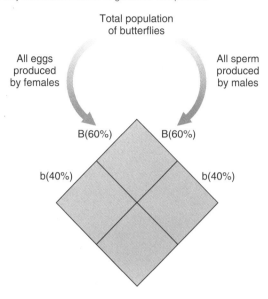

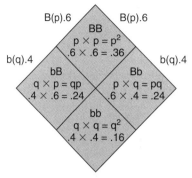

Dominant genotype frequency
= p × p = .6 × .6 = .36

Heterozygote frequency
= p × q + p × q = 2pq = 2(.24) =.48

Recessive genotype frequency
= q × q = q² = 0.4 × 0.4 = .16

The frequency of all genotypes can be added:
p² + 2pq +q² = 1
.36 + 2(.24) + .16 = 1

reproduce slowly, such methods are impractical. How, then, can we calculate gene frequencies in real populations? Can we, for example, measure the frequency of a recessive allele such as albinism in a human population?

These questions are important ones because studying the frequencies of harmful genes in human populations has useful medical applications. Fortunately, geneticists have developed techniques that enable them to investigate gene frequencies more easily.

# THE HARDY-WEINBERG LAW

Early in the twentieth century, two investigators, Godfrey Hardy and Wilhelm Weinberg, independently developed a way to estimate dominant and recessive allele frequencies. Their method is called the **Hardy-Weinberg law.** As part of their method, they proposed using *p* and *q* to denote dominant and recessive allele frequencies. To see how their law works, let's use the example of wing spot color in butterflies.

We found that the dominant allele frequency (*p*), for blue wing spots, was 0.6 and that the recessive allele frequency (*q*), for white wing spots, was 0.4. Another way of saying this is that 60 percent of the eggs produced by females in the population and 60 percent of the sperm produced by males will have the B allele (fig. 16.2). Logically, then, 40 percent of the eggs and sperm produced in the population will have the b allele.

The Hardy-Weinberg law can be derived by considering the way alleles combine at fertilization. When the eggs and sperm in this population combine, the proportion of each genotype produced will be determined by the relative numbers of each allele. Figure 16.3 shows how the allele frequencies are multiplied to produce the three different genotypes. This method of calculation is the same technique we used in

chapter 15 to indicate the combination of gametes from two individuals. The difference in figure 16.3 is that we're showing the combination of gametes in an *entire population*.

When alleles in a population are combined, we can discover the frequency of offspring with a given genotype by multiplying the allele frequencies (Thinking in Depth 16.1). In other words, if we know the allele frequencies, we can calculate the genotypes. But how do we discover individual allele frequencies?

# CALCULATING GENE FREQUENCIES

The information presented in figure 16.3 provides an indication of how we can calculate gene frequencies. Notice in the figure that $q^2 = 0.16$. In a population, $q^2$ is equal to the frequency of the recessive phenotype and is easy to measure. We can find the value of $q$, the recessive allele frequency, by taking the square root of $q^2$.

$$q = \sqrt{q^2} = \sqrt{0.16} = 0.4$$

Now that we know the value of $q$, we can also find the value of $p$. We know that $p$ and $q$ added together must equal 1, since the frequency of all the alleles added together will be equal to 1.

$$p + q = 1$$

Using a simple rule of algebra, we find that

$$p = 1 - q$$

Thus,

$$p = 1 - 0.4 = 0.6$$

# USING PROBABILITY TO CALCULATE ALLELE FREQUENCIES

Another way of viewing the combination of alleles illustrated in figures 16.2 and 16.3 is with the demonstration shown in figure 1. Suppose 100 marbles are placed in container 1. Each marble represents an allele contributed by a male. Sixty marbles are blue (for the blue spot allele) and 40 are white (for the white spot allele). These numbers represent the relative proportions of blue and white spot alleles in the population. An identical container represents the alleles contributed by females.

If one marble is drawn from container 1, the chance that it is blue is 60 out of 100, or 0.6, since 60 blue marbles are present. The chance of drawing a white marble is 40 out of 100, or 0.4. These same probabilities hold true for container 2.

The chance of drawing a blue marble from both containers at the same time can be determined by multiplying the two probabilities: $0.6 \times 0.6 = 0.36$ (fig. 1a). The product of the two probabilities is the chance of producing a homozygous dominant individual. The chance of producing a homozygous recessive individual (two white marbles) is $0.4 \times 0.4 = 0.16$ (fig. 1b).

The calculation for heterozygotes is slightly more complicated since there are two ways to form a heterozygote: a white marble from container 1 and a blue marble from container 2, or a blue marble from container 1 and a white marble from container 2. Since there are two ways to form a heterozygote, we must add the two probabilities.

| Container 1 | | Container 2 | | |
|---|---|---|---|---|
| white (0.4) | × | blue (0.6) | = | 0.24 |
| blue (0.6) | × | white (0.4) | = | 0.24 |
| | | | | 0.48 |

Thus the chance of forming a heterozygote is 0.48. These numbers are the same as the ones we calculated in figure 16.3 and are derived from simple rules of probability.

**FIGURE 1** Allele combinations. Marbles represent individual alleles for blue and white spot color. The probabilities of drawing a specific marble color from each container are multiplied to give the probability of forming a specific genotype.

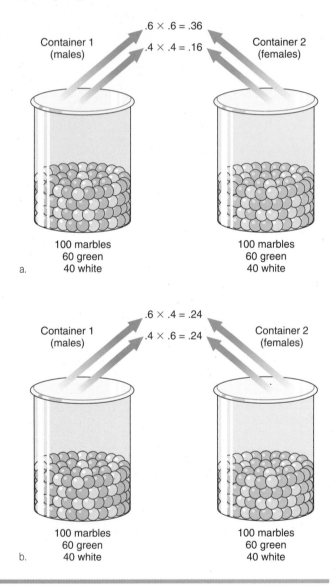

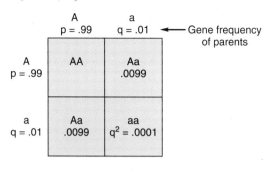

FIGURE 16.4    Gene frequency for albinism. If the gene frequency for albinism is 0.01 in the parental generation, the *same* frequency is found among the offspring.

To see how this equation is applied to real populations, let's look again at albinism in humans. Earlier we said that one in 10,000 individuals is an albino. In other words, albinism occurs with a frequency of 0.0001. This frequency is the same as $q^2$. Once we know $q^2$, we can find $p$.

$$q^2 = 0.0001$$
$$q = \sqrt{q^2} = \sqrt{0.0001} = 0.01$$
$$p = 1 - q$$
$$p = 1 - 0.01 = 0.99$$

Thus the allele frequency of $p$ is 0.99 and the allele frequency of $q$ is 0.01.

What happens when this population interbreeds? Will the gene frequency change? Figure 16.4 shows the allele frequencies of a parent generation and the genotypes and allele frequencies of their offspring. Notice that the frequency of aa is the same as in the parent generation! This is one of the more surprising outcomes of the Hardy-Weinberg law: *Gene frequencies remain constant from generation to generation.* A population in which gene frequencies remain constant is said to be in **genetic equilibrium.** As we will see later, certain conditions must be met for genetic equilibrium to occur.

Could $q$ ever be greater than $p$ for any allele in a population?

Another somewhat surprising result of the Hardy-Weinberg law is the relatively high frequency of heterozygotes in a population when there are so few homozygotes.

TABLE 16.2    Frequency of Albinism by Ethnic Group

| Population | Frequency |
|---|---|
| U.S. African-American population | 1 in 15,000 |
| U.S. white population | 1 in 37,000 |
| Hopi and Navajo populations | 1 in 200 |

For example, in figure 16.4, the heterozygotes are denoted by Aa. We can determine their number with the following equations:

$$
\begin{aligned}
\text{heterozygotes} &= qp + qp \\
&= 2pq \\
&= 2(.99)(0.01) \\
&= 0.0198 \\
&= \text{about 1 in 50 individuals}
\end{aligned}
$$

Even though only 1 in 10,000 albinos are present in the population, 1 in 50 individuals carry a recessive albino allele! This is far more than we might expect given the small number of albinos.

We have been looking at albinism in the entire human population. If we study individuals in a smaller population, however, they will not necessarily exhibit the same gene frequencies. The frequency of albinism varies widely by ethnic background, for example, as shown in table 16.2. Other factors can affect gene frequency as well, as we will see in the next section.

What would cause the frequency of albinism in the Hopi and Navajo Indian population to decrease?

In a population, three genotypes are present: AA, Aa, and aa. The genotypic frequencies are

AA = 0.04      Aa = 0.32      aa = 0.64

What will the gene frequencies of A and a be in the second generation?

## DEVIATIONS FROM HARDY-WEINBERG

The Hardy-Weinberg law is extremely useful in analyzing population genetics, but it represents an ideal situation where certain assumptions are made. Real populations only approximate this ideal and usually deviate from it to at least some extent. In the following sections we will discuss various deviations from the Hardy-Weinberg ideal.

### Random Mating

One assumption made by Hardy-Weinberg is **random mating,** which means that any female is equally likely to mate with any male. In other words, sperm carrying the A allele

**FIGURE 16.5** Genetic effect of migration. The frequency of the allele for B blood type gradually decreases from 25–30 percent in western Asia to a low of 0–5 percent in parts of western Europe. These data have been proposed as evidence that migration can change gene frequency.
*Source: Data from A.E. Moarant, et al.,* The Distribution of the Human Blood Groups and Other Polymorphisms, *1976, Oxford University Press.*

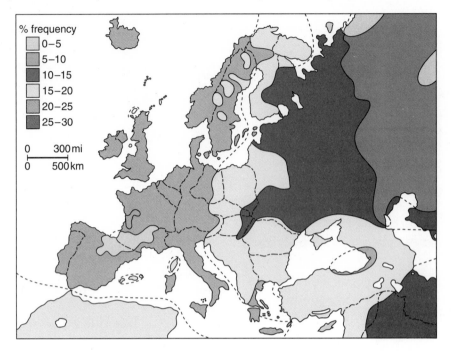

will combine proportionally with eggs carrying the A allele and sperm carrying the a allele. Does random mating actually occur? Probably not. Many factors can cause deviations from randomness. Individuals with certain characteristics may be attracted to others with like characteristics. Or opposites may attract. Such factors are extremely difficult to determine, but most likely mating will not be random in every respect. When mating is not random, its effect must be taken into account when calculating gene frequencies.

## Genetic Drift

In small populations, a condition called **genetic drift** may occur in which gene frequencies change simply by chance because variations in mating, reproductive success, or survival of offspring cause differences from generation to generation. In large populations, genetic drift is unlikely. In order for the Hardy-Weinberg law to apply, the population under study must be large or have no genetic drift.

## Migration

**Migration,** the movement of individuals into or out of a population, might also change gene frequencies. Such movement can lead to **gene flow** from one location to another. Do migration and gene flow significantly affect gene frequencies? To find out, let's examine the data in figure 16.5. How do you interpret these data?

The figure shows the frequency of the B blood type allele in Europe and western Asia. The frequency of this blood type gradually decreases from almost 30 percent in Asia to less than 5 percent in parts of western Europe. A variety of studies have indicated that large numbers of people have migrated from Asia to Europe. Figure 16.5 is usually interpreted as evidence that migration has affected the gene frequencies for the B blood type allele across Europe. For gene frequencies to remain constant, then, Hardy-Weinberg assumes no migration.

Suppose the gene frequencies for a population are $p$ and $q$ = 0.5. Individuals with what genotype could migrate without changing gene frequencies?
What assumptions are we making in our interpretation of figure 16.5?

## Mutation

Another way gene frequencies can change is through **mutations.** As we saw in chapters 14 and 15, mutations occur when changes in DNA lead to altered proteins. Since these changes occur in DNA, they are inherited. Thus, the Hardy-Weinberg law assumes no mutation.

**Back-mutations** occur when mutated genes mutate back to their original form. How would back-mutations affect changes in gene frequency?

*Reproduction and Inheritance*

**FIGURE 16.6** Mutation. If a trait initially has only a single allele, the gene frequency will be 1.0 (at generation 1). Using $1.4 \times 10^{-5}$ as an estimate of the mutation rate, the new allele resulting from mutation will decrease the frequency of the original allele as the new allele increases in frequency. *Source: Data from Michael Cummings,* Human Heredity: Principles and Issues, *1988, West Publishing Company, St. Paul, MN.*

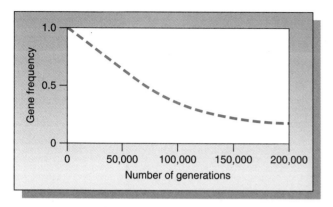

**FIGURE 16.7** Effects of selection on gene frequency. The change in gene frequency vs. number of generations is shown. Line X—all alleles equally fit, thus gene frequency does not change. Line Y—aa fitness = 1; AA and Aa do not reproduce. Line Z—AA and Aa are 90 percent as fit as aa. Line Q—aa is 90 percent as fit as AA and Aa. Notice that line Z changes more rapidly than line Q.

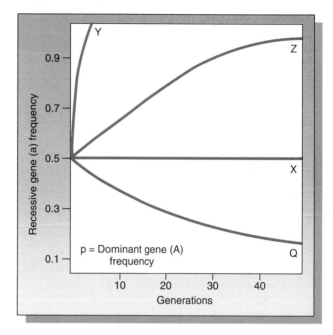

How often do mutations occur? How seriously can they affect gene frequencies? A simple way to measure the rate of mutation in a population is to observe the appearance of a new dominant trait. One such trait is achrondroplasia, a form of dwarfism that causes short arms and legs. One survey revealed seven achrondroplastic individuals in 242,257 births. Thus the mutation rate is $7/242{,}257 = 1.4 \times 10^{-5}$. Figure 16.6 shows how slowly the gene frequency will change if a mutation appears in a population at this rate. If one generation represents about 20 years, then fifty generations will cover 1,000 years. As figure 16.6 shows, even in 20,000 generations there would be almost no change in gene frequency.

Why is it important that the mutation to achrondroplasia occurs on a dominant allele? If the gene for achrondroplasia were recessive, how would this complicate measurement of the mutation rate?

## Selection

Finally, the Hardy-Weinberg law assumes that all genotypes are equally likely to survive and to reproduce. This means that there is no **selection** for one genotype over another. When selection *does* occur, one genotype is more likely to have an advantage and to survive than others. In reality, some genotypes *are* likely to be more successful in mating and in overcoming environmental stresses.

Selection can be considered an expression of the **fitness** of a particular genotype. Fitness is a measure of how well a genotype succeeds in reproducing its alleles in the next generation. If the fitness of a genotype equals zero, then individuals of that genotype will not reproduce because of early death or failure to mate or some other factor. If the fitness of a genotype equals one, then it will be the *only* genotype to

survive and reproduce. All other genotypes will not reproduce. Fitness, then, can range from zero to one. The most fit genotypes will be selected for survival.

To what extent and how quickly can selection change gene frequency? Figure 16.7 shows calculated changes in gene frequency when fitness varies. We can consider several possibilities. When all genotypes (AA, Aa, aa) are equally fit, then gene frequencies remain constant as expected (line X). This is the assumption made by the Hardy-Weinberg law.

Suppose the recessive allele (a) is more fit than the dominant allele (A). In the most extreme case, the genotypes AA and Aa will fail to reproduce and the A allele will be eliminated from the population in one generation (line Y). If the AA and Aa phenotypes are 90 percent as fit as aa (line Z), the A allele will be removed slowly. About 1 percent of A will remain in the gene pool after fifty generations.

If, however, the A allele is more fit—say aa is 90 percent as fit as AA and Aa—then the a allele is removed. Removal is slower, with 18 percent remaining after fifty generations (line Q).

Why is removal slower for the recessive allele? Since it is recessive, the a allele can be "hidden" in the heterozygous genotype and protected from removal to some extent. Even so, removal of more than 80 percent of the a alleles in just fifty generations represents a rapid change in gene frequency when compared to the effects of mutation, migration, and genetic drift.

If aa had a fitness of zero, would the a allele be eliminated from the gene pool as rapidly as A when the fitness of AA and Aa were zero? How would you graph this scenario in figure 16.7?

Although the Hardy-Weinberg law requires several, often unrealistic assumptions, it is an extremely useful way to view inheritance in populations. The recognition that gene frequencies *should* remain constant allows us to focus on the effects that are operating when gene frequencies change in real situations. Such effects represent deviations from the assumptions of the Hardy-Weinberg law. Each of the deviations has a different effect on gene frequencies. Genetic drift can produce significant changes in small populations but will not be serious in large populations. Migration can have major effects but will not over a short period unless large numbers of individuals migrate. Mutation can lead to major changes in individual alleles, but mutation rates are usually very low.

Unlike the other forces, selection can lead to rapid and major changes in gene frequency, as we have seen. But do such changes occur in the real world? If so, what specific environmental and genetic factors affect selection? In chapter 17 we will examine selection in more detail.

## SUMMARY

*Population genetics applies the laws of Mendelian genetics to groups of individuals. One of the most important aspects of population genetics is determining gene frequencies. One method of determining gene frequency is the Hardy-Weinberg law, which assumes random mating and no genetic drift, migration, mutation, or selection. As you review concepts in population genetics, keep in mind the following key points.*

1. The genes in an interbreeding population can be considered a **gene pool.** The **gene frequency** of a particular allele is defined as the number of alleles for a specific trait divided by the sum of alleles for all expressions of the trait.
2. The sum of allele frequencies for all expressions of a trait must be equal to 1.
3. The **Hardy-Weinberg law** predicts that gene frequencies will remain constant from generation to generation if certain specific assumptions are valid.
4. Hardy-Weinberg assumes there is no **genetic drift,** or no change in gene frequencies by random fluctuation.
5. **Random mating** means that matings between all genotypes and phenotypes are equally likely, a condition assumed by Hardy-Weinberg.
6. Hardy-Weinberg assumes that there is no **migration,** or movement of individuals in or out, in populations.

7. Hardy-Weinberg assumes that there are no **mutations,** changes in DNA that lead to altered proteins.
8. Hardy-Weinberg assumes no **selection,** in which one genotype has an advantage and is more likely to survive than others.
9. At **genetic equilibrium,** $p^2 + 2pq - q^2 = 1$, where

   $p^2$ = frequency of homozygous dominant individuals

   $2pq$ = frequency of heterozygous individuals

   $q^2$ = frequency of homozygous recessive individuals

10. In a population at equilibrium, gene frequencies can be calculated with the following two equations:

    $q = \sqrt{q^2}$ = frequency of homozygous recessives

    $p = 1 - q$

## REFERENCES

Cummings, Michael R. 1988. *Human Heredity: Principles & Issues* (St. Paul: West Publishing).

Wallace, Bruce. 1981. *Basic Population Genetics* (New York: Columbia University Press).

## EXERCISES AND QUESTIONS

1. Consider a population with the following genotypes,

| Genotype | AA | Aa | aa |
|---|---|---|---|
| Number of individuals | 60 | 20 | 20 |

Which of the following most accurately represents the allele frequencies in this population (page 198)?

|  | Frequency | |
|---|---|---|
|  | A | a |
| a. | 0.6 | 0.2 |
| b. | 0.7 | 0.3 |
| c. | 0.8 | 0.4 |
| d. | 1.2 | 0.4 |
| e. | None of the above | |

2. A biologist had 31 dark mice and 31 light mice. When bred, a dark and a light mouse were always mated. All offspring were dark. Which of the following are the gene frequencies for the dark gene (D) and the light gene (d) in the parents and offspring?

|  | Parents | | Offspring | |
|---|---|---|---|---|
|  | D | d | D | d |
| a. | 1.0 | 1.0 | 2.0 | 0 |
| b. | 0.5 | 0.6 | 1.0 | 0 |
| c. | 1.0 | 1.0 | 1.0 | 1.0 |
| d. | 0.5 | 0.5 | 0.5 | 0.5 |
| e. | Gene frequencies cannot be determined from the information given. | | | |

3. Could $q$ ever be greater than $p$ for any allele in a population (page 201)?
   a. No, the recessive allele cannot be more frequent than the dominant allele.
   b. Yes, whenever there are equal numbers of homozygous dominant and recessive individuals in a population.
   c. Yes, if all individuals are heterozygous.
   d. No, $p$ is by definition the more frequent allele.
   e. Yes, whenever the recessive allele is more frequent.

4. What would cause the frequency of albinism in the Hopi and Navajo Indian population in table 16.2 to decrease (page 201)?
   a. The migration of Hopis and Navajos out of the population.
   b. More frequent marriages between albino Hopis and Navajos.
   c. The intermarriage of Hopis and Navajos with members of the U.S. white population.
   d. The deaths of equal numbers of albino and normal children.
   e. More frequent marriages between Navajos and Hopis.

5. In a population three genotypes are present: AA, Aa, and aa. Each has equal reproductive success. The genotypic frequencies are
   AA = 0.04      Aa = 0.32      aa = 0.64
   What will the allele frequencies of A and a be in the second generation (page 201)?
   a.   A = 0.6        a = 0.4
   b.   A = 0.5        a = 0.5
   c.   A = 0.4        a = 0.6
   d.   A = 0.3        a = 0.7
   e.   A = 0.2        a = 0.8

6. A student studying a population of butterflies wanted to determine the gene frequencies for wing color. She knew that in this butterfly wing color is exhibited as either yellow or black and that black is dominant. She also knew that half the butterflies in this population have yellow wings and half have black wings.
   Which of the following possibilities would she most likely reject?
   a. There are more yellow genes than black genes.
   b. There are more black genes than yellow genes.
   c. There are twice as many yellow genes as black genes.
   d. There are equal numbers of black and yellow genes.
   e. There are three times as many yellow genes as black genes.

7. Suppose the allele frequencies for a population are $p = 0.5$ and $q = 0.5$. Individuals with which of the following genotypes could migrate without changing frequencies (page 202)?
   a. Homozygous dominant
   b. Individuals that are haploid
   c. Heterozygous
   d. Individuals that are diploid
   e. Homozygous recessive

8. The interpretation of figure 16.5 discussed on page 202 most likely makes which of the following assumptions?
   a. Gene frequencies in Europe are now different from those in the distant past.
   b. Migration can affect gene frequencies.
   c. Individuals migrate into a population but none migrate out.
   d. Mutation in the blood type gene can affect gene frequencies.
   e. Gene frequencies vary from Asia to western Europe.

9. Back-mutations occur when mutated genes mutate back to their original form. How would back-mutations most likely affect changes in gene frequency (page 202)?
   a. Since more mutations are occurring, gene frequency would change more rapidly.
   b. More individuals would be heterozygous.
   c. Gene frequency must be calculated using more than two alleles.
   d. Gene frequency would change more slowly.
   e. More individuals would be homozygous for the mutation.

10. If the gene for achrondroplasia were recessive, how would this most likely complicate measurement of the mutation rate (page 203)?
    a. The dominant allele would be more frequent.
    b. The gene frequency for achrondroplasia would be lower than expected from measuring only recessive phenotypes.
    c. Recessive alleles would not be apparent in heterozygotes.
    d. Homozygous recessive individuals would not survive long enough to be counted.
    e. The recessive allele would more likely be decreased due to selection.

11. If the aa genotype in figure 16.7 had a fitness of zero, would the a allele be eliminated from the gene pool as rapidly as A when the fitness of AA and Aa were zero (page 204)?
    a. Yes. If the allele has a fitness of zero, it will not survive.
    b. No. The recessive allele in heterozygotes would not be eliminated.
    c. Yes, if both the a allele and the A allele can be present in the same individual.
    d. No, since the aa genotype can contribute alleles to the next generation.
    e. Yes. No a allele will be present in the second generation if fitness is zero.

*The exercises presented in this section provide an opportunity for you to explore relationships between biology and other academic disciplines such as law, literature, and psychology. There are no "right" answers to the questions posed in each exercise. Your response may be a short viewpoint or an in-depth research paper, depending on your teacher's instructions. References are provided with each exercise, should you want or need to refer to them as part of your assignment.*

## Political Science
### LYSENKOISM

Even though Mendelian genetics has provided a successful explanation of inheritance, other unscientific explanations have sometimes held a strong emotional appeal. Perhaps the most striking (and damaging) example of this continued appeal is the influence of geneticist Trofim Denisovich Lysenko on Soviet genetics from the 1930s until the 1960s. Apparently, Lysenko saw many parallels between unscientific explanations of inheritance and the Marxist philosophy that was the basis for the Soviet state until the 1990s. In the words of one observer:*

> [Lysenko's views] have strong analogies with Marxist political dogma. The existence of genes is simply denied. Mendelians are obviously "idealists" since they study something that doesn't exist. Heredity is transmitted by every particle of the body (just as every worker in Russia contributes toward the future of the state). When a plant is suddenly given new environmental conditions there is a "shattering"

of its heredity (like a political revolution). This shattering is a kind of shock treatment. It makes the plant peculiarly plastic to change. The new environment then produces desirable changes in the plant which are transmitted permanently to all later generations.

Soviet geneticists who opposed Lysenko's views were banished or imprisoned. For a generation, genetics research in the Soviet Union was severed from the successful experimental research going on in the rest of the world.

How can conflicts between scientific research and political systems be resolved or prevented?

Which is more important in the national interest, political ideology or "good science"?

*M. Gardner, 1957, *Fads and Fallacies in the Name of Science* (New York: Dover Publications).

## Literature
### ATTITUDES TOWARD SCIENCE

Lewis Wolpert* cites the following rather disturbing views of science by well-known writers.

> Knowledge has killed the sun, making it a ball of gas with spots. . . . The world of reason and science. . .this is the dry and sterile world the abstracted mind inhabits.
>
> D. H. Lawrence

> Modern science. . .abolishes as mere fiction the innermost foundations of our natural world: it kills God and takes his place on the vacant throne so henceforth it would be science that would hold the order of being in its hand as its sole legitimate guardian and so be the legitimate arbiter of all relevant truth. . . .People thought they could explain and conquer nature—yet the outcome is that they destroyed it and disinherited themselves from it.
>
> Vaclav Havel

Wolpert goes on to state that

> [Science] is viewed with a mixture of admiration and fear, hope and despair, seen both as the source of many of the ills of modern industrial society and as the source from which cures for these ills will come.

Within the context of the last few chapters, would you view molecular biology and genetics as a beautiful story of great power and explanation or as knowledge that can be dangerous and destructive?

What view would D. H. Lawrence take?

*L. Wolpert, 1993, *The Unnatural Nature of Science* (Cambridge, Mass.: Harvard University Press).

## PATENTS ON LIVING ORGANISMS

Governments routinely issue patents to inventors who develop devices or processes that are new, useful, and not obvious to others. Such patents protect the inventor from imitation and provide a limited monopoly on the product so that profits will yield a reward for the risk the inventor took to develop the new product.

In general, this patent system works reasonably well, but with the techniques of biotechnology scientists can develop new and unique living organisms that present special patent problems. For example, living organisms reproduce. Suppose a new gene were inserted into the genetic material of an existing organism. If the geneticist sells one of these organisms, does the buyer have the right to sell the offspring, or would that be a violation of the geneticist's patent rights?

Consider a case in which the gene sequence of an organism is altered to produce a slightly different protein. This protein is patented. Suppose a competitor develops a protein with very similar properties and characteristics but not exactly the same amino acid sequence. Would this be a patent violation?

John H. Barton, 1991, "Patenting Life." *Scientific American* 264, no. 3: 40–46.

## *Ethics*

## IN VITRO FERTILIZATION

In 1978, the first baby grown from an egg fertilized outside the human body was born. Since then, this procedure, called in vitro fertilization, has become routine. Eggs and sperm are removed from the woman and man and are maintained under laboratory conditions that simulate the normal environment of a woman's body. When eggs and sperm are mixed, fertilization occurs and the fertilized egg (or the multicelled embryo resulting from several cell divisions) is implanted in the womb. In about 14 percent of the cases, a normal pregnancy and birth results.

While this procedure is straightforward and allows many women to bear children who might otherwise not be able to, it poses several emotional and ethical problems.* If eggs or sperm are donated by family members or close friends of the mother, the child may later become involved in serious family conflicts. Whose child is it—the donors of the eggs and sperm, or the mother who gave birth to the child?

The identity of the donors and recipients are much easier to conceal if the participants are unknown to each other. But donors are difficult to recruit. Is it ethical to offer financial incentives to potential donors?

These difficult questions have led to much controversy and emotional distress. How would you resolve them?

*Robert M. L. Winston and Alan H. Handyside, 1993, "New Challenges in Human In Vitro Fertilization," *Science* 260: 932–936.

# PART IV

# EVOLUTION AND ECOLOGY

CHAPTER
17

SELECTION

**D**oes selection lead to significant changes in gene frequency? In the last chapter we learned that selection is one deviation to the Hardy-Weinberg law. We saw that it *could* change gene frequencies, but does evidence show that it does in real life situations? What drives selection? And if genetic change can occur this way, how much change is possible?

To begin answering these questions, let us consider the data on insecticide resistance shown in figure 17.1. Can we interpret these results in terms of selection? Notice that over the years American farmers have increased their pesticide use, the number of insect species resistant to pesticides has also increased. The most widely accepted interpretation of these data is that when insects are exposed to these toxic substances, the ones that are resistant to it survive and reproduce; the ones that are not resistant die. If resistance is inherited, then later generations will have a greater frequency of the insecticide-resistant genes. The **selecting factor** in this case is exposure to the insecticide.

Why was resistance to insecticides so low in the 1950s? If all exposure to insecticides were to be eliminated in 1995, how would resistance change in subsequent generations?

A question about the insects is now apparent. What caused the insecticide resistance? One explanation is that the insecticide acts on the insects' genes to create mutations that confer resistance. Insects that have acquired resistance by exposure will survive. A second possibility is that resistance is already present in a few individuals in the insect population; when these survive and reproduce, they increase the gene frequency of resistance.

## LEDERBERG'S EXPERIMENTS

These two hypotheses were investigated by U.S. biologist Joshua Lederberg. He examined bacteria and their increasing resistance to antibiotics, which was similar to that observed in insects. Lederberg performed the experiment diagrammed in figure 17.2.

First, Lederberg grew bacteria in a solid sheet on a nutrient agar medium. Then he used the surface of a velvet pad to pick up bacteria from the original agar plate and transfer the bacteria to agar plates containing penicillin. By maintaining a marked orientation of the velvet pad (indicated by the arrows), he deposited bacteria on the fresh plates, called the replica plates, in the same positions they held on the original plate. As figure 17.2 shows, only a few penicillin-resistant colonies survived on the plates containing penicillin. Lederberg surmised that these colonies grew from single resistant bacteria on the original plate. Was the observed penicillin resistance generated in the bacteria by

**FIGURE 17.1**    Insecticide resistance. Exposure of insect populations to insecticides increased rapidly from the 1940s until the 1980s. At the same time, more and more insect species exhibited resistance to these insecticides.    *Source: Data from D. C. Gunn and J .G. H. Stevens, editors, "Pesticides and Human Welfare" in* Science, *226:1293, 1976, Oxford University Press, Oxford England.*

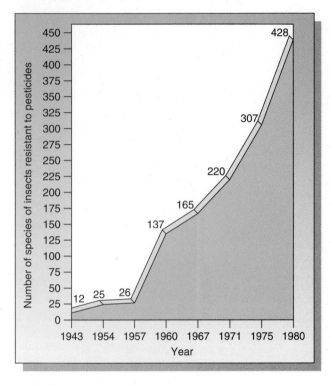

**FIGURE 17.2**    Replica plating experiments. Bacteria are grown on the original plate to form a continuous sheet. Next, a velvet pad on a cylindrical block is pressed on the surface of the original plate picking up bacteria from the surface. Then the pad is briefly pressed on the surface of the replica plates and transfers bacteria to the agar surface.

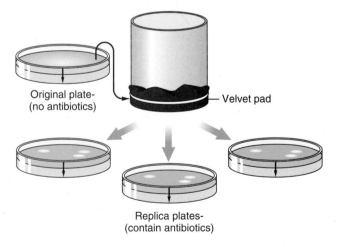

exposure to penicillin, or did it already exist in a few bacteria in the original plate? Since resistant colonies appeared at exactly the same positions on the replica plates, it appeared that resistance existed in bacteria located at these positions on the original plate before they were exposed to penicillin.

*Evolution and Ecology*

To further investigate this possibility, Lederberg returned to the original plate and removed a few bacteria from the spot where the resistant bacteria were obtained. He grew these bacteria in a culture containing *no* penicillin. When he transferred some of these cells to an agar plate containing penicillin, all of them were penicillin-resistant, *even though they had no prior exposure to penicillin.* Lederberg argued that his results indicated that penicillin resistance existed prior to any exposure and is not induced by contact with penicillin.

What results would you predict if bacteria were removed from the original plate at a position where bacteria were *not* found to be penicillin-resistant on the replica plates and then placed in a penicillin medium?

According to Lederberg's interpretation, penicillin resistance must be a genetic trait that existed in a few bacteria even though the bacteria population had never been exposed to penicillin. This interpretation suggests that resistance must have originated as a random, chance mutation in the genetic material of the bacteria. If there was no selection either for or against this mutation, the allele would be retained in subsequent generations at a very low frequency. When exposed to penicillin, however, the resistant allele would have an enormous advantage over the nonresistant allele and its gene frequency would increase dramatically.

See figure 16.7 in the previous chapter. Which curve would correspond to an experiment in which a bacteria colony was exposed to penicillin for the first time?

## SELECTION IN THE PEPPERED MOTH

In another study of selection, biologists investigated various insects in the British Isles. In the 1890s, they observed that many species existed in both dark-colored and light-colored forms. One particular species, the peppered moth, was found to be distributed unevenly throughout the island. The distribution of light- and dark-colored forms is shown in figure 17.3.

Biologists observed the dark form more frequently in central and southeastern England, while the light form predominated in other areas. What caused this difference in distribution? Could it be due to selection?

The areas where the dark form predominated were heavily industrial areas where air pollution such as soot and smoke had darkened the leaves and trunks of trees. Could the pollution have affected the color of the moths? In the 1950s, British biologist H. B. D. Kettlewell hypothesized that

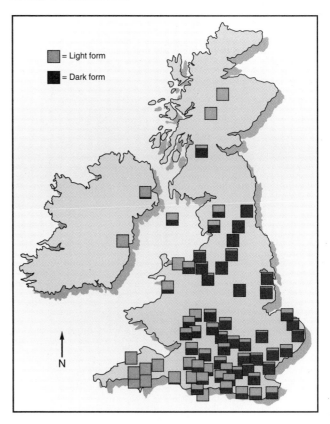

**FIGURE 17.3** Peppered moth distribution. This map of the British Isles shows the distribution of peppered moths. The portion of a circle colored black indicates the percent of dark-colored moths found in that area. For example, a circle half black indicates that 50 percent of the moths are of the dark form. The highest percent of dark moths are found in the urban and industrial areas. *Source: Data from H. B. D. Kettlewell, "Further Selection Experiments on Industrial Melanism in the Hepidoptera" in Heredity, V. 10, p. 287, 1956.*

the coloration of the moths provided protective camouflage so that they were less easily detected by predators (fig. 17.4). In industrial areas where the trees were darkened, the dark form was camouflaged; in unpolluted areas where the trees were a natural color, the light form was camouflaged.

Kettlewell predicted that the dark-colored form of the moth was more likely to survive in industrialized areas, while the light-colored form was more likely to survive where there was less pollution. He tested his hypothesis by marking equal numbers of dark and light moths with a small spot of paint under their abdomens. He released these moths in a nonpolluted area of southwestern England and in an industrial area around Birmingham in central England. Later he captured moths in these same two areas and recorded the number of marked light and dark forms. His data are shown in table 17.1

As Kettlewell predicted, a greater percentage of light moths survived in the unpolluted area, while a greater percentage of dark moths survived in the polluted area. He argued that these data supported his hypothesis.

FIGURE 17.4 Peppered moth camouflage. (a) The light and dark form of the peppered moth on the trunk of a tree darkened with soot and other industrial pollution. (b) Light and dark forms are shown on a lighter tree trunk in an unpolluted area.

a.

b.

The interpretation of the data in table 17.1 makes what assumption?

Breeding experiments between light-colored and dark-colored peppered moths show that the allele for dark color is dominant. Does this result affect Kettlewell's hypothesis?

| TABLE 17.1 | Light- and Dark-Colored Moths in Southwest and Central England |
|---|---|

| | Marked Moths Captured as Percent of Those Released |  |
|---|---|---|
| | Southwest England (unpolluted) | Central England (polluted) |
| Light form | 12.5 | 13.0 |
| Dark form | 6.3 | 27.5 |

Kettlewell's data and observations show how environment affected the forms of the peppered moths, but they give no indication of *why* survival of the two forms differed in the two areas. As we mentioned, however, Kettlewell hypothesized that difference in color offered protection from predators. To investigate this question, he placed moths of both forms on dark-colored and light-colored tree trunks in their natural settings. He set up blinds so that he could unobtrusively observe and photograph the moths. Kettlewell observed that birds tended to consume the uncamouflaged moths in both sites. His data are consistent with the data in table 17.1 and support his hypothesis.

In another experiment with peppered moths, caterpillars of the light form and dark form were fed either soot-covered leaves from polluted regions or clean leaves from unpolluted regions. The food source had no effect on the color of the resulting adult moths. What was the purpose of this experiment?

## BALANCED POLYMORPHISM

From Kettlewell's experiments with peppered moths it is apparent that selection and changes in gene frequency depend both on the alleles present in the population and on specific environmental conditions (i.e., selection factors). The allele for light color has a **selective advantage** under some conditions, but the allele for dark color has advantages under other conditions. Selection, then, may sometimes help maintain two different alleles in a population because of variations in the environment. This maintenance of two or more phenotypically distinct forms in a population is called **balanced polymorphism.**

Another example of balanced polymorphism is found in the British land snail, *Capacea,* where the shells may be banded or unbanded (fig. 17.5). These snails are found in a wide variety of environments, including rocky and grassy areas and woodlands. Biologists have observed the distribution of these snails and have gathered the data summarized in table 17.2.

Why is the distribution different in the two environments? A primary predator of the snails is the song thrush. When a thrush captures a snail, it carries the snail to nearby rocks where it can break open the shell and expose the flesh. By examining the shells found near these rocks, biologists can estimate the numbers of banded and unbanded snails consumed by the birds. These numbers are compared with the numbers of live snails found in a uniform grassy environment (table 17.3).

214

*Evolution and Ecology*

FIGURE 17.5   Polymorphism in snails. (a) Banded and unbanded
forms of one species of snail. (b) Song thrush breaking open captured
snails on a rock serving as an "anvil."

a.

b.

| TABLE 17.2 | Distribution of Banded and Unbanded Land Snails | |
| --- | --- | --- |
| | **Distribution in Percent** | |
| **Environment** | **Banded Snails** | **Unbanded Snails** |
| Uniform background (short grassland) | 10 | 90 |
| Broken or mottled background (woodlands or brushy areas) | 90 to 60 | 10 to 40 |

| TABLE 17.3 | Living vs. Captured Snails in a Uniform Environment | | | |
| --- | --- | --- | --- | --- |
| | **Number of Snails in Uniform Environment** | | | |
| | **Banded** | **Unbanded** | **Total** | **Percent Banded** |
| Living | 264 | 296 | 560 | 47.1 |
| Killed by thrushes | 486 | 377 | 863 | 56.3 |

In this uniform environment, more banded snails were consumed by the birds, indicating that the camouflage of the snails' shell pattern provides a selective advantage.

What hypothesis was being tested in the experiment summarized in table 17.3?

A student examined the data in table 17.3 and argued that it was unnecessary to collect living snails because the shell fragments gathered near the rocks was sufficient evidence. How would you respond to this student's statement?

## SELECTION AT THE MOLECULAR LEVEL

The variations we have described in these examples are phenotypic differences we can see just by looking at the organisms in question. Some variations are harder to detect, such as variations in enzymes that might be influenced by environmental conditions. For example, scientists studying the fruit fly have discovered several forms of the enzyme alcohol dehydrogenase. Some of the forms are more resistant to heat than others. A logical hypothesis is that selection has resulted in fruit flies that live in warmer climates having the heat resistant form of the enzyme. These flies would presumably have a selective advantage over flies without the heat resistant form.

When biologists measured the forms of alcohol dehydrogenase in flies from warm climates and flies from cool climates, they found that, as they had predicted, the heat-resistant enzyme was more prevalent in flies from warm climates. These results clearly suggest that selection can act at the molecular level to influence the genetic makeup of individuals.

What additional experimental data would add support to the hypothesis that selection affects enzyme form?

FIGURE 17.6 Distribution of sickle-cell anemia in Africa. The percentage of the population in each area suffering from sickle-cell anemia is shown by the indicated shading. *Source: Data from A. C. Allison, "Protection Afforded by Sickle-Cell Traits Against Subtertian Malarial Infection" in* British Medical Journal, *1:290-301, 1954.*

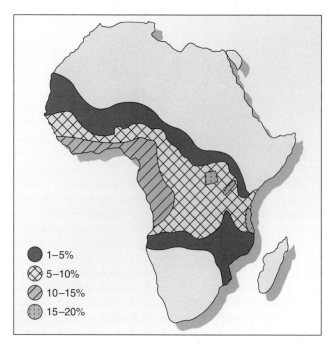

1–5%
5–10%
10–15%
15–20%

Another example of how selection works at the molecular level is provided by sickle-cell anemia. In chapter 15, we saw that this disease is the result of a single amino acid substitution in the hemoglobin molecule (see fig. 15.8). This single amino acid substitution causes a variety of destructive effects that are usually fatal to sufferers of the disease at an early age.

If sickle-cell anemia is so destructive, why would the allele that carries it continue to exist? We would expect the allele to be rapidly eliminated since affected individuals are not likely to reproduce and pass on the trait to subsequent generations.

In fact, sickle-cell anemia does exist and is particularly common in parts of Africa and the Middle East (fig. 17.6). These parts of the world also happen to be areas where malaria is prevalent. Careful investigation has shown that individuals who are heterozygous for the sickle-cell trait are resistant to malaria, whereas normal homozygotes are not. The heterozygous individual has a major advantage over both homozygous sickle-cell genotypes (who suffer the destructive effects of the disease) and homozygous normal genotypes (who are highly

susceptible to malaria). As a consequence, the heterozygotes are most likely to survive and preserve the sickle-cell allele in the population. Biologists have discovered such **heterozygote superiority** in a number of traits.

## VARIABILITY

In all the examples we have covered in this chapter, selection requires the existence of a variety of alleles. If all the genes in a population were homozygous, selection would not take place. Selection, then, depends on **variability,** the amount of genetic variation or the number of different kinds of genes found in the population. How much variability exists in populations? This is an important question when it comes to estimating the amount of change that selection can cause.

Estimating variation can be a difficult task. If we simply examine phenotypic variation, we will not discover recessive alleles in heterozygotes. Moreover, many alleles for traits such as antibiotic resistance will remain hidden unless the organisms that carry the gene are exposed to specific selection factors.

The task of estimating variation has been made a great deal easier by techniques that permit the separation and analysis of proteins. Since a single protein chain is the product of a particular sequence of nucleotides in DNA and represents a single gene, a change in the gene will produce an altered protein chain. If amino acid substitutions in the protein result in a change in the electrical charge on the protein molecule, then the original and the altered protein molecules can be separated by **electrophoresis,** a technique for separating molecules in an electric field. Electrophoretic methods are so sensitive that scientists can detect differences of only one amino acid in several hundred.

Of course, it is not realistic, even with electrophoretic methods, to examine every gene in an organism. What biologists need is a reasonable estimate of the variation. They make such estimates by measuring variability in about twenty genes. The results are commonly reported as the **heterozygosity** of the organism, or the average proportion of genes at which an individual in a population has two different alleles for the trait.

What do these measurements reveal? Geneticists have measured the heterozygosity of a variety of organisms, some of which are listed in table 17.4.

Geneticists regard heterozygosities of this magnitude as quite large. For example, humans have about 100,000 genes. A heterozygosity of 6.7 percent means that 6,700 genes in a typical individual are heterozygous. This much diversity would theoretically allow the individual to produce more

| TABLE 17.4 | Heterozygosity in Major Groups of Plants and Animals |
| --- | --- |

| Organism | Percent Heterozygosity |
| --- | --- |
| Plants | 17 |
| Insects | 15 |
| Marine invertebrates | 12 |
| Fish | 8 |
| Birds | 4 |
| Mammals | 5 |
| Humans | 6.7 |

than $10^{2,000}$ genetically different gametes, a number greater than the estimated number of atoms in the universe!

Recall from chapter 16 that the mutation rate in most organisms is fairly low. It is not possible that mutation could be creating this much variability in one generation. Apparently, the genetic diversity is carried along from generation to generation. As mutations do occur, many are retained and become part of the genetic variability. Selection, then, acts on a sizable amount of genetic variation and in many circumstances creates a major change in gene frequency.

## SUMMARY

*Selection in plants and animals leads to significant changes in gene frequency. Selection may be caused by environmental factors, and it relies on variability in the genetic makeup of species. As you review the information presented about selection, keep in mind the following key points.*

1. **Selection** occurs when one genotype can reproduce and survive better than others; it results in changes in gene frequency.
2. **Selecting factors** in the environment, such as the presence of an insecticide, antibiotics, or predators, can act on the phenotype of organisms to provide a **selective advantage** of one phenotype over another.
3. Lederberg's replica plating experiment transferred bacteria to agar plates containing penicillin. Resistant colonies appeared at the same position on each replica plate, suggesting that penicillin resistance existed in a few bacteria before any exposure to the antibiotic.
4. Lederberg cultured bacteria from positions on the original plate where resistant bacteria had been obtained. All the bacteria from the culture were penicillin-resistant, indicating that favorable genetic mutations exist before exposure to selection.

5. Moth coloration appears to serve as protective camouflage. Moths with protective coloration exhibit higher survival rates than moths without protective coloration.
6. **Balanced polymorphism** is the maintenance of two or more phenotypically distinct forms in a population; it can preserve variability in populations.
7. **Heterozygote superiority** occurs when the heterozygous genotype has a survival advantage over both the homozygous dominant and recessive forms. This effect tends to preserve both alleles in the population.
8. Variability in a population can be measured as **heterozygosity,** the average proportion of genes in an individual that are heterozygous.
9. **Heterozygosity** ranges from 17 percent to 4 percent in living organisms. These figures represent a high genetic variability.

## REFERENCES

Allison, Anthony. August 1956. "Sickle Cells and Evolution." *Scientific American* 195, no. 2: 87–94.
Ayala, Francisco J. September 1978. "The Mechanisms of Evolution." *Scientific American* 239, no. 3: 56–69.

Hamilton, Terrell H. 1967. *Process and Pattern in Evolution* (London: Macmillan).
White, Michael J. D. 1978. *Modes of Speciation* (San Francisco: W. H. Freeman).

## EXERCISES AND QUESTIONS

1. Observe figure 17.1 (page 212). Which of the following would most likely explain why insect resistance to insecticides was so low in the 1950s (page 212)?
   a. No genes for insecticide resistance existed in the population before 1960.
   b. Selecting factors had acted on the population to decrease the insecticide resistance gene.
   c. The graph shows that insect populations were low in the 1950s.
   d. Little exposure to insecticide had provided selection favoring the resistant gene.
   e. Exposure to the insecticide is necessary to create mutations for resistance.
2. If all exposure to insecticides was eliminated in 1995, how would resistance change in subsequent generations (page 212)?
   a. The level of resistance would not change.
   b. Resistance would decrease rapidly in one or two generations to the levels found in the 1950s.

    c.   Resistance would later be found in even more insects in the population.

    d.   Resistance would decrease gradually in ten to fifteen generations to the levels found in the 1950s.

    e.   Resistance would reach an equilibrium with susceptibility so that about 50 percent of the insects would be resistant and 50 percent susceptible.

3.   Which of the following results would you most likely predict if bacteria were removed from the original plate in figure 17.2 at a position where bacteria were *not* found to be penicillin-resistant on the replica plates and then placed in a penicillin medium (page 213)?

    a.   Bacteria would grow, but not as well as bacteria from the original resistant colonies.

    b.   These bacteria would be resistant to other antibiotics but not to penicillin.

    c.   These bacteria could not be cultured in a medium that does not contain penicillin.

    d.   After exposure to the penicillin, the bacteria would grow but not before exposure.

    e.   The bacteria would not survive in the penicillin medium.

4.   Suppose Lederberg's replica plating experiment (fig. 17.2) had yielded the results shown in the figure above. Which of the following is the best interpretation of these data?

    a.   Penicillin has no significant effect on bacteria from the original plate.

    b.   Bacteria that are penicillin-resistant can grow on the replica plates, but other bacteria cannot.

    c.   Exposure to penicillin induces resistance because replica colonies appear at different positions.

    d.   Most bacteria on the original plate are resistant to antibiotics other than penicillin.

    e.   The experimenter changed the orientation of the velvet pad when making replica plates.

5.   See figure 16.7 in chapter 16. Which curve would correspond to an experiment in which a bacteria colony was exposed to penicillin for the first time (page 213)?

    a.   X

    b.   Y

    c.   Q

    d.   Z

6.   The interpretation of the data in table 17.1 makes what assumption (page 214)?

    a.   Marking the moths with a spot of paint makes them easier to recapture.

    b.   Greater numbers of dark moths than light moths survive in polluted areas.

    c.   Both light and dark moths are found in polluted and unpolluted areas.

    d.   The number of captured moths is representative of the number of surviving moths in the environment.

    e.   Coloration of the moths provided camouflage protecting them from predators.

7.   In another experiment with peppered moths, caterpillars of the light form and dark form were fed either soot-covered leaves from polluted regions or clean leaves from unpolluted regions. The food source had no effect on the color of the resulting adult moths. This experiment was done to investigate which of the following possibilities (page 214)?

    a.   The selecting factor is the color of caterpillars rather than predation by birds.

    b.   Beneficial mutations or characteristics can be generated by environmental conditions.

    c.   Alleles providing a selective advantage are found in caterpillars as well as adult moths.

    d.   Caterpillars can migrate from polluted to unpolluted areas and from unpolluted to polluted areas.

    e.   Soot-covered leaves may provide less nutrition than clean leaves.

8.   Breeding experiments between light-colored and dark-colored peppered moths show that the allele for dark color is dominant. Does this result affect Kettlewell's hypothesis (page 214)?

    a.   No. Kettlewell's hypothesis referred to moth coloration, not genetics.

    b.   Yes. The recessive allele will be eliminated more quickly by selection.

    c.   No. Selection acts in the same way on both dominant and recessive alleles.

    d.   Yes. This explains why there are more dark moths in polluted areas.

    e.   No. The hypothesis predicts that dark is dominant and light is recessive.

9.   A current biology textbook states that "most new mutations are recessive" but offers no evidence to support the statement. Which of the following assumptions would be the most important in an argument supporting the book's statement?

    a.   Most individuals are heterozygous for most genes.

    b.   Most individuals are homozygous for most genes.

c. Most existing genes are recessive.

d. Most existing genes code for active enzymes.

e. Mutations are rare events.

10. Which of the following hypotheses was most likely being tested in the experiment with banded and unbanded snails summarized in table 17.3 (page 215)?

   a. Snail shell pattern provides protective camouflage from predators.

   b. Individual snails camouflaged in both woodlands and short grasslands escape predation.

   c. There are more dead snails than live snails in any given area.

   d. Predatory birds require rocks to break open the snail shells.

   e. Woodlands are a more suitable environment for snails than grasslands.

11. A student examined the data in table 17.3 and argued that it was unnecessary to collect living snails because the shell fragments gathered near the rocks was sufficient evidence to test the hypothesis. Which of the following is the best response to this student's statement (page 215)?

   a. The student's assessment of the results in table 17.3 is accurate.

   b. The data on shell fragments did not test the hypothesis; only living snails should be collected.

   c. If the age of the shells were known, then data from shell fragments would be sufficient.

   d. Either data on shell fragments or living snails would have been sufficient.

   e. Data on living, as well as dead, snails provide additional support for the hypothesis.

12. Which of the following experimental investigations would add the most support to the hypothesis that selection has resulted in fruit flies from warm climates having a higher frequency of heat-resistant enzymes than fruit flies from cool climates (page 215)?

   a. Measure the frequency of heat-resistant alcohol dehydrogenase in organisms from warm climates other than fruit flies.

   b. Expose fruit flies with heat-resistant enzymes to warm temperatures and see if they survive better than flies without the heat-resistant form.

   c. Extract the alcohol dehydrogenase and measure more precisely the temperature range in which the enzyme is active.

   d. Repeat the original experiment to see whether the same results are obtained.

   e. Measure the amount of alcohol dehydrogenase present in heat-resistant and non-heat-resistant flies.

13. In the polluted areas of England the gene frequencies for the dark and light alleles of insects are almost constant even though intensive selection is taking place. Should this cause doubts about the reliability of the Hardy-Weinberg law?

   a. No. It is assumed that selection will lead to equilibrium of gene frequencies.

   b. Yes. When selection is taking place it is not possible for gene frequencies to remain constant.

   c. No. The Hardy-Weinberg law does not prohibit constant gene frequencies when selection is occurring.

   d. Yes. If selection is taking place, gene frequencies will be constantly changing.

   e. No. The effects of selection will most likely be offset by migration and mutation.

14. Consider an isolated population living in a uniform environment and subjected to intensive selection. Which of the following will happen to the heterozygosity of the population?

   a. Heterozygosity will increase.

   b. If heterozygosity is originally more than 10 percent, it will increase.

   c. Heterozygosity will not change.

   d. If heterozygosity is originally less than 10 percent, it will decrease.

   e. Heterozygosity will decrease.

# SPECIATION

How much change can occur in a population due to selection? Can selection bring about **speciation,** the development of separate species from a single original population? A **species** is a group or population of organisms that will not normally interbreed with other species. Can selection bring about such separation?

In the last chapter we saw that environmental differences will lead to distinct variations in characteristics. For example, pollution affected coloration in peppered moths and climate affected the enzymes present in fruit flies. Environments, however, usually differ from one another in more than just one respect. Living in different environments and being exposed to many selection factors would likely cause organisms to be different from one another in many ways.

## RACES AND SUBSPECIES

To investigate the nature and amount of variation possible in populations, U.S. botanists Jens Clausen, David Keck, and William Hiesey conducted careful experiments with the plant *Potentilla*. This plant grows in California in an area that ranges from near the seacoast to high elevations in the Sierra Nevada Mountains. The conditions vary from very mild near sea level to a harsh, alpine climate (fig. 18.1).

*Potentilla* plants grown at different altitudes exhibit significantly different appearances. The size and shape of the leaves can vary as well as the height of the plant. Are these differences caused by the plants' environments, or are they genetic differences due to selection?

Clausen, Keck, and Hiesey examined this question by collecting plants at three different locations: Stanford, near sea level; Mather, at middle altitudes; and Timberline, near 10,000 feet in the Sierras. They cut each plant into three parts and replanted one part at each of the three locations. The responses of the plants at each location are shown in figure 18.2.

In most cases the plants grew best in their native locations. In the other locations growth was less vigorous. The one exception was the plants collected at Timberline; they exhibited greater growth at Mather. How can we interpret these observations?

Apparently, selection has produced plant populations at each site that are genetically most "fit" for the conditions at that site. This process, termed **adaptation,** results in groups or populations of organisms that are genetically different from one another. The entire population of *Potentilla,* from sea level to the mountains, represents a single gene pool in which plants are *capable* of interbreeding but do not normally do so because of the separation of distance. Plants at each location, however, have developed into different **races** that are genetically adapted to survive and flourish in their local environments.

Why did Clausen, Keck, and Hiesey divide one plant from each site into three cuttings rather than collect three different plants?

If plants from Stanford, Mather, and Timberline were genetically identical, what results would you expect in the experiment?

**FIGURE 18.1** Growing range of *Potentilla*. This cross section of California shows the sites and elevations where *Potentilla* plants were collected at native growing locations and the sites where plants from all three locations were grown.

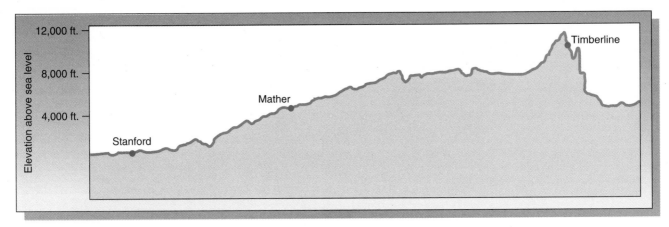

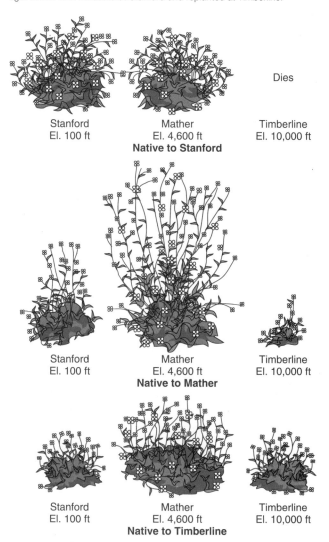

**FIGURE 18.2** Response of *Potentilla* grown at different locations. The site of collection is shown at left, while the site where plants were grown is shown at bottom. For example, the plant shown in the lower right corner was collected and grown at Timberline, while the plant in the upper right corner was collected at Stanford and replanted at Timberline.

Stanford
El. 100 ft

Mather
El. 4,600 ft

Timberline
El. 10,000 ft

Dies

**Native to Stanford**

Stanford
El. 100 ft

Mather
El. 4,600 ft

Timberline
El. 10,000 ft

**Native to Mather**

Stanford
El. 100 ft

Mather
El. 4,600 ft

Timberline
El. 10,000 ft

**Native to Timberline**

This kind of variation in populations that are spread over wide geographic areas, in which organisms in one location will develop distinct characteristics, is commonly observed in a great many species. When the differences among the races become more obvious, the populations are considered **subspecies.**

Consider the common song sparrow found throughout North America (fig. 18.3). Biologists recognize more than thirty subspecies of this bird. Even though all song sparrows are very similar in most ways, individual birds from separate regions, such as the Eastern, Desert, and Kenai subspecies, exhibit clear differences in their appearance and their songs.

In spite of their distinctive and separate characteristics, these subspecies are capable of interbreeding and producing offspring. In other words, when subspecies come into contact they can exchange genes. Apparently, the distinctive characteristics of the subspecies are maintained because the great majority of the individuals in a subspecies are isolated and do not come in contact with members of other subspecies.

A student argued that if subspecies differed in some characteristics, then they should not be able to interbreed. What was this student assuming?

## ISOLATION AND SPECIES FORMATION

We have seen that subspecies can exchange genes when they come in contact with one another. However, many groups of organisms that are obviously similar and live in the same environment do not interbreed. One example is the Lincoln sparrow and the song sparrow (fig. 18.4). These birds, though very similar, do not interbreed as subspecies do. The same is true of sugar maples and red maples (fig. 18.5). Species like sugar and red maples that are similar in appearance but do not interbreed are called **sympatric** if they live in the same geographic area. If they are geographically separated, they are called **allopatric.**

How could organisms have developed into such similar species but not interbreed? Were these groups originally subspecies that changed so much genetically that interbreeding is no longer possible? How could such genetic change occur if as subspecies the individuals continued to exchange genes? Interbreeding assures that groups of organisms will remain part of the same gene pool. Even if mutation and selection act to change part of a population, exchange of genes tends to offset such changes. We would expect the interbreeding population to remain a single species.

According to the argument in the previous paragraph regarding gene exchange in a population, why can't a single, continuous, interbreeding population split into two noninterbreeding groups?

One hypothesis is that part of a population becomes separated from the rest of the population and cannot exchange genes; this is called **isolation.** If mutation and selection occur in this isolated group, they result in increasing changes in the genetic makeup of the group. This hypothesis

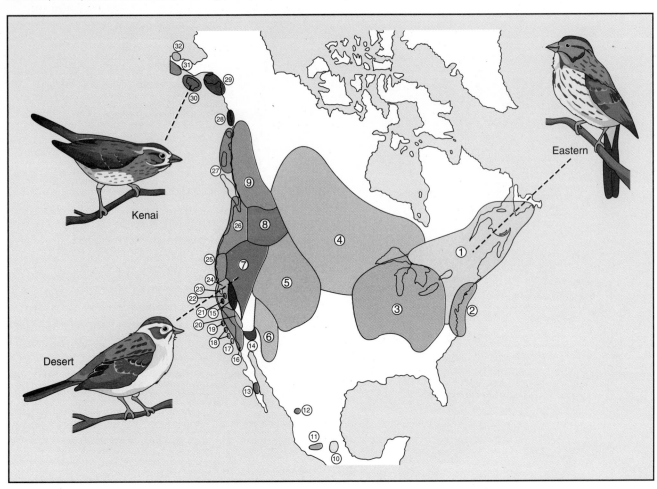

FIGURE 18.4    Sympatric sparrows that do not interbreed. Even though Lincoln and song sparrows occupy the same range and are very similar in appearance, they do not interbreed.

Lincoln sparrow                    Song sparrow

is illustrated in figure 18.6. Since the isolated population is separated from the parent population by distance or physical barriers such as mountains or a river, this scenario is called **geographic isolation.** If this isolated group with its new genetic makeup reunites with the original parent population (fig. 18.6*d*), interbreeding may or may not occur. If the isolated population has changed to such an extent that interbreeding is not possible, then a new **species** has developed. The two groups of organisms, now living in the same area, are incapable of exchanging genes under natural conditions. This is called **reproductive isolation.**

Are new species actually formed in this way? Certainly biologists know of a great many natural populations that correspond to each part of figure 18.6. We have considered only a few. Figure 18.6*b* corresponds to the example of song sparrows. Figure 18.6*c* possibly corresponds to the situation with *Potentilla* in the Sierras. Figure 18.6*d* represents the breeding dynamics of red and sugar maples and Lincoln and song sparrows.

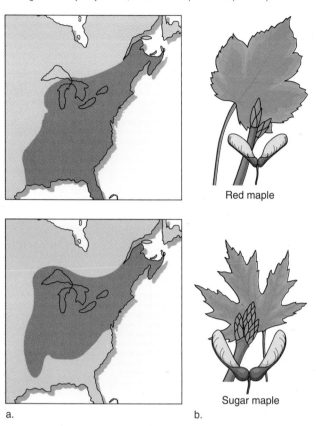

**FIGURE 18.5** Sympatric but separate maple species. (a) The ranges of red and sugar maples overlap over almost the entire areas of occupation. (b) Many similarities are apparent in leaves, buds, and seeds. Although obviously very similar, these two maples are separate species.

Red maple

Sugar maple

a.                    b.

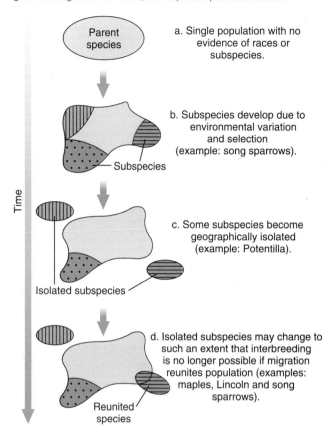

**FIGURE 18.6** Hypothesized model of speciation. (a) A single homogeneous population. (b) Environmental differences and selection act on an extended population to produce races and subspecies. (c) Some parts of the population may become isolated. Any genetic characteristics cannot be exchanged between populations. (d) If populations reunite and gene exchange does not occur, then separate species are evident.

Parent species

Time

a. Single population with no evidence of races or subspecies.

b. Subspecies develop due to environmental variation and selection (example: song sparrows).

Subspecies

c. Some subspecies become geographically isolated (example: Potentilla).

Isolated subspecies

d. Isolated subspecies may change to such an extent that interbreeding is no longer possible if migration reunites population (examples: maples, Lincoln and song sparrows).

Reunited species

The extremely long time periods required for genetic change to accumulate in isolated groups prevents biologists from directly observing the development of reproductive isolation. Most likely it takes millions of years to isolate a population and allow sufficient genetic change to develop a new species. So many examples, however, seem to be explained by the model in figure 18.6 that the supporting arguments are very persuasive. Indeed, most scientists consider geographical isolation a necessity for the origin of almost all species.

## REINFORCING SPECIATION

One question we might ask about the model in figure 18.6 is, how can two closely related but separate species remain separate species once they occupy the same area? Wouldn't the two species change over time to become even more similar and merge again into one species?

We can investigate this question by examining similar species in which populations of the species overlap in some areas but not in others. For example, two species of lizards, *Typhlosaurus lineatus* and *Typhlosaurus gariepensis,* live in the Kalahari Desert of Africa. Both species prey primarily on

termites, and their ranges overlap considerably. Biologists measured certain characteristics of the lizards, including body length, head length, and size of prey consumed. These data are shown in figure 18.7.

The data show that head and body length and prey size are larger for *T. lineatus* lizards found in sympatric areas than for those found in allopatric areas. Moreover, all variables are larger for *T. lineatus* than for *T. gariepensis.* How can we interpret these data? Why are *T. lineatus* lizards different in the two areas?

Contrary to what we might expect, these sympatric species become more dissimilar rather than more alike where their ranges overlap. Why? In this case, biologists argue that if both species consumed the same-sized prey, there would be less food for both of them. Apparently, selection acts to minimize competition for the same food source by adapting organisms to consume different-sized prey. Selection favors larger *T. lineatus* consuming larger prey in sympatric areas, so there is *less* competition for the same food source. This effect is called **character displacement.** No such selection acts in allopatric areas.

FIGURE 18.7 Character displacement. Results from two species of lizards (*T. lineatus* and *T. gariepensis*) studied in the Kalahari Desert are shown. In the sympatric area, *T. lineatus* exhibits a larger head and body size and consumes larger prey than it does in the allopatric area.

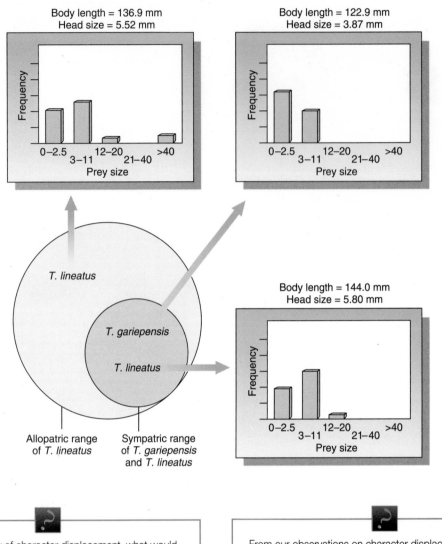

Body length = 136.9 mm
Head size = 5.52 mm

Body length = 122.9 mm
Head size = 3.87 mm

Body length = 144.0 mm
Head size = 5.80 mm

Based on the theory of character displacement, what would most likely happen to *T. gariepensis* if *T. lineatus* were removed from the environment?

From our observations on character displacement in frog calls, what can you conclude about hybrids formed by interbreeding between the two species?

Another study compared the mating calls of two closely related frog species in allopatric and sympatric areas. In allopatric areas, the calls of the two frogs were found to be very similar, but in sympatric areas the calls were markedly different. Since mating calls appear to be the primary method male and female frogs use to locate one another during breeding season, different mating calls would serve to prevent interbreeding. Conversely, if mating calls in sympatric areas were similar, then extensive interbreeding would likely occur between the two species.

# THE SPECIES CONCEPT

An important concept in biology is that species are reproductively isolated groups. Such genetic isolation ensures relative stability in the characteristics of the species—characteristics that have led to that species' successful survival. Without reproductive isolation, less adaptive genes could enter the gene pool and decrease the survival rate of the species. Thus, maintaining genetic isolation is a primary survival advantage.

Earlier we defined a species as a group of organisms incapable of exchanging genes with another species *under natural conditions*. Genes often can be exchanged under artificial conditions, however. For example, mallard ducks and pintail ducks, two separate species, will interbreed in captivity to produce **hybrids,** offspring of two different subspecies or species. In their natural environment they do not interbreed. Many species of plants can be artificially pollinated with pollen from other species to produce viable hybrid seed. Such cross-pollination, however, does not normally occur in nature.

## ISOLATING MECHANISMS

If some species can interbreed artificially, what prevents them from interbreeding under natural conditions? Some mechanism must be at work that maintains the isolation of species.

As we have seen, selection results in differences between species. We would expect such differences to include mechanisms that prevent interbreeding. One such difference we have already seen is variety in the mating calls of frogs. This is an example of **behavioral isolation,** which is common in animals. Many animals exhibit species-specific courtship rituals that are a prerequisite for mating.

In plants, where behavior is not a factor in reproduction, we observe other mechanisms. Consider the example of scarlet and black oaks. These species are found in the same regions and for the most part remain separate. In a few areas, however, they interbreed and produce hybrids. Why should this interbreeding occur? Further observation reveals that scarlet oaks thrive in poorly drained or swampy areas. Black oaks, on the other hand, thrive in well-drained, dry areas. Apparently, this **habitat isolation** is sufficient to maintain the distinctness of the two species. Interbreeding occurs only where swamps and dry land are found close together.

Biologists have hypothesized many isolating mechanisms and have observed in nature almost all of them. A sample of these isolating mechanisms are listed in table 18.1.

## POLYPLOIDY

In our discussion so far, speciation appears to be a very slow process involving geographic isolation possibly followed by migration that reunites isolated groups. Character displacement can create further species differences. Observations of chromosome numbers in some species, however, suggest that new species, particularly in plants, sometimes may originate much more rapidly.

Consider, for example, the chromosome numbers observed in several species of chrysanthemum. These separate species exhibit diploid chromosome numbers of 2n = 18, 36, 54, 72, and 90. Interestingly, each of these numbers is equal to the previous number plus 18. For example, 54 = 36 + 18. It seems as though a new species arises by the addition of another "set" of 18 chromosomes.

**TABLE 18.1  Isolating Mechanisms**

I. Mechanisms that prevent contact between gametes
   A. *Behavioral:* Different mating behaviors prevent interbreeding (frog calls)
   B. *Habitat:* Species live in the same areas but occupy different habitats (scarlet and black oaks)
   C. *Seasonal:* Species breed at different times of the year
II. Mechanisms that allow interbreeding but result in offspring with a low survival advantage
   A. *Hybrid sterility:* Hybrid offspring produced are healthy but cannot reproduce
   B. *Hybrid inviability:* Hybrid species are produced but are less "fit" than either parent species

How could chromosome numbers occur in such a regular sequence? How could they increase in such a systematic manner? To investigate this question we must reconsider events in mitosis and meiosis. When an organism such as chrysanthemum reproduces, it produces gametes with the haploid chromosome number. At fertilization, haploid male and female gametes fuse and restore the diploid number. Subsequent growth of the organism by mitotic cell division maintains the diploid chromosome number. What could possibly change this sequence of events to increase the chromosome number?

Apparently, the chromosome number is changed by some disruption in cell division. One well-accepted explanation is illustrated in figure 18.8. Chromosome duplication and separation are completed in mitosis, but then complete cell division fails to occur. This results in a cell with double the number of chromosomes. If this cell now divides normally, tissue will form where the cells have the double chromosome number. Moreover, if this part of the organism can reproduce asexually (which occurs commonly in plants but not in animals), it will produce a complete new organism with a double, or tetraploid (4n) chromosome number.

In the next generation, the tetraploid, or 4n, organism generates gametes, which are now 2n. These gametes, however, must unite with gametes from an organism that has the same chromosome number. Otherwise, the two gametes would contribute different chromosome complements. Chromosomes would not pair properly in meiosis and subsequent spore and gamete formation would be disrupted.

The chance of two different organisms doubling their chromosome number and then meeting to join gametes may seem so unlikely as to be impossible. Considering the vast numbers of organisms and gametes produced, however, there is reasonable possibility that two such gametes will unite to form a new species. These new species formed by the multiplication of chromosome number are called **polyploids.**

Although apparently rare in sexually reproducing animals, polyploidy accounts for many new plant species. Indeed, scientists estimate that as many as 40 percent of all flowering plant species are polyploids.

**FIGURE 18.8** Polyploidy is one possible way cells can increase in chromosome number. (a) A cell with a normal diploid number of four divides mitotically, but it does not complete separation into two cells, leaving a cell with 8 chromosomes. (b) Further mitotic divisions, (c) and (d) generate tissue with 2n=8.

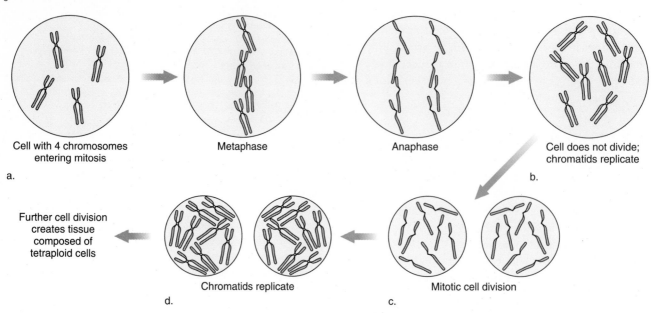

Cell with 4 chromosomes entering mitosis

a.

Metaphase

Anaphase

Cell does not divide; chromatids replicate

b.

Further cell division creates tissue composed of tetraploid cells

Chromatids replicate

d.

Mitotic cell division

c.

## SUMMARY

*Biologists can explain how new species are generated and maintained in terms of selection, geographic isolation, and character displacement. Mutation, population genetics, and molecular biology provide the foundation for our understanding of speciation. Thus, a complete explanation of speciation relies on concepts ranging from the molecular level to the level of entire populations. As you review this chapter on speciation, keep in mind the following key points.*

1. Selecting factors acting on individuals of a population in different environments may produce **races** that are genetically **adapted** to these various environmental conditions.
2. As differences in races become greater, a race can be considered a **subspecies.**
3. While races and subspecies can interbreed, **species** are groups of organisms incapable of exchanging genes with other groups under natural conditions.
4. **Speciation** is the development of separate species by mutation, isolation, and selection.
5. Species are **reproductively isolated** from other organisms. Under some unusual natural circumstances or artificial breeding conditions, species or subspecies can interbreed to produce **hybrids.**
6. **Geographical isolation** is generally considered necessary for a separate species to form. Part of a population becomes isolated from the rest of the population, and then mutation and selection in the separate groups lead to increasing genetic differences.

If reunited, the two groups may differ to such an extent that interbreeding does not occur.

7. Species occupying the same areas are called **sympatric;** those that occupy different areas are called **allopatric.**
8. When very similar species occupy the same area, **character displacement** can create further divergence between the species. Such divergence apparently leads to decreased competition between species.
9. Preservation of species as genetically isolated groups is considered highly beneficial because it prevents less adaptive genes from entering the species gene pool.
10. A variety of **isolating mechanisms** generally prevent interbreeding between species. These include behavioral and habitat isolation and hybrid sterility.
11. **Polyploidy** can give rise to new species in a single generation by increasing the chromosome number in offspring. Most likely, the chromosome number increases due to malfunctions in mitosis when chromosomes duplicate but the cell fails to divide.

## REFERENCES

Grant, V. 1981. *Plant Speciation,* 2d ed. (New York: Columbia University Press).

Lewontin, R. C. 1978. "Adaptation." *Scientific American* 239, no. 3: 213–230.

Stebbins, G. L. 1971. *Chromosomal Evolution in Higher Plants* (Reading, Mass.: Addison-Wesley).

———1971. *Processes of Organic Evolution,* 2d ed. (Englewood Cliffs, N.J.: Prentice-Hall).

## EXERCISES AND QUESTIONS

1. Why did Clausen, Keck, and Hiesey divide one plant from each site into three cuttings rather than collect three different plants (page 222)?
    a. To provide controls as well as experimental plants at each site.
    b. To treat all collected plants in the same way.
    c. As shown in figure 18.2, plants had a different appearance, but initially they should appear identical.
    d. Native plants at each site should not be removed from the soil.
    e. To ensure that the plants were genetically identical.

2. If plants from Stanford, Mather, and Timberline were genetically identical, what results would you expect in the experiment (page 222)?
    a. The response of cuttings from one site would be the same at all sites (e.g., plants collected at Timberline would grow the same way at Timberline, Mather, and Stanford).
    b. At each site, the response of the three cuttings planted would be the same.
    c. No conclusions could be drawn from the experimental results.
    d. The response of *all* plants at *all* sites would be the same.
    e. Plants would not grow at Timberline.

3. A student argued that if subspecies differed in some characteristics, then they should not be able to interbreed. Which of the following was this student most likely assuming (page 223)?
    a. Subspecies differ in at least some characteristics.
    b. Subspecies may be geographically isolated from each other.
    c. Mutations affecting all biological processes occur with the same frequency.
    d. Selection will act on any phenotypic variation in subspecies.
    e. Subspecies exhibit more differences in reproduction than do races.

4. According to the argument regarding gene exchange in a population, why can't a single, continuous, interbreeding population split into two noninterbreeding groups (page 223)?
    a. Genetic change will not necessarily spread throughout the population.

    b. If environments are different in the population's range, the subspecies should become more different.
    c. Different subspecies may come in contact in only part of their ranges.
    d. Continuous exchange of genes will keep the population the same throughout.
    e. Isolating mechanisms can develop that prevent gene exchange.

5. A botanist collected plants from California and Texas. These plants were identical in appearance, but when the biologist attempted to interbreed them, they produced no seeds. Which of the following most likely accounts for these observations?
    a. Plants from California and from Texas are widely separated by distance and cannot interbreed.
    b. Whether a plant does or does not produce seed is not as important as the number of seed that germinate and produce offspring.
    c. Genetic differences could prevent fertilization but have no effect on the plants' appearance.
    d. If plants appear identical, they should interbreed and produce fertile seeds.
    e. Environmental conditions are possibly very different in Texas and California.

6. A type of frog lives in an area stretching from Georgia to Maine. The frogs apparently form a continuous population with no visible barriers. It has been noticed, however, that although the frogs can interbreed with their nearby neighbors, frogs from the extremes of the range (i.e., Georgia and Maine) will not interbreed, even when brought together in captivity. This example would cause you to most seriously question which of the following?
    a. Gene frequencies in a continuous population will remain constant unless there is migration of individuals with different gene frequencies.
    b. Mutations may occur in a population that can change the genotype of some members of the population.
    c. Genes can move or flow from one part of a continuous gene pool to all other parts.
    d. In a continuous population stretching over a large area, environmental factors may be different in different parts of the area.

7. Suppose you observe two populations of plants that resemble each other closely and that occupy ranges that do not overlap. You form two hypotheses:
   a. The two populations were once one.
   b. The two populations were originally different but became similar.
   You observe that individuals from the two populations can interbreed and that 10 percent of the hybrids are fertile. Your biology instructor claims that this evidence favors hypothesis 1. Which of the following assumptions is your instructor most likely making?
      A. In order to produce fertile offspring, two organisms must have many genes in common.
      B. The same mutations are not likely to occur in two different populations.
   a. A only
   b. B only
   c. Both A and B
   d. Either A or B but not both
   e. Neither A nor B

8. Based on the theory of character displacement, which of the following would most likely happen to *T. gariepensis* if *T. lineatus* were removed from the environment (page 226)?
   a. Body and head size would decrease in subsequent generations.
   b. A greater variety of prey would be consumed.
   c. Head size would increase but body size would decrease.
   d. Body and head size would increase in subsequent generations.
   e. No change in body size or food type would be observed.

9. From our observations on character displacement in frog calls, which of the following would you most likely conclude about hybrids formed by interbreeding between the two species (page 226)?
   a. They would occupy a much smaller part of the range than either parent species.
   b. They could out-compete either parent species for food.
   c. They would be unable to interbreed with either parent species.
   d. Their mating call would be the same as that of one of the two parent species.
   e. They would be less likely to survive or reproduce than either parent species.

10. Consider two similar birds, A and B, that live along the northern United States from Maine to Washington state. In Washington, only A is found and it breeds in early July. In Maine, only B is found and it also breeds in early July. In Minnesota both birds are found; A breeds in late June and B breeds in late July. Which of the following predictions is most likely to be true?
   a. A and B can interbreed and the hybrids are fertile.
   b. A and B can interbreed but the hybrids are sterile.
   c. A and B will not interbreed.
   d. The area of the range in which A and B overlap will most likely increase.
   e. The survival of baby birds is lower in Minnesota than in other parts of the range.

11. Sometimes similar kinds of plants interbreed and produce offspring that exhibit "hybrid vigor." These hybrids grow taller, are physically stronger and more resistant to disease, and have other characteristics superior to the parent plants. You might expect these hybrid plants to become more common than either of the parent plants, but this seldom occurs. Which of the following best explains why?
   a. The hybrid plants cannot produce offspring.
   b. Mutation may occur more frequently in the parent plants.
   c. Hybrids have some characteristics of one parent and some characteristics of the other.
   d. The hybrid plants reproduce asexually.
   e. For genetic change to occur in two similar groups of organisms, they must be separated by some type of barrier.

12. "Hybrids are often sterile because they cannot produce functional gametes." Which of the following best supports this statement?
   a. Inability of hybrids to mate may be due to isolating mechanisms that prevent union of gametes.
   b. Meiosis occurs much less frequently than mitosis in cell division.
   c. In many organisms, such as ferns and mosses, meiosis does not form gametes.
   d. Chromosomes from parents of different subspecies or species cannot pair in meiosis.
   e. If hybrids can reproduce asexually, lack of functional gametes will not prevent reproduction.

*Evolution and Ecology*

CHAPTER
19

EVOLUTION

an selection and speciation lead to change in organisms over long periods of time? Could these mechanisms give rise to groups as diverse as algae, giraffes, and oak trees? While it seems fairly reasonable for similar species to have come from the same parents, it is quite another to visualize a common origin for organisms that are vastly different. As we will see in this chapter, however, evolution is capable of producing a great diversity of organisms from just a few species.

## DARWIN AND WALLACE

Biologists have long been intrigued by the incredible diversity among the earth's living creatures and have wondered how such variety came to be. More than 200 years ago scientists proposed hypotheses and gathered observations to try to understand how species developed. The high point of these early efforts was the publication in 1859 of Charles Darwin's *On the Origin of Species by Means of Natural Selection,* in which he explained his theory of evolution through natural selection. At the same time, another scientist, Alfred Russel Wallace, proposed ideas very similar to Darwin's.

In this context, **evolution** means change over time (usually a very long time) in the diversity and adaptation of biological organisms. The key to Darwin's arguments is **natural selection,** the idea that the environment acts on the variation in populations and "selects" the most fit. This is basically the concept presented in chapters 16 and 17. Darwin and Wallace had no knowledge of modern genetics, molecular biology, or biochemistry. Instead, Darwin based his conclusions on his observations of diversity within and between species, the adaptation of organisms to their environment, and the struggle of organisms for survival and reproduction.

A second important part of Darwin's explanation was **common descent,** which proposed that all present-day organisms have evolved in a slow, gradual process (i.e., descended) from at most a few early, primitive organisms (fig. 19.1). In other words, animals and plants as different as roses and iguanas have one common ancestor. Darwin's concept of common descent was strongly influenced by evidence from the fossil record and new information about the age of the earth. Fossils of animals and plants suggested that dramatic change occurred in organisms over exceptionally long periods of time. Evidently, living organisms were continually changing, some becoming extinct and new ones appearing. According to common descent, all this change began with a few simple organisms.

Many scientists realized that the process of evolution was slow, but if the earth was only a few thousand years old, as many people thought, how could evolution have created the incredible diversity scientists observed? Geologists helped answer this question by discovering evidence that the earth was far older than anyone previously realized. Their findings helped validate Darwin's and Wallace's arguments.

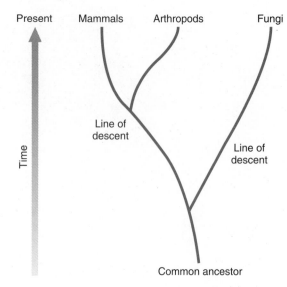

**FIGURE 19.1** Common descent. Lines of descent are used to illustrate proposed relationships between the evolution of organisms. This example indicates that organisms as diverse as fungi, mammals, and anthropods (such as insects) have descended from common ancestors over millions of years.

These two key points of Darwin's explanation of diversity—evolution through natural selection and common descent—represented major departures from explanations popular at the time. Previously, the hypothesis of **inheritance of acquired characteristics** had been a major influence on explanations regarding the diversity of species. According to this hypothesis, offspring could acquire characteristics formed by their parents' conscious interactions with the environment. For example, giraffes had developed long necks over many generations because each generation had stretched their necks to reach leaves on taller and taller trees (fig. 19.2).

According to the inheritance of acquired characteristics hypothesis, organisms develop certain traits because they "need" or "want" them. In 1809, a French naturalist named Jean Lamarck published an explanation of evolution based largely on these ideas. His views were reasonably consistent with evidence available at the time and were widely accepted.

Darwin, on the other hand, emphasized extensive variability that arose due to chance variation and completely independent of any environmental factors. In his hypothesis, environmental selection acts on this variation so that those organisms most fit will survive and reproduce. In short, Darwin's arguments proposed that selection, as we have studied in previous chapters, *could* lead to the observed diversity in living organisms.

Many species of flowering plants have developed flower shapes that can be pollinated by only one species of insect. How would Lamarck explain this development? How would Darwin explain it?

*Evolution and Ecology*

**FIGURE 19.2** Natural selection vs. acquired characteristics. Natural selection proposes that giraffes, for example, exhibit a variation in neck length. (*a*) Those that cannot reach the leaves serving as their food source will not survive. (*b*) The surviving long-necked giraffes reproduce and thus pass on the trait to their offspring. (*c*) Acquired characteristics proposes that giraffes with short necks stretch to reach leaves and, as a result, offspring have longer necks.

Natural selection (Darwin-Wallace Theory)

a.            Time

Acquired characteristics (Lamarck Theory)

b.            Time

Darwin's hypothesis has proven remarkably consistent with twentieth-century discoveries in genetics and molecular biology in which evidence indicates that mutations occur at random in DNA and give rise to variation in genetic characteristics. Lamarck's hypothesis has received no significant support from experimental investigations. For example, recall that Joshua Lederberg's experiment in chapter 17 investigated the possibility that the environment could create beneficial mutations. His results, however, showed that any beneficial mutations existed *before* exposure to selecting factors. Lederberg's results contradict Lamarck's hypothesis of evolution and support Darwin's.

Darwin's explanation of evolution through natural selection proposes that selection and speciation have given rise to the great variety of organisms on the earth. Is Darwin's theory correct? Many scientists over the years have tested it. What do their studies show? Does evidence indicate that the diverse organisms on earth are related by common descent

FIGURE 19.3

FIGURE 19.3 Similarities among organisms. The branch points in the diagram indicated by the arrows show where common ancestry diverged and developed into different populations that would no longer interbreed, according to the hypothesis. There would be more similarities between mammals and anthropods than between fungi and mammals.

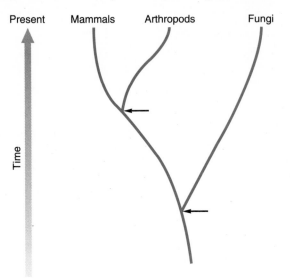

and have developed through natural selection? In the following sections we will examine the evidence for Darwin's hypothesis of evolution.

## EVIDENCE FOR EVOLUTION

In their search for evidence to test Darwin's explanation, scientists have been guided by a hypothesis:

> *Hypothesis* If organisms have arisen by common descent through natural selection, then they should have many similar characteristics. Moreover, the more recent the divergence from a common ancestor, the greater the similarities; the more distant the divergence, the greater the differences.

This hypothesis is illustrated in figure 19.3. To test it, scientists have drawn on their knowledge of many aspects of science, including DNA coding, protein structure, hybrid DNA, comparative morphology, and the fossil record.

## DNA Coding

To determine whether species have evolved from a common ancestor, it is logical to look for similarities among the species. The best place to look is the genetic code carried by nucleic acids. How similar is this code in different organisms? Recall from chapter 14 that sets of three nucleotide monomers in DNA code for specific amino acids. There is no reason to expect the code to be similar from organism to organism. After all, *any* three nucleotides should serve as well as any other to match with tRNA.

| TABLE 19.1 | Amino Acid Substitutions in Cytochrome C |
| --- | --- |

| Organism | Number of Differences from Humans |
| --- | --- |
| Chimpanzee | 0 |
| Monkey | 1 |
| Rabbit | 9 |
| Cow | 10 |
| Pigeon | 12 |
| Bullfrog | 20 |
| Fruit fly | 24 |
| Yeast | 42 |

If organisms have different origins (i.e., there is no common descent), then we would expect their genetic codes to be vastly, if not completely, different. Experimental evidence indicates, however, that *all* organisms, both plants and animals, have basically the same genetic code! This finding is strong support for the hypothesis that all species have a common ancestor.

Suppose DNA were extracted from chicken cells and injected into a frog egg nucleus. What would you expect to observe about the proteins synthesized by the frog egg?

## Protein Structure

Another test scientists have developed to determine how closely related species are is to examine the sequence of amino acids in proteins. If the sequences in the same kind of protein are similar from species to species, we can be more confident that the theory of common descent is correct. The best known of these investigations studied the amino acid sequence in cytochrome C, a protein in the electron transport pathway (see chapter 6). The cytochrome C molecule has 104 amino acids.

What similarities and differences would you predict in the amino acid sequence? Table 19.1 shows the number of amino acid differences between human cytochrome C and cytochrome C from a variety of other organisms. For example, cytochrome C from bullfrogs differs from human cytochrome C at 20 of the 104 amino acid positions. Even between organisms as different as humans and yeast there are only 42 differences!

What kind of data would you expect in table 19.1 if the data contradicted the hypothesis?
A student argued that if the genetic code is the same in all organisms, then the amino acid sequence in cytochrome C should be the same in all organisms. How would you refute this student's argument?

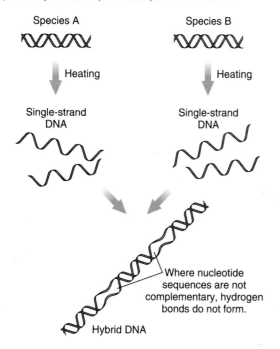

FIGURE 19.4    Hybrid DNA. SIngle-stranded DNA is prepared by heating purified double-stranded DNA. When single strands from different species are mixed, the strands will pair where nucleotide sequences are complementary but will not pair in other parts of the strands.

| TABLE 19.2 | DNA Hybrid Formation |
| --- | --- |
| **Organism** | **Percent of the DNA Sequence Complementary to Human DNA** |
| Human | 100 |
| Monkey | 88 |
| Lemur (a primate) | 47 |
| Mouse | 21 |
| Hedgehog | 19 |
| Chicken | 10 |

Biologists who conducted the tests on amino acid sequences in cytochrome C argued that after lines of descent diverged (fig. 19.1), mutations resulting in amino acid substitutions accumulated in each line. Thus, the number of amino acid differences is a measure of the amount of time that has passed since the two lines diverged. This certainly seems to be the case based on the data in the table, since yeast and humans show far more differences than rabbits and humans, for example. Scientists have used the degree of difference in amino acid sequences to propose time scales of when various species diverged.

In the proposed line of common descent leading to chimpanzees, it is argued that bullfrogs diverged from this line twice as long ago as cows. This is because there are twice as many amino acid differences between bullfrog and chimpanzee cytochrome C as between cow and chimpanzee cytochrome C. What does this argument assume?

## Hybrid DNA

Another way to investigate similarities among species is to examine the sequence of nucleotides in DNA. The argument here is similar to that for comparing amino acid sequences: The more recent the divergence of the lines of descent, the greater the similarity between the DNA of the two species. If the data show nucleotide sequence similarity between species that are similar in physical characteristics, it will support the hypothesis.

Establishing nucleotide sequences as biologists did with amino acid sequences would be a daunting task. Fortunately, scientists can investigate similarities in DNA sequences by forming DNA "hybrids" from separate species. In chapter 18, the term "hybrid" referred to the offspring from interbreeding organisms. Here it refers to the combination of DNA from different species. All DNA is composed of the same four nucleotide monomers and is formed by two complementary strands of nucleotides bonded together by hydrogen bonds between base pairs (chapter 14). If these strands are separated and mixed with single strands from another species, then new, hybrid double strands can form (fig. 19.4). The double strand will form, however, only in areas where the molecules are complementary. In areas where the nucleotide sequences are not complementary, the two strands will remain unpaired. Biologists can measure the degree of pairing by determining the amount of energy required to separate paired strands (Thinking in Depth 19.1). The results obtained from such investigations are shown in table 19.2.

As our hypothesis predicted, DNA from related species is more similar than DNA from less related species.

What interpretations could you make based on the data in table 19.2?

These compelling results from comparisons of the genetic code, amino acid sequences, and DNA nucleotide sequences strongly support the hypothesis that species have a common ancestor. The arguments for common descent and divergence appear to be well justified.

Other studies, many conducted earlier than the molecular experiments we have just described, investigated the evidence for evolution. These studies compared body structure among a variety of organisms, both living and extinct.

## HYBRID DNA

One of the useful experiments scientists have developed to test hypotheses of evolution is creating hybrid DNA. The results of this simple procedure provide clues to the ancestry of species.

Biologists separate purified double-stranded DNA into single strands by heating them to about 100°C to break the hydrogen bonds between nucleotides (chapter 3). They then mix single-stranded DNA from two species and incubate the strands for a time sufficient to form hybrid DNA molecules (fig. 19.4).

In parts of the strands where nucleotide sequences are complementary, bases will pair; in other areas, hydrogen bonds will not form. Once the base pairs have formed, the hybrid DNA solution is slowly heated. As the temperature increases, more hydrogen bonds break and the double strands separate. The greater the number of hydrogen bonds that have formed, the greater the amount of heat energy required to break all the bonds and allow the strands to separate. Hybrid DNA from species that are very similar will have more complementary sequences and would be expected to separate at higher temperatures. Conversely, hybrid DNA from very different species would be expected to separate into single strands at lower temperatures.

As the hybrid DNA solution is heated, the amount of single-stranded DNA is periodically measured. Figure 1 shows data from three species of fruit fly. Single strands of DNA from *D. melanogaster* were mixed with single strands from one of the other two species. The three experiments are indicated by the three curves. Below 60° C, all DNA exists in the double-stranded form. When the solution reaches 100° C, all DNA has separated into the single-stranded form.

In the curve to the far right labeled *D. melanogaster,* the added single strands are from the same species. In other words, *D. melanogaster* strands were mixed with *D. melanogaster* strands. All areas of the molecules are complementary, of course, and the maximum amount of heat is required to separate the strands.

The horizontal line through the middle of the graph shows the temperature at which 50 percent of the strands in each experiment separate; it is used to compare the three curves. For the *D. melanogaster* curve, the 50 percent point is 78° C. For *D. simulans* the 50 percent point is 75° C. These figures indicate that there is a 3 percent difference in the DNA nucleotide pairs of *D. melanogaster* and *D. simulans.*

**FIGURE 1**   Separation of hybrid DNA strands. The three curves measure the amount of energy required to separate DNA. Curves represent strands from (a) *D. melanogaster* and *D. melanogaster,* (b) *D. simulans* and *D. melanogaster,* and (c) *D. funebris* and *D. melanogaster.* Energy is measured as the temperature at which 50 percent of the strands have separated.

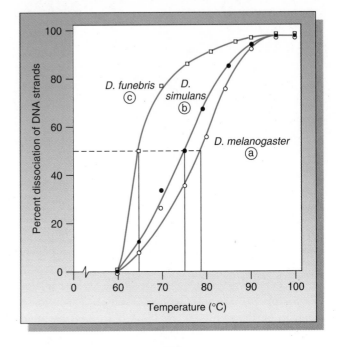

In an experiment comparing other fruit fly species to *D. melanogaster,* would it be possible to find a 50 percent DNA separation point higher than 78° C?

The 50 percent separation point for *D. funebris* is 64° C. Thus, there is a 14 percent difference in the DNA nucleotide pairs of *D. melanogaster* and *D. funebris.*

## Comparative Morphology

One of Darwin's lines of evidence was based on studies of vertebrate embryos. An embryo is the developmental stage before birth or hatching. As Darwin discovered, embryos from different species are very similar, especially at early stages of development (fig. 19.5).

The similarities among embryos of such different species as humans and turtles include gill slits and a prominent tail. These observations are taken as evidence for common descent because structures common to species before they diverged are apparent in embryos, even though they do not exist in adult organisms.

FIGURE **19.5** Comparative embryology. Three developmental stages are shown for six different organisms. In the first stage (a), the embryos appear very similar. In the second stage (b) obvious differences appear but gill pouches are still apparent even in the human embryo. By the last stage (c), distinctive characteristics of each species are apparent.

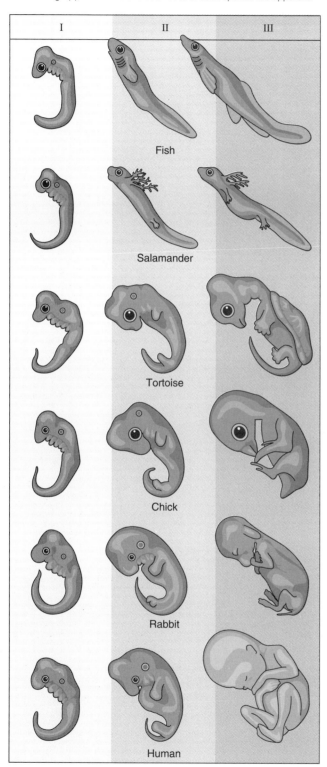

| I | II | III |
|---|---|---|

Fish

Salamander

Tortoise

Chick

Rabbit

Human

Biologists have also compared the skeletal structures of various species (fig. 19.6). From the early reptile forelimb structure, they have identified similarities in the number and position of bones. Again, this evidence from skeletal structures is taken as an indication of common descent and divergence.

Does figure 19.6 represent an observation or an interpretation?

## *Fossil Record*

Perhaps one of the most valuable sources of information on evolution is the **fossils** or remains of organisms in the earth's crust. Early geologists recognized that layers of sediment were built up by the deposition of material eroded from parts of the earth's surface. The position of these layers is an indication of the time when they were deposited. Generally, the deeper the layer, the earlier it was formed.

Sometimes plants and animals are trapped in the sediment as a layer is formed, leaving a preserved impression or form of the organism. Analysis of these remains can indicate the structure of the organism as well as the age of the fossil. Sometimes these fossils are amazingly detailed (fig. 19.7).

The kind of information that can be determined from fossils is illustrated in figure 19.8, which shows the evolution of the modern horse from earlier species. Progressive changes in the structure of teeth and legs can be traced through the succession of fossils.

Fossils have revealed that about 250,000 species have lived on earth over the past 600 million years. This number is impressive considering that there are frequent gaps in the fossil record and that remains are often fragmentary or deformed. Nevertheless, biologists have identified many evolutionary sequences, such as that for the horse, as well as the remains of numerous extinct organisms.

The support for evolution is both impressive and abundant. As we have seen, extensive evidence has been provided by molecular analysis, comparative morphology, and fossil analysis. The theory of evolution has been subjected to rigorous tests, and in every serious line of investigation the results have supported, and thus strengthened, the hypothesis. The general features of evolution are accepted by scientists, many of whom continue to investigate specific evolutionary details (Thinking in Depth 19.2).

## CHALLENGES TO EVOLUTION

Most scientists accept evolution as the best explanation for the diversity of species on earth. Other explanations have been offered, however, and some individuals and groups strongly disagree that evolution adequately explains the origin of species.

**FIGURE 19.6** Comparative skeletal structures. Broad arrows indicate proposed lines of descent. Numbers indicate corresponding structures developed from the early reptile. For example, structures in the early reptile have changed through evolution into the structures labeled 1 to 5 in the other organisms.

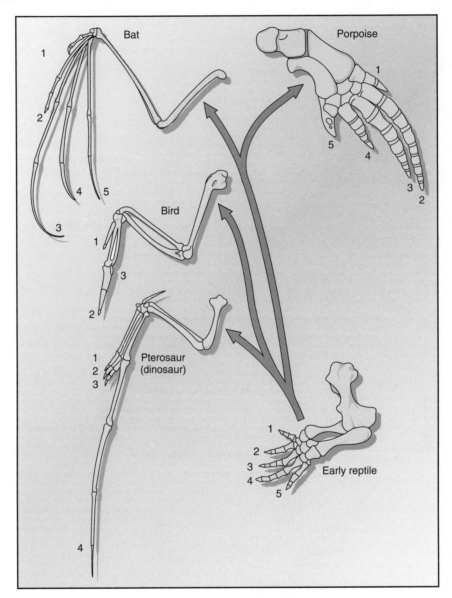

Generally, objections to evolution are raised on religious or moral grounds. Several groups cite religious texts, such as the Christian Bible and the Muslim Koran, as evidence that all species were created within a short span of time and have changed very little throughout the earth's history. In some cases, the disagreement between scientists and religious groups has led to legal challenges and alterations in textbooks and teaching methods.

Objections to evolution have received a measure of popular support, which many scientists find strange and alarming. They believe that theories should be judged on the basis of the scientific method and the weight of evidence. Dorothy Nelkin, (see References), a U.S. science historian, has argued that two factors contribute to the objections to evolution: a perceived threat to "traditional values" and a resentment of what is seen to be scientific dogmatism (Thinking in Depth 19.2). Religious groups object to the teaching of evolution on the grounds that the theory of common descent equates humans with animals, a view they believe undermines the correct relationship between humans

## IS EVOLUTION A FACT?

What is a fact? If we accept the definition given in chapter 1 that a fact is a well-supported hypothesis, then we can consider evolution a fact. If, however, we define a fact as a "truth" that can never be disproven, then we cannot call evolution a fact.

The strong support for evolution from a wide variety of experiments and observations leads scientists to accept it as an exceptionally well-supported hypothesis to the extent that the basic mechanisms of evolution are well established. Indeed, scientists sometimes express their confidence in evolution in such strong terms that they are accused of "scientific dogmatism." In other words, they appear to believe that evolution is the absolute truth. In fact, the mechanisms of evolution are continually subjected to experimentation. While the basics are established, scientists may disagree on details such as the rate of evolution and the effect of variability in species.

Such disagreement is sometimes misinterpreted as uncertainty about and lack of confidence in the fundamental theory of evolution. While this is not the case, critics of evolution have used these disagreements as a basis for arguments against it.

FIGURE 19.7    Fossil remains. This fossil of an early ancestor of the modern horse recovered from 50-million-year-old shale deposits is so well preserved that contents of the digestive system can be analyzed. In this case, the gut contains fossilized leaves. Fossils of other organisms revealed fur and tiny ear bones.

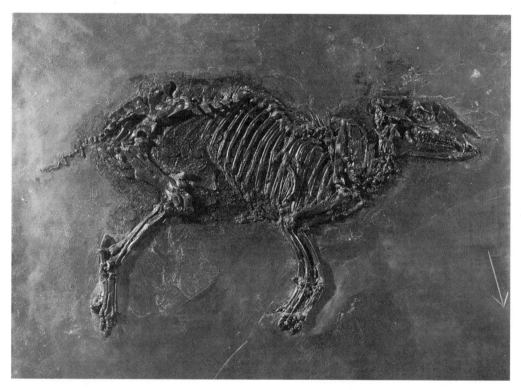

and other organisms and leads to a devaluing of human life. Nelkin points out that these objections represent a serious misunderstanding of both evolutionary theory and the scientific process.

Despite these challenges, the theory of evolution can be regarded as one of the primary principles of biology. The supporting evidence is impressive and is produced by a variety of experimental approaches, thus providing an exceptionally high degree of confidence in the interpretations. There is little doubt that evolution is one of the most important principles in modern biology.

**FIGURE 19.8** Evolution of the horse. Fossils of the skull, forelimb, hindlimb, and teeth of horses from different geological ages are shown. Geological strata from which fossils were obtained are shown on the left. The succession and relative age of fossils has permitted a reconstruction of progressive change in the horse.

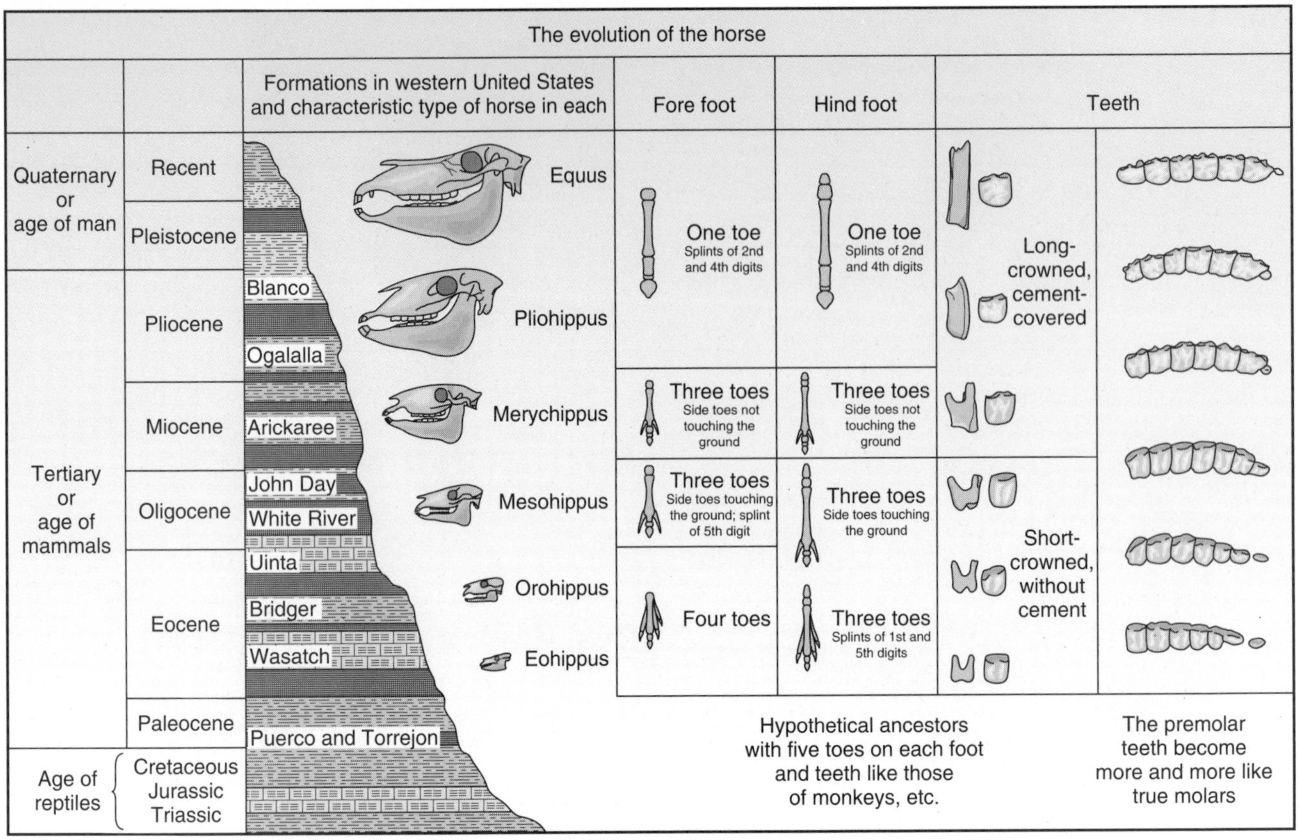

The evolution of the horse

## SUMMARY

*Numerous and diverse investigations have repeatedly tested Darwin's and Wallace's explanation of evolution through natural selection and common descent. These investigations of DNA coding, protein structure, DNA hybridization, comparative morphology, and fossil evidence continue to provide strong support for their explanation. Other hypotheses, such as the inheritance of acquired characteristics, are not supported by the evidence.*

*Together with selection (chapter 17) and speciation (chapter 18), the concepts and evidence presented in this chapter provide a strong foundation for our understanding of the incredible diversity and variety of life on our planet. As you integrate the information from this chapter with other chapters in the book, keep in mind the following key points.*

1. Charles Darwin proposed evolution through **natural selection** and **common descent** in 1859 to explain the origin of the enormous diversity in living organisms. At about the same period, Alfred Russel Wallace developed a similar explanation.

2. Darwin and Wallace were strongly influenced by observations of biological diversity, evidence from fossil plants and animals, and interpretations of the age of the earth.

3. The **inheritance of acquired characteristics** is the proposed transmission of characteristics to offspring that parents have developed through conscious interaction with the environment. Such views were widely accepted before Darwin's explanation and were supported by French naturalist Jean Lamarck.

4. Careful experiments have failed to support Lamarck's views. For example, Joshua Lederberg's experiment in chapter 17 contradicts Lamarckian interpretations.

5. The same triplet code, matching amino acids with the sequence of nucleotides in mRNA, is found in all living organisms. This strongly suggests a common origin for all living organisms.

6. When amino acid sequences from a protein in closely related organisms are compared, they show a high degree of similarity. As increasingly different organisms are compared, they show an increasing number of amino acid substitutions.

7. Single strands of DNA extracted from different organisms can be mixed. Hydrogen bonds will form where nucleotides are complementary to form **hybrid DNA**. The amount of base pairing is interpreted as a measure of the relatedness between organisms.

8. **Comparative morphology** of embryo development and skeletal structures indicates similarities between organisms that is interpreted as evidence of common descent.

9. Fossils preserve a record of animals and plants that shows a change in characteristics over time and supports arguments for common descent.

10. Although the evidence for evolution by natural selection and common descent is compelling, there have been numerous challenges in which evolution is perceived as "scientific dogmatism" or a threat to traditional values.

## REFERENCES

Dobzhansky, R., F. J. Ayala, G. L. Stebbins, and J. W. Valentine. 1977. *Evolution* (San Francisco: W. H. Freeman).

Mayr, E. 1978. "Evolution." *Scientific American* 239, no. 3: 46–55.

National Association of Biology Teachers. 1978. *A Compendium of Information on the Theory of Evolution and the Evolution-Creationism Controversy* (Reston, Va.: NABT).

Nelkin, D. 1976. "The Science-Textbook Controversies." *Scientific American* 234, no. 4: 33–39.

## EXERCISES AND QUESTIONS

1. Many species of flowering plants have developed flower shapes that can be pollinated by only one species of insect. According to natural selection, which of the following best explains this development (page 232)?
   a. Subspecies of the plant producing the flower likely exhibit different shapes in different environments.
   b. Even though only a specific insect pollinates the flower, many other insects pollinate other plants.
   c. Many plants are wind pollinated and do not require insects for fertilization.
   d. Many diverse flower shapes develop by chance over time and those that are successfully pollinated survive and reproduce.
   e. If pollination did not take place, the plant would not reproduce.

2. Consider the example described in question 1. According to the inheritance of acquired characteristics, which of the following best explains this flower development (page 232)?
   a. The fossil record has revealed that flower shapes have changed dramatically over time.
   b. The plants developed a flower shape so that successful pollination would be possible.
   c. The inheritance of acquired characteristics argument has been used in cases other than that described in question 1.

   d. If the flowers are not pollinated, then reproduction would not be possible.
   e. Many plants are wind pollinated and do not require insects for fertilization.

3. Suppose a mother and father exercised regularly to build up their strength and stamina before they had any children. They stated that they did this so that their children would be stronger and have more stamina. Which of the following best describes their reasoning?
   a. They have not yet had any children so they can't make an argument.
   b. It is based on natural selection.
   c. It is a valid argument and reasonable prediction.
   d. They would have to exercise for many years to have any effect on their children.
   e. It is Lamarckian.

4. A biologist examined an old copper mine site where excess metal ions in the earth dumped from the mine were poisonous to plants. While no plants were growing on the dumped earth, the biologist predicted that plants would later move in and begin growing in this area. Which of the following best states the hypothesis on which the prediction is based?
   a. Plants will develop mutations for resistance to the toxic metal ions so they can grow on the dumped earth.
   b. No plants are known that are resistant to the toxic effects of metal ions.

c. Some plants may have preexisting mutations giving them resistance to the toxic metal ions and thus can grow in the dumped earth.

d. Competition between plants will create mutations for resistance to the toxic metal ions.

e. One kind of plant would be expected to grow in the toxic area, but not more than one.

5. Suppose DNA were extracted from chicken cells and injected into a frog egg nucleus. What would you expect to observe about the proteins synthesized by the frog egg (page 234)?

a. Proteins would be synthesized but the amino acid sequence would be different from that normally found in either frogs or chickens.

b. Some proteins that are characteristic of chicken cells but not frog cells would be synthesized.

c. No protein characteristic of chicken cells would be synthesized because the frog tRNA could not match with mRNA.

d. Proteins synthesized would have half a molecule characteristic of chicken protein and half characteristic of frog protein.

e. Messenger RNA could not be formed on the chicken DNA and thus no protein could be synthesized.

6. Which of the following best describes how the data in table 19.1 would be different if the data contradicted the hypothesis regarding amino acid substitutions (page 234)?

a. The same differences would be found between chimpanzees and all other organisms.

b. There would be no differences between cows, monkeys, and rabbits.

c. A greater difference would be found between fruit flies and yeast.

d. A different protein was examined but the same amino acid differences were observed.

e. At least one amino acid difference between chimpanzees and humans would be found.

7. A student argued that if the genetic code is the same in all organisms, then the amino acid sequence in cytochrome C should be the same in all organisms. Which of the following best refutes this student's argument (page 234)?

a. Amino acids are coded by sequences of three nucleotides in both DNA and mRNA, but only tRNA "matches" with the code in mRNA.

b. If the genetic code changed, then even greater differences would be expected between the cytochrome C amino acid sequences from different organisms.

c. The data in table 19.1 should have included data on the cytochrome C differences from a greater variety of organisms.

d. Even if the genetic code *did* change between different kinds of organisms, the amino acid sequence could be the same in these different organisms.

e. Amino acids are coded by a sequence of three nucleotides, so a nucleotide can change in mRNA, resulting in an amino acid substitution without a change in the code.

8. In the proposed line of common descent leading to chimpanzees, it is argued that bullfrogs diverged from this line twice as long ago as cows. This is because there are twice as many amino acid differences between bullfrog and chimpanzee cytochrome C as between cow and chimpanzee cytochrome C. Which of the following does this argument most likely assume (page 235)?

a. The rate of amino acid substitution is the same in cytochrome C from bullfrogs and from cows.

b. The data in table 19.1 shows twice as many differences between bullfrogs and chimpanzees as between cows and chimpanzees.

c. Bullfrogs diverged from the line of descent leading to chimpanzees twice as long ago as did cows.

d. There is twice as much cytochrome C in the body of a cow as in the body of a bullfrog.

e. Amino acids could be substituted at almost any point in the cytochrome C molecule in both cows and bullfrogs.

9. In a hybrid DNA experiment similar to the one described in Thinking in Depth 19.1 comparing other fruit fly species to *D. melanogaster,* would it be possible to find a 50 percent DNA separation point higher than 78° C (page 236)?

a. Yes. Since complete separation does not occur until 100° C, other hybrid fruit fly DNA could be found between 78° C and 100° C.

b. No. At temperatures higher than 78° C more than 50 percent of the DNA strands have separated.

c. Yes. As long as the energy input (i.e., temperature) is increased, more hydrogen bonds will be broken and further separated.

d. No, because 78° C represents the energy to separate two DNA strands from the same species where all areas are complementary.

e. Yes. The temperature at the 50 percent separation point depends on the length of the DNA molecule (i.e., number of hydrogen bonds). If the molecule is longer, the 50 percent temperature will be higher.

10. Based on the data in table 19.2, which of the following is the best interpretation (page 235)?
    a. Mouse DNA is more similar to chicken DNA than to human DNA.
    b. Differences in DNA are greater between a human and a chicken than between a human and a mouse.
    c. The earlier an organism diverges from a line of descent, the greater the differences in DNA from more recently divergent organisms.
    d. Humans have a greater amount of DNA than do any of the other organisms shown in table 19.2.
    e. If DNA from a horse were used in this experiment, the complementary DNA would be between 47 percent and 21 percent.

11. Does figure 19.6 represent an observation or an interpretation (page 237)?
    a. Interpretation. Several different species are compared but other comparisons could still be made.
    b. Observation. The bone structures of the limbs are shown as they would be observed in the organisms.
    c. Interpretation. Similarities between bones of different species are indicated to explain the relationships between the species.
    d. Observation. The figure does not explain *how* the structures evolved, only that they are similar.
    e. Interpretation. Only part of each organism is shown (the limb), not the whole organism.

# POPULATIONS AND COMMUNITIES

How do organisms interact with other organisms and with their surrounding physical environment? What influences their survival and reproduction?

In the past few chapters, we examined several simple ways in which organisms and their environments interact. For example, we considered environmental effects on peppered moths and land snails only in terms of their predators (birds) and the background vegetation that camouflages them. As you might imagine, these organisms are influenced by a great many other environmental factors.

In another example, we looked at how *Potentilla* adapts to different environments, but we examined only the results of the adaptations, not the specific interactions of *Potentilla* with its environment. In other words, many factors probably shaped *Potentilla's* adaptation, but we did not identify or examine them. The interactions between organisms and their environment is the focus of this chapter.

The study of organisms' relationship to their environments and to each other is called **ecology.** The variety and complexity of these relationships is enormous. In this chapter, we will look at just a few of the most important interactions that make up ecology. Though we will be examining a fraction of the earth's ecological relationships, we are likely to be impressed with the incredible complexity of these relationships.

To begin, let's consider the most obvious interaction between organisms—that between predator and prey. Figure 20.1 illustrates relationships between predators and prey in an aspen parkland environment in Canada. The arrows indicate which organisms are consumed by other organisms. If we trace a single sequence of arrows, such as grasses and helianthus (bottom of figure) to prairie vole to great horned owl, we can see the relationships among these organisms. This sequence is called a **food chain.** All the food chains in this parkland environment together form a **food web,** which represents the food sources and feeding patterns of all the organisms in the environment.

## COMMUNITIES

Another way to view relationships among organisms is to consider the kinds of plants and animals that live in a defined area, or **habitat.** Such assemblages of interacting organisms are called **communities.** Together, communities and their nonliving environments function as **ecosystems.** Figure 20.1 indicates several communities, such as the mature poplar community, the willow community, and the prairie community. Notice that organisms in one community may be food for organisms in another community. When communities are interdependent in this way, we call them minor communities. When they are large enough to provide food for all members, we call them major communities.

What are the characteristics of communities? Are they organized in specific ways? Do all communities have features in common? We can answer these questions in several ways. As a start, let's look at one way biologists categorize the organisms found in a single community.

## TROPHIC LEVELS

A convenient way to categorize organisms in a community is by food source. For example, plants such as grass are consumed by insects, which are consumed by small birds, which are consumed by hawks. Biologists call these categories **trophic levels.** Figure 20.2 shows the trophic levels in two typical communities.

The first trophic level includes all organisms that do not consume other organisms. These **producers** obtain food from inorganic materials, usually by photosynthesis. In the aquatic community the producers are phytoplankton; in the terrestrial community the producers are grass.

Organisms that consume producers are called **herbivores** or **primary consumers.** Zooplankton and grasshoppers are primary consumers. The next trophic level, perch and insect-eating birds, are called **secondary consumers.** Hawks and bass, which consume the secondary consumers, are called **tertiary consumers.**

Could a single species occupy more than one trophic level? A cave can be considered a habitat. Could a cave community support all the trophic levels listed above? If not, how could the organisms in the cave community survive?

One way to examine relationships in a community is to count the number of organisms in each trophic level. Figure 20.3 shows the number of organisms in two different ecosystems, a grassland and a temperate forest. In the grassland, the producers are individual grass plants; the herbivores are grass-consuming insects and larvae. As you can see from the data, the numbers decrease as we move from producers to tertiary consumers, forming a **pyramid of numbers.**

In the temperate forest community we see a different pattern. The number of producers is considerably fewer than the number of primary and secondary consumers. Why is this so? In a forest, many of the producers are large trees, which can support a great many insects. Thus, only a few large producers are needed to produce food for many small herbivores.

**FIGURE 20.1**  Food web in an aspen parkland of Canada. Arrows indicate consumers and their food sources. For example, grasshoppers consume grass. Organisms are grouped as members of separate communities where they are generally found, but organisms can move from one community to another.

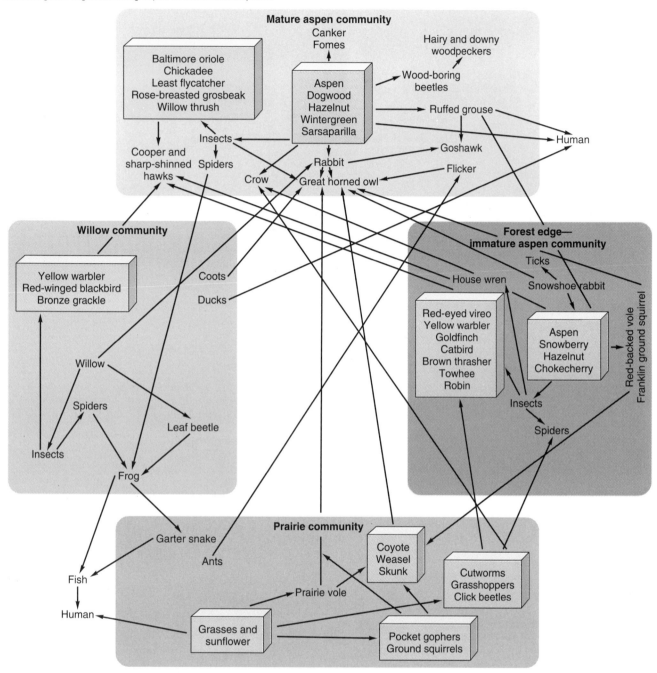

How would you expect the pyramids in figure 20.3 to change in winter?

Could the numbers of organisms be equal in all trophic levels?

## BIOMASS PYRAMIDS

The number of organisms in each trophic level of a community can be misleading, because producers can be tiny or very large, as we have seen. We can get a more accurate picture of the relationships between organisms in a community by examining the total weight, or **biomass,** in each trophic level. Biomass equals the number of organisms multiplied by their average weight.

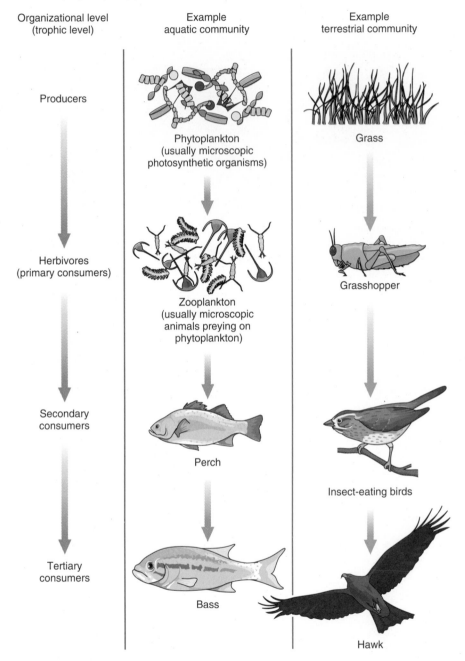

We can present the data on biomass in the same way we presented the data on numbers, forming a **pyramid of biomass** from producers to tertiary consumers (fig. 20.4). The two pyramids often appear similar, because large numbers of producers also have a large biomass. In both aquatic (Wisconsin lake and coral reef) and terrestrial (old field and tropical forest) communities, biomass is usually greatest at the producer level and progressively decreases in higher trophic levels.

Notice the fourth biomass pyramid shown in figure 20.4, that of a tropical forest. This pyramid includes the biomass of organisms called **decomposers,** which break down dead organic matter into forms producers can use. Decomposers include bacteria and fungi that feed on dead material. Their biomass can make up a significant bulk of the community.

Why do communities have more producer than consumer biomass? The most obvious explanation is that since herbivores obtain all their food material (and therefore all their body weight) from producers, the total weight of herbivores could not be greater than the total weight of

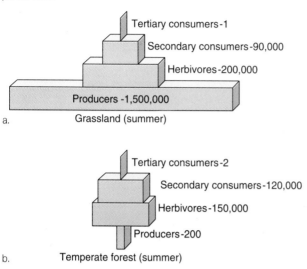

FIGURE 20.3    Pyramid of numbers. (a) The total number of organisms found in each trophic level in a typical grassland community in the summer is shown. (b) Data on the total number of organisms in a temperate forest community in summer shows the number of consumers per 0.1 acre.

FIGURE 20.4    Pyramid of biomass. Data show the grams (dry weight) of organisms per square meter in each trophic level. This biomass in the trophic levels exhibits a pyramid shape similar to the pyramid of numbers.

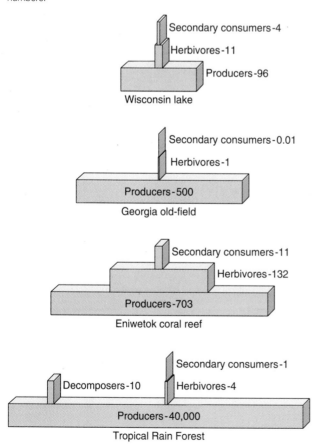

FIGURE 20.5    Exceptions to the typical biomass pyramid. A comparison of biomass in each trophic level for a freshwater aquatic community in summer and winter is shown. Data measure the milligrams of dry weight per cubic meter of water for primarily microscopic producers, herbivores, and consumers.

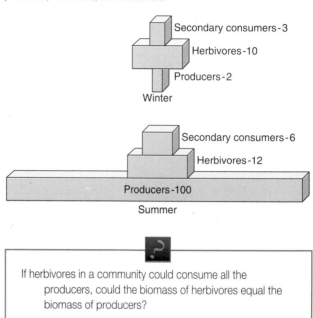

If herbivores in a community could consume all the producers, could the biomass of herbivores equal the biomass of producers?

producers. The same reasoning is used for the relative amounts of biomass in other trophic levels.

Table 20.1 compares the numbers and biomass of an aquatic community and a terrestrial community. In both cases, the individual producers are small but they are numerous and so constitute a large biomass. The other groups of organisms include both herbivores and consumers in each category and are *not* organized as trophic levels. For example, "fish" include fish that are only herbivores as well as fish that are secondary and tertiary consumers.

A large, air-tight glass cage contained grass, herbivorous insects, and two field mice. One of the mice was fed grain with starch containing $C^{14}$. (See chapter 4 on tracing metabolic pathways.) $C^{14}$ would ultimately be found in which other organisms in the cage?

In most communities, the biomass pyramid model holds true, but there are exceptions. Could producer biomass ever be less than herbivore biomass? How? In aquatic communities, such as the one for winter shown in figure 20.5, the producers are small organisms with short life spans. At particular times, in this case winter, the biomass of these producers may be small. As they are consumed, they are rapidly replaced. Over a period of time, the biomass of producers is greater than that of herbivores even though a single measurement could indicate the opposite.

**TABLE 20.1** Comparison of Numbers and Biomass in Aquatic and Terrestrial Ecosystems

| | Open-water Pond | | | Meadow of Old Field | |
| Assemblage | No/m² | Gm Dry wt/m² | Assemblage | No/m² | Gm Dry wt/m² |
|---|---|---|---|---|---|
| Algae (producers) | $10^2$–$10^8$ | 5.0 | Grasses (producers) | $10^2$–$10^3$ | 500.0 |
| Microscopic animals | $10^5$–$10^7$ | 0.5 | Insects and spiders | $10^2$–$10^3$ | 1.0 |
| Insects, mollusks, and crustaceans | $10^5$–$10^6$ | 4.0 | Soil organisms | $10^5$–$10^6$ | 4.0 |
| Fish | 0.1–0.5 | 15.0 | Birds and mammals | 0.01–0.03 | 0.3 |

## COMMUNITY INTERACTIONS

Usually each trophic level in a community or ecosystem includes many different species. These species interact with each other and with species in other trophic levels in a multitude of ways. How do these interactions affect the species involved?

We have discussed the type of interaction in which one organism serves as a food source for another organism. Foxes feeding on rabbits and birds consuming insects are examples of the **predator-prey relationship.** In both these examples, the prey is killed. Some predators, however, feed off their prey without killing them. For example, mosquitoes feed on the blood of animals without causing more harm than a mosquito bite. Caterpillars eat the leaves of trees without destroying the trees.

How do predators and prey affect one another? Figure 20.6 shows data from studies of sawfly populations. Small mammals, in this case shrews, prey on sawfly cocoons, and insect-consuming birds prey on the sawfly larvae. In both examples, the predator population grows when the prey population grows. The obvious interpretation is that a large prey population can produce a bigger food supply for predators and thus support a large predator population.

In another study, the population of mites was compared to that of their predators. Figure 20.7 shows changes in these populations. Notice that the increases and decreases in the predator population clearly lag behind increases and decreases in the prey population. These fluctuations are called **predator-prey oscillations.** When the predator population is low, prey species are more likely to survive and reproduce. Increasing prey provide a greater food supply, thus increasing the predator population. But as predators increase, more individuals are eliminated from the prey population, leading to a decrease in numbers of prey.

**FIGURE 20.6** Predator-prey relationships. (a) The number of small mammals (shrews) compared to the number of sawfly cocoons. The sawfly cocoons serve as a primary food source for the shrews. (b) The number of sawfly larvae-consuming birds is compared to the number of sawfly larvae. *(a) Source: Data from C. C. Holling, "The Components of Predation as Revealed by a Study of Small Mammal Predation of the European Pine Sawfly" in Can. Entomologist, 91:293-320, 1959. (b) Source: Data from C. H. Buckner and W. J. Turnock, "Avain Predation on the Larch Sawfly,* Prisphora Erischsonii (Htg), *in Ecology, 46:223-236, 1965.*

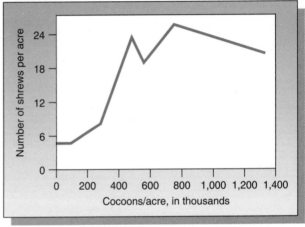

a.

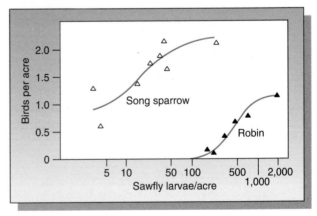

b.

A student argued that a large predator population would sometimes eliminate all prey individuals. What is the student most likely assuming?

What factors might increase the lag time between population growth of prey and population growth of predators?

*Evolution and Ecology*

FIGURE 20.7 Predator-prey oscillations. (a) Data show the fluctuating sizes of predator-prey populations. The prey is a mite, *E. sexmaculatus*. The rise and fall of prey population leads to the corresponding changes in predator population. (b) Idealized fluctuations in population are interpreted as follows: The prey increases because predators are few, but as prey increases, predators more easily capture prey and reproduce more rapidly in response to greater food source. *Source: Data from C. B. Huffaker, "Experimental Studies on Predation: Dispersion Factors and Predator-Prey Oscillations" in Hilgardia, 27:343–383, 1958.*

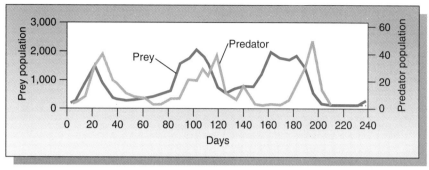

a.

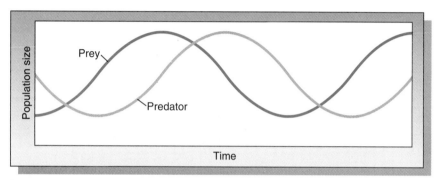

b.

## COMPETITION

In a different type of interaction between species in a community, organisms may rely on the *same* resources instead of using each other as resources. Such a relationship is called **competition.** What happens when two species compete? Do their populations change?

It seems reasonable to predict that competition between species would lead to a decrease in population levels, because neither species would have all the resources it would have without competition. In one of the earliest and most revealing studies of competition, ecologist G. F. Gause, working in the early part of this century, investigated two species of paramecium. These single-celled organisms live in aquatic environments and feed on bacteria and yeast. Figure 20.8 shows the data Gause obtained when he grew two species, *Paramecium caudatum* and *Paramecium aurelia,* in separate cultures and together in the same culture.

In separate cultures, each species exhibits generally the same growth. They grow rapidly at first and then reach a stable population size, which is characteristic of a normal growth pattern (Thinking in Depth 20.1). This stable population size, which occurs after several days, represents the **carrying capacity** of the environment, which is the maximum sustained population that the environment can support.

**FIGURE 20.8** Paramecium growth. (a) When paramecium are grown in separate cultures, population growth is almost identical. (b) When grown in the same culture vessel, however, one species does not survive. (c) Paramecium are single-celled organisms that feed on smaller organisms such as bacteria and yeast. *Source: Data from G. F. Gause, The Struggle for Existance, 1934, Williams & Wilkins Publishers, Baltimore, MD.*

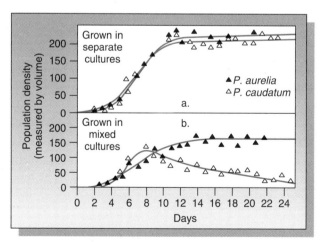

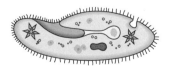

c.

## POPULATION GROWTH

When a population grows, say, for example, a population of cells, the number of cells in that population increases from few to many in a characteristic pattern. The graph in figure 1 illustrates how the growth increases with time. The increase occurs slowly at first then becomes more rapid. Why should the growth curve have this shape?

To visualize why growth occurs in this way, consider what happens when one cell divides repeatedly.

1 cell → 2 cells → 4 cells → 8 cells → 16 cells

Change =1  Change =2  Change =4  Change =8

As the population of cells doubles with each generation, the change in cell number increases exponentially. If these cell numbers (i.e., the population size) are graphed against time, then we see the type of curve shown in figure 1a.

What assumption is being made in the example of population growth in which the cell or population number doubles with each generation?

Populations cannot, however, continue to grow indefinitely. At some point, they are limited by the availability of resources in the environment. When this happens, growth slows to a steady state, as shown in figure 1b. At this point, cells or organisms are reproducing as fast as they are dying and the birth rate equals the death rate. In practice, populations usually fluctuate just above and below their carrying capacity because birth and death rates are not exactly equal at all times.

**FIGURE 1** Growth curves. (a) There is an initial rapid increase in population size when individuals or cells have no constraints on growth. (b) As population grows and resources may become limited, growth can slow and reach a steady state representing the carrying capacity of the environment. Notice that the initial stages of (b) are identical to (a).

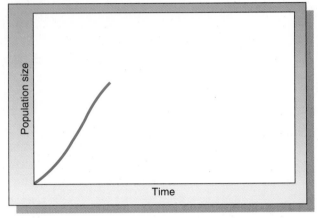

a.

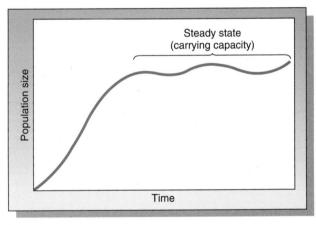

b.

When the two species of paramecium were grown together, *P. caudatum* did not survive. This is an example of **competitive exclusion:** one species survived at the expense of the other. These results have been interpreted as evidence that *P. aurelia* is more successful at competing for food under the conditions in Gause's experiment.

What different interpretations could you make of the data in figure 20.8?

In a second competition study, Gause used another paramecium species called *P. bursaria*. In this experiment, the two species were able to survive when they were grown together (fig. 20.9). This type of interaction is called **coexistence.** Both species survived, but the size of each population was decreased. Further observations of the mixed culture revealed that one species tended to feed primarily on bacteria suspended in the culture solution while the other fed primarily on yeast growing on the bottom of the culture vessel. Biologists have viewed this observation as evidence that *P. caudatum* and *P. bursaria* can coexist because their food sources are slightly different, thus decreasing competition for food.

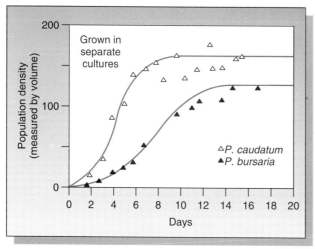

**FIGURE 20.9** Coexistence. (a) Two species of paramecium grown in separate cultures exhibit similar but not identical growth curves. (b) When grown together in the same culture, both population decreases but neither is eliminated. *Source: Data from G. F. Gause, The Struggle for Existance, 1934, Williams & Wilkins Publishers, Baltimore, MD.*

a.

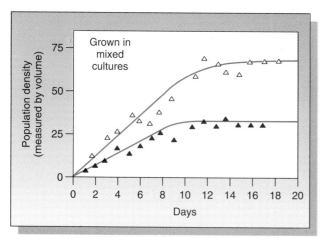

b.

In what way do the data in figure 20.9*b* suggest that one species of paramecium does not feed *only* on bacteria and the other species *only* on yeast?

In another study of competition, biologists examined the growth of clover. Figure 20.10 shows the results when two species of clover were grown separately (pure) and together (mixed). In the mixed culture, the leaf area of both species was decreased, but they both survived. Yet the two species were competing for the same light and water resources. How did they manage to coexist?

The data in figure 20.10 show that *T. repens* grows faster and reaches a maximum leaf density sooner than *T. fragiferum.* Biologists observed, however, that *T. fragiferum*

**FIGURE 20.10** Coexistence in populations of clover. When grown without the other species present (that is, in a pure culture), the growth is shown by the broken line. *T. repens* grows faster but then decreases. When grown together, both species exhibit decreased growth but both survive. *Source: Data from J. L. Harper and J. N. Chatworthy,"The Comparative Biology of Closely Related Species, VI Analysis of the Growth of Trifolium Repens and Fragifesum in Pure and Mixed Populations" in J. Exp. Bot., 14:172–190, 1963.*

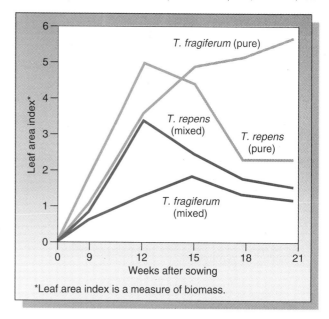

has longer stems and higher leaves. This species grows more slowly, but it is able to grow higher and avoid being completely shaded by *T. repens.* The timing of the growth of each species and differences in their height allow them to coexist.

Based on these examples, we can see that coexistence is possible if competing species differ, even slightly, in the way they utilize environmental resources. If different species use the same resources in the same way, as with *P. caudatum* and *P. aurelia,* then one will not survive.

This interpretation of the data makes sense if you recall character displacement, which we discussed in chapter 18. If two similar species inhabit the same environment, competition and selection can lead to greater differences between the species. The lizard species in the Kalahari Desert is a case in point. Competition for the same food source (termites) resulted in differences in size and body structure, which allowed the two species to feed on different size termites.

## MUTUALISM AND COMMENSALISM

Another type of interaction is illustrated by the relationship between insects, such as bees, and flowering plants. Both organisms benefit from the interaction. The flowers are pollinated and thus can reproduce, and the bees obtain nectar and pollen. Such mutually beneficial interactions are examples of **mutualism.**

In another type of interaction, called **commensalism,** one species benefits and the other neither benefits nor is harmed. A typical example is the relationship between birds

FIGURE 20.11   Succession. Changes in an abandoned cultivated field over 200 years are shown. Grasses and shrubs initially occupy the site but are gradually replaced by shrubs and trees.

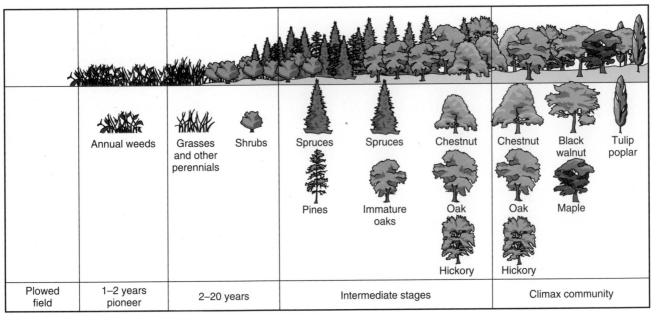

| Plowed field | 1–2 years pioneer | 2–20 years | Intermediate stages | Climax community |

200 years (variable)

and the trees in which they nest. The birds obtain shelter and protection from the trees, but the trees receive no apparent benefit from the birds.

While these examples of species interactions are not exhaustive, they serve to indicate the complexity of the relationships among organisms in a community. All of these interactions can significantly influence community structure and dynamics.

## SUCCESSION

Given the complexity of species interactions and the diversity of organisms in most communities, would you expect communities to be stable? Suppose you examined a community at twenty-year intervals. Would you observe differences?

Consider an area where relatively few organisms are growing, such as a recently abandoned field. Would this area change over time? Biologists have studied community changes in abandoned fields in great detail. Figure 20.11 summarizes these changes. Over a period of 150 to 200 years, abandoned fields tend to undergo a typical sequence of changes as different organisms occupy the site at different times. The grasses initially growing in the old field give way to shrubs, then pine trees, and finally oaks and hickory. This progression from one community type to another is called **succession.**

Of course, biologists were not able to observe a single site continuously for more than 150 years. Instead, they examined many fields that had been abandoned for varying

periods of time. In this way, they developed a description of continuous change from "snapshots" of several fields.

What assumption are biologists making when they determine a scheme of succession by observing many separate sites?

Based on our study of communities and species interactions in this chapter, it seems reasonable to propose that organisms in a community modify their physical environment over an extended period of time. They change the environment, enabling other, different organisms to become established, which leads to a further stage of succession. Some modifications are fairly obvious. Shrubs provide shade and protection for seedlings of other species. The debris that accumulates from dead plant material can change soil characteristics to suit new plants. Given these kinds of physical changes, succession appears to be a natural and logical occurrence in communities.

Although community interactions are extremely complicated, we have been able to observe a few relatively simple patterns in this chapter. Based on these patterns, we can make further generalizations about how energy and materials cycle through communities and ecosystems. In the next several chapters we will discuss these cycles, gaining a greater understanding of ecosystems and communities in the process.

## SUMMARY

*Organisms interact with one another and with their physical environments in many ways. These interactions affect their survival and reproduction. The interactions of populations in communities determine the structure and dynamics of those communities, and can even change the physical environment. As you read further about community interactions, keep in mind the following key points from this chapter.*

1. **Ecology** is the study of the relations of organisms to their environment and to each other.

2. A **food web** and **food chains** indicate the food consumption relationships among organisms.

3. An assemblage of interacting organisms in a defined area is called a **community,** while an **ecosystem** defines the integrated functioning of communities and their nonliving environments.

4. Organisms in a community can be divided into four **trophic levels: producers, herbivores** or **primary consumers, secondary consumers,** and **tertiary consumers.**

5. The **pyramid of numbers** represents the numbers of organisms in each trophic level.

6. The **pyramid of biomass** represents the total weight of organisms in each trophic level.

7. Interactions between organisms in a community can influence the kinds, numbers, and biomass of organisms in each of the trophic levels. Common types of community interactions include the **predatory-prey relationship, competition, mutualism,** and **commensalism.**

8. When a small population begins to grow in an environment with ample resources, growth is rapid at first. Soon the population stabilizes at the **carrying capacity,** the maximum, sustained population size that the environment can support.

9. **Succession** is the orderly progression from one community type to another.

## REFERENCES

May, R. M. 1978. "The Evolution of Ecological Systems." *Scientific American* 239, no. 31: 160–75.

Miller, G. T. 1985. *Living in the Environment* (Belmont, Calif.: Wadsworth Publishing).

Putman, R. J., and S. D. Wratten. 1984. *Principles of Ecology* (Berkeley: University of California Press).

Smith, R. L. 1990. *Ecology and Field Biology, 4th ed.* (New York: Harper & Row).

## EXERCISES AND QUESTIONS

1. Which of the following organisms could occupy more than one trophic level (page 246)?
   a. A mosquito that feeds on rabbit and deer blood
   b. A black bear that eats berries and fish
   c. A bird that eats several kinds of beetles
   d. A deer that consumes green twigs as well as leaves
   e. A hawk that eats several species of small birds

2. Caves can be considered a habitat that supports populations of organisms. Could a cave community support all the trophic levels, from producers to tertiary consumers (page 246)?
   a. No. Caves are only a small part of the space available to living organisms and the biomass would be small compared to that found in other communities.
   b. Yes. Any community must have all the trophic levels to be considered a complete community.
   c. No. There would not be sufficient biomass to support larger animals such as deer or bears.
   d. Yes. As long as a measurable amount of biomass can be found in the community, this biomass could be divided among trophic levels.
   e. No. Producers could not survive inside the cave, but organic material could be carried into the cave to support other trophic levels.

3. How would the pyramids of numbers in figure 20.3 change in winter (page 249)?
   a. There would be a greater decrease in forest herbivores than in grassland herbivores.
   b. Secondary consumers would increase in both grassland and forest.
   c. Producers in the grassland would decrease.
   d. In at least some trophic levels, the numbers would increase.
   e. In the grassland, herbivores would decrease more than producers.

4. In which of the following cases would the numbers of organisms in all trophic levels most likely be equal (page 249)?
   a. Producers are primarily large trees.
   b. Only one species is found in each trophic level.
   c. When there are no more than three trophic levels.
   d. The tertiary consumers are very small organisms.
   e. Organisms in all trophic levels are about the same size.

5. If herbivores in a community could consume all the producers, could the biomass of herbivores equal the biomass of producers (page 249)?

 a. No. Since producers are constantly dying as well as reproducing, the producer biomass could decrease.

 b. Yes, if the herbivores were smaller and more numerous than the producers.

 c. No. Some consumed material would be excreted as waste and would not be part of herbivore biomass.

 d. Yes. Some herbivores will probably be consumed by secondary consumers.

 e. No. Producers are usually dependent on photosynthesis to produce biomass from carbon dioxide and water.

6. A large, air-tight glass cage contained grass, herbivorous insects, and two field mice. One of the mice was fed grain with starch containing $C^{14}$. $C^{14}$ would ultimately be found in which of the following organisms in the cage (page 249)?

 a. Only the grass and one field mouse

 b. The grass, insects, and one field mouse only

 c. Both field mice only

 d. The insects and one field mouse only

 e. All the organisms

7. A student argued that the large predator population in figure 20.7 would sometimes eliminate all prey individuals. Which of the following is the student most likely assuming (page 250)?

 a. None of the prey organisms could hide from and avoid predators.

 b. The predators could eliminate all the prey individuals.

 c. The predators are larger in size than any of the prey individuals.

 d. There are more predators than there are prey.

 e. The size of the predator and prey populations oscillates.

8. Which of the following factors would likely increase the lag time between population growth of prey and population growth of predators in figure 20.7 (page 250)?

 a. Larger size of prey individuals

 b. Slower reproduction by the predators

 c. The size of the geographical area occupied by the community containing predators and prey

 d. Larger size of predator individuals

 e. Amount of food available as prey

9. The figure below shows the increase in the number of eggs produced by female predators as the number of prey individuals increases in the environment. These data most likely support which of the following hypotheses?

 a. Predators will outnumber prey individuals if prey are numerous.

 b. Female predators will consume more prey than male predators will consume.

 c. Predators can consume almost all prey under some conditions.

 d. Greater food consumption increases population size of the predators.

 e. The larger the size of individual predators, the more rapid their production.

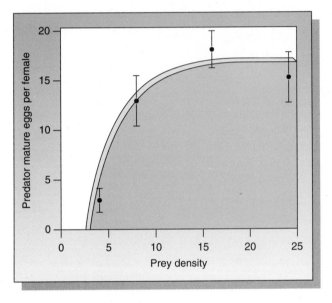

Source: Data from A. G. F. Dixon, "An Experimental Study of the Searching Behavior of the Predatory Coccinellid Beetle Adalia Decempunctata (L). J. Anim. Ecol., 28:259-281, 1959.

10. In Thinking in Depth 20.1, what assumption is being made in the example of population growth in which the cell or population number doubles with each generation (page 252)?

 a. Birth and death rates are equal.

 b. The cell or population number doubles with each generation.

 c. No cells or individuals die.

 d. The population will reach a stable size.

 e. A generation can reproduce in a few days.

11. Which of the following is an acceptable interpretation of the data in figure 20.8 (page 252)?

a. *P. aurelia* releases a waste product poisonous to *P. caudatum* but not to itself.

b. Food organisms had been eliminated from the culture after about twenty-four days.

c. *P. caudatum* dies out in the culture after about twenty-four days.

d. *P. aurelia* is a larger organism than *P. caudatum*.

e. When grown in separate cultures, the two species grow equally well.

12. In what way do the data in figure 20.9*b* suggest that one species of paramecium does not feed *only* on bacteria and the other species *only* on yeast (page 253)?

a. If that were the case, then one of the species should have died out.

b. If so, then both species should have produced larger populations in mixed culture than in separate cultures.

c. If that were the case, then both species should have died out in the mixed culture.

d. If so, then both bacteria and yeast would have been eliminated from the mixed culture.

e. If so, growth in mixed culture should have been the same as in separate cultures.

13. What assumption are biologists making when they determine a scheme of succession by observing many separate sites (page 254)?

a. Succession will occur in the same way at all the sites where observations are made.

b. Observations should be made at a variety of sites to detect differences in succession.

c. Succession is an orderly sequence of changes in the community types at a single location.

d. A single scheme of succession can be concluded from observations at many separate sites.

e. Succession can occur in a variety of ways at different sites.

# CHAPTER 21

# ENERGY AND COMMUNITIES

n the last chapter we saw how the biomass and numbers of organisms in each trophic level in a community are related. We saw that organisms in one trophic level feed on organisms in lower trophic levels. This suggests that energy is being passed from one trophic level to another. Are the relationships between energy use in various trophic levels similar to the relationships between biomass and numbers in various levels? How much energy is available in each trophic level? How much energy moves from one level to another?

These are important questions because *all* organisms in *all* trophic levels require energy for a variety of purposes. Energy is necessary to synthesize molecules (chapter 3), to control the movement of material across membranes (chapter 5), to generate heat (chapter 6), to obtain and eliminate material (chapter 9), and to effect movement and communication (chapter 11).

A general overview of energy flow, shown in figure 21.1, indicates that in most communities producers capture solar energy in molecular bonds through photosynthesis. They use some of this energy and pass the rest on to organisms in other trophic levels. Those organisms release the energy via cellular respiration. We will examine this model of energy flow in communities, called an **energy flow diagram,** in more detail in this chapter. *How much* energy is captured in communities? Are some communities more efficient at capturing energy than others?

## ENERGY MEASUREMENT

The amount of light energy coming from the sun to the earth has been measured in a variety of ways. Scientists estimate that it averages between 3,000 and 4,000 kilocalories per square meter per day (kcal/m$^2$/day) for most parts of the United States. The general equation for photosynthesis indicates how biologists can measure the amount of energy captured by producers.

$$6\ CO_2 + 6\ H_2O \xrightarrow{\text{light}} C_6H_{12}O_6 + 6\ O_2$$

We can use either the amount of organic material ($C_6H_{12}O_6$) produced or the amount of oxygen produced to estimate how much energy is captured by plants.

We know that while plants are capturing energy through photosynthesis, other organisms such as animals are using oxygen and organic material in respiration. What scientists are actually measuring in communities, then, is the **net result** of photosynthesis and respiration (fig. 21.2). This result, also called the **net production,** represents the total amount of energy captured minus the amount released in respiration. We can express this relationship as an equation:

net production (NP) = gross production (GP) − respiration (R)

This equation will be true for all the producers in an ecosystem measured together, as well as for a single organism.

---

FIGURE 21.1   Simple energy flow. Energy enters the community from an outside source, usually the sun, and is captured in photosynthesis. Much of this photosynthetic energy is used as respiration directly by producers. The remainder may be consumed by herbivores or represent decay or storage. Herbivores and other consumers exhibit respiration and loss of energy to decay or storage.

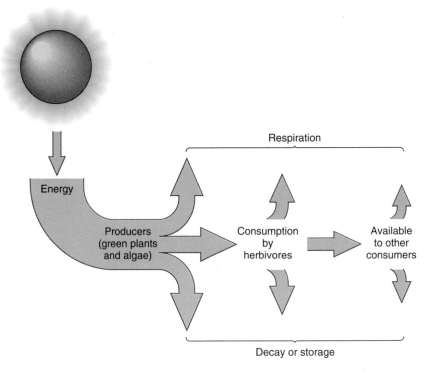

FIGURE 21.2 Production. The amount of energy captured in the chemical bonds of material produced in photosynthesis represents *gross production*. Some of this material is used in respiration by the producers. The remainder represents net production, which may be consumed by herbivores, stored as plant growth, or decay.

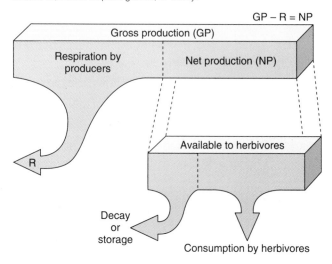

$$GP - R = NP$$

Using gross production, respiration, and net production, what best represents the plant growth harvested from a community?

How do biologists measure gross production? In simple terrestrial communities, such as grasslands, they can harvest new plant growth and measure the energy content of this material. Recall from chapter 3 that we can determine the energy content of organic material by burning it in a bomb calorimeter. The number of calories in the material is a measure of the amount of energy in the producers. We can estimate respiration rates by using a respirometer, as we described in chapter 6. The studies described in this chapter make use of these techniques to study energy flow through typical communities.

In aquatic communities it is usually more convenient to measure energy as oxygen production rather than as the production of organic material, or biomass production. One of the most useful ways to make these measurements is with the light and dark bottle method, shown in figure 21.3.

Samples of water that include algae, other green plants, and consumers are placed in a transparent bottle and in a "dark bottle" that does not allow light to enter. These bottles are placed in the pond or stream from which the water samples were taken. Researchers periodically measure the oxygen content of the two bottles (fig. 21.4). During daylight hours, oxygen in the light bottle increases. In the

**FIGURE 21.3** Light bottle–dark bottle experiment. (*a*) Water from the aquatic community is placed in a light bottle (transparent and allowing transmission of light) and a dark bottle (allowing no entry of light). (*b*) The two bottles are sealed and placed in the pond, lake, or river comprising the aquatic community. Oxygen content of the two bottles is monitored and is used to calculate gross production.

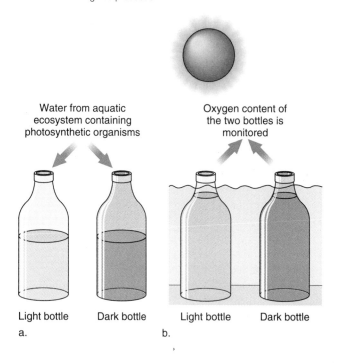

dark bottle, however, oxygen content decreases because respiration is occurring while photosynthesis is not. Thus the dark bottle provides a measurement of respiration. In the light bottle, photosynthesis and respiration are taking place simultaneously, so the oxygen measured represents the net production (NP) of the two processes. Knowing the respiration rate and the net production, biologists can calculate the gross production.

What changes would occur in oxygen concentration in light and dark bottles at night?
In addition to producers, the light and dark bottles would contain microscopic consumers. How would these consumers affect the measurement of gross production and net production?

## EFFICIENCY AND PRODUCTIVITY

Biologists have collected data on efficiency and productivity from a variety of communities, both terrestrial and aquatic. When we examine these data, shown in table 21.1, we can detect several general features. First, the efficiency of natural

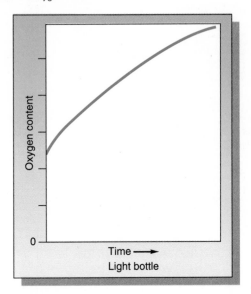

FIGURE 21.4    Light bottle–dark bottle data. The hypothetical data expected from the light bottle–dark bottle experiment is shown. In the light bottle, photosynthesis exceeds respiration and the oxygen content increases. In the dark bottle, no photosynthesis is possible and respiration decreases the oxygen content.

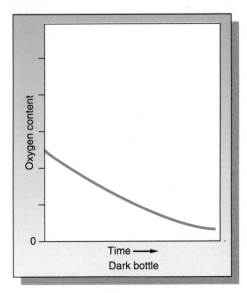

ecosystems (a temperate forest, a spring, and a lake) is around 1 percent or less, while the efficiency of cultivated systems (sugarcane and cornfields) is considerably higher. In both cases we can see that only a small portion of the energy in sunlight is captured.

Measures of productivity, shown in the second half of table 21.1, indicate that a large part of the gross production is expended as respiration, producing energy for growth and maintenance of the producers. (You may want to refer to chapter 6 where the production of energy as ATP by respiration was described in detail.) The amount remaining as net production is the energy available for herbivores to consume.

The highest rates of production efficiency, that is, the ratio of net production to gross production, are found in the cultivated communities (alfalfa field and pine plantation). These figures are somewhat misleading, however, as are the figures for efficiencies in the first half of the table, because cultivated fields require high energy inputs in the form of applied insecticides, operation of farm machinery, and selection of genetically superior seed. It is interesting to note that what we might consider the most productive community, the rain forest, retains the smallest percentage of its gross production as net production (28.9 percent).

A student knew that respiration decreases when temperature decreases. He argued that the ratio of net production to gross production (NP/GP) could be increased if plants were grown at lower temperatures. What was this student assuming?

## HERBIVORE CONSUMPTION

If net production represents all the energy available to herbivores, how much of that energy do herbivores actually consume, and how do they use it? Figure 21.5 shows how a typical herbivore, a deer, uses energy. About 25 percent of the deer's food is undigestible and passes through the animal as feces. Additional material is excreted as urine and digestive

| TABLE 21.1 | Efficiency and Productivity |

| A. Efficiency of Solar Energy Capture | | | | | |
|---|---|---|---|---|---|
| Community | Temperate Forest (northern USA) | Spring (Silver Springs, Florida) | Lake (Lake Mendota, Wisconsin) | Sugarcane Field (Hawaii) | Cornfield (Israel) |
| Percent efficiency* | 0.9 | 1.2 | 0.4 | 7.6 | 6.8 |

| B. Productivity (kcal/m²/year) | | | | | |
|---|---|---|---|---|---|
| | Alfalfa Field (U.S.A.) | Young Pine Plantation (England) | Oak/Pine Forest (New York) | Large Spring (Silver Springs, Florida) | Rain Forest (Puerto Rico) | Coastal Waters (Long Island, New York) |
| Gross production (GP) | 24,400 | 12,200 | 11,500 | 20,800 | 45,000 | 5,700 |
| Respiration (R) | 9,200 | 4,700 | 6,400 | 12,000 | 32,000 | 3,200 |
| Net production (NP) | 15,200 | 7,500 | 5,100 | 8,800 | 13,000 | 2,500 |
| Ratio** (NP/GP) | 62.3% | 61.5% | 43.5% | 42.3% | 28.9% | 43.8% |

*Calculated as percent of solar radiation captured by photosynthesis.

** Indicates the production efficiency, or the proportion of material or energy remaining as net production after respiration is subtracted from gross production.

FIGURE 21.5 Herbivore utilization of food. Of the food energy consumed by a typical herbivore, almost 50% is expended in heat production and 35% is eliminated as waste. Less than 20 percent represents body weight gain. *Source: Data from R. L. Cowan, "Physiology of Nutrition as Related to Deer" in First National White-Tailed Deer Disease Symp., p. 1–8, 1962.*

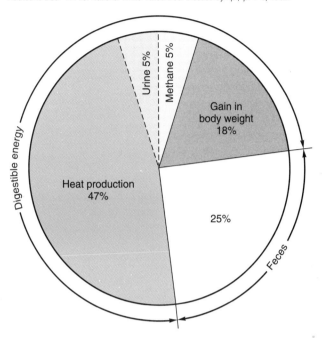

FIGURE 21.6    Food webs. (a) Aquatic community, about 25 percent of the net production is diverted to the detritus food chain. (b) In the forest community, over 60 percent of the net production enters the detritus food chain.

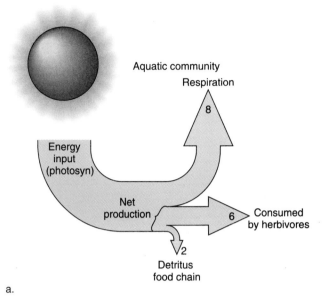

\* Numbers represent energy flow in kcal/m$^2$/day.

system gas (methane). More than 45 percent of the ingested energy is dissipated in heat production. Less than 20 percent results in net production by the herbivore and is available for consumption by secondary consumers. These percentages regarding energy use are similar for secondary and tertiary consumers.

---

A farmer proposed raising chickens to feed to minks. She would use the mink pelts for mink coats and grind up the remaining mink carcasses for chicken feed. The farmer would have to buy only enough grain to equal the weight of the mink pelts and feed this grain with the mink carcass material to the chickens to maintain a full supply of food for the minks. What is wrong with the farmer's plan?

---

We have seen that herbivores consume some of the energy produced by plants and other producers. Another avenue of consumption is via the decay of dead organic material by bacteria and fungi. This is called the **detritus food web.** How much energy is utilized by decay organisms? Are trophic levels associated with the decay pathway?

Figure 21.6 shows examples of typical detritus food webs in aquatic and terrestrial communities. As you can see, there are significant differences between them. In the forest community, most net production from producers flows to the detritus food web, while in the aquatic community the opposite is true.

As we saw in figure 21.5, only a small portion of the energy consumed by herbivores, or primary consumers, is

---

The data in figure 21.6 are observations on energy flow in two communities. What interpretations can you draw from these observations?

---

available for growth. This pattern of energy use is also typical of secondary and tertiary consumers. If this is so, then how much energy is actually transferred from one trophic level to another? To answer this question, biologists have conducted careful studies of numerous ecosystems and have recorded the data shown in table 21.2.

| | | Percent of Gross Production Consumed | |
|---|---|---|---|
| **Trophic Level** | **Cedar Bog Lake, Minnesota** | **Lake Mendota, Wisconsin** | **Silver Springs, Florida** |
| Herbivores (primary consumers) | 13.3 | 8.7 | 16.0 |
| Small carnivores (secondary consumers) | 22.3 | 5.5 | 11.0 |
| Large carnivores (tertiary consumers) | Not present | 13.0 | 6.0 |

**TABLE 21.2**  Trophic Level Energy Intake Efficiency

**FIGURE 21.7**   Energy flow in a forest community. Energy flow is shown as grams/m² year. In the absence of large herbivores, little energy is consumed in grazing.

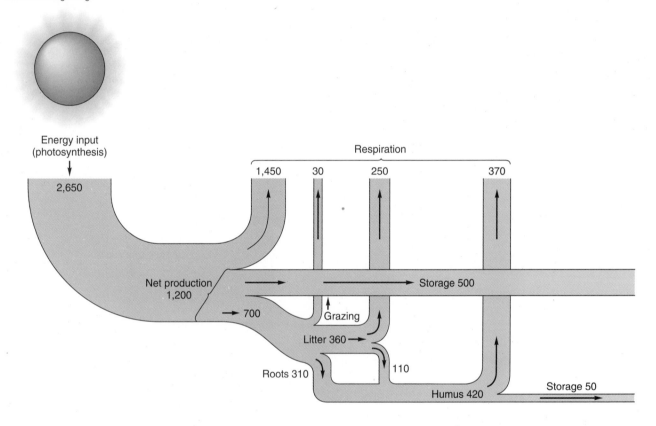

These data show that a relatively small portion of the energy consumed by one trophic level is passed on to or consumed by the next trophic level. For example, in Cedar Bog Lake, herbivores consume only 13.3 percent of the gross production by producers, and secondary consumers (small carnivores) consume only 22.3 percent of the energy consumed by herbivores. These figures are comparable to those for the deer in figure 21.5. As a general rule, about 10 percent of the gross production of energy is transferred from one trophic level to the next.

Using the general rule of 10 percent energy transfer, how much energy would be consumed by tertiary consumers if gross production of producers was 1,000 kcal/m²?

We have covered the basic features of how energy is captured, used, and transferred in communities. To apply these concepts, let us trace the flow of energy through two sample communities.

## A FOREST COMMUNITY

One of the most detailed studies of a forest community has been done on an oak-pine forest on Long Island, New York. The flow of energy in this system is shown in figure 21.7. Energy is measured in grams of dry matter per square meter per year (g/m²/yr). This forest is somewhat protected and does not contain large herbivores, such as deer. Most of the plant material in this system is consumed by insects and a few small mammals.

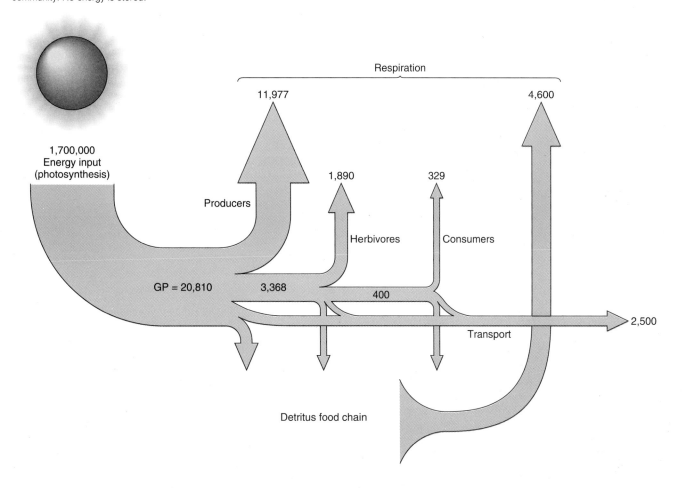

**FIGURE 21.8** Energy flow in aquatic community. Over 10 percent of the energy is carried away by the stream flow as particulate organic matter. Respiration and transport account for all the energy entering the community. No energy is stored.

Why is energy flow measured as grams of dry matter rather than kilocalories?

Most (55 percent) of the gross production of 2,650 g/m²/yr is lost as cellular respiration, leaving a net production of 1,200 g/m²/yr. In the absence of major herbivores, only about 30 g/m²/yr is consumed by grazing, representing a transfer of 1.1 percent. Litter (dead leaves, decaying wood) accounts for 360 g/m²/yr. Storage, as new tree growth, accounts for a large portion of the net production.

In this community, the detritus food web apparently accounts for a major part of the consumption of net production. In the absence of larger consumers, much of the energy from producers is directed to the decay of litter and material in humus (organic soil material).

If large herbivores were present in this community, which component of energy flow would most likely decrease?

This detailed analysis of a forest community illustrates several of the major features of energy flow: Most of the solar energy captured in photosynthesis ultimately leaves the community as respiration; net production is divided between herbivore consumption, storage, and decay; and, in general, the amount of energy leaving the community as respiration and retained as stored plant growth is balanced by the gross production.

## AN AQUATIC COMMUNITY

Our second example is an aquatic community, which differs a great deal from the forest community. An extensive investigation of a large spring at Silver Springs, Florida, has revealed the energy flow pattern shown in figure 21.8. As in the forest community, most (63.4 percent) of the gross production of 20,810 kcal/m²/yr is lost as plant respiration, while 3,368 kcal/m²/yr is consumed by herbivores (insect larvae, snails, and small herbivorous fish). In contrast to the forest community, the spring maintains a large population of herbivores, as well as secondary and tertiary consumers. At every stage, about 20 percent of the energy is diverted to decomposers.

These organisms are primarily bacteria and crayfish and are major contributors to community respiration.

While none of the energy apparently is retained as storage, about 12 percent (2,500 kcal/m²/yr) is carried out of the community as particulate material by streams flowing away from the spring.

As in the forest community, the greatest part of the energy entering the aquatic community as gross production leaves the system as respiration. Respiration and export of particulate material are equal to gross production. Again, we see that energy entering the community is equal to energy leaving the community.

Comparing energy flow in these two sample communities has indicated several similarities. One of the most important observations is that the data on energy flow constitutes an **energy balance sheet**—energy entering the community is equal to energy "leaving" the community in the form of respiration and storage or export. Equally important, we can see that energy flows *through* the system. Once this energy is expended in respiration, it cannot be recaptured. The energy source, usually the sun, must continually supply energy to the community. In contrast, matter, such as carbon and nitrogen, is recycled by communities and used again and again. This cycling of matter in communities is the focus of the next chapter.

What kind of occurrence in the forest community would be comparable to the export of particulate matter out of the aquatic community?

## SUMMARY

*Communities obtain energy from the sun, usually in the form of photosynthesis by producers, and pass that energy through herbivores, secondary consumers, and tertiary consumers. Organisms in each trophic level use only a part of this energy for growth, eliminating the rest through respiration. As you continue reading about communities, keep in mind the following key points about energy flow we have studied in this chapter.*

1. The amount of solar radiation striking the earth's surface averages 3,000 to 4,000 kcal/m²/day. A portion of this energy is captured in photosynthesis. The amount of energy captured is the **gross production (GP)** and can be measured as the amount of organic material produced or the amount of oxygen released in photosynthesis.

2. **Net production (NP)** is the biomass that remains after respiration (R).

3. The relationship between gross production, net production, and respiration is expressed as NP = GP – R.

4. In terrestrial communities, energy production can be estimated by collecting new plant growth and burning it in a bomb calorimeter. Respiration can be estimated by determining $CO_2$ production.

5. In aquatic systems, the changes in oxygen levels in light and dark bottles is a measure of respiration and net production.

6. The solar capture efficiency of most natural ecosystems is around 1 percent; the efficiency of intensively cultivated systems can be 6 to 8 percent.

7. The net production from producers can be consumed by herbivores, stored in the community, or enter the detritus food web where material undergoes decomposition and decay.

8. Only a small portion of the material consumed by herbivores, secondary consumers, and tertiary consumers is used to produce new biomass. Only this material is available for consumption by the next trophic level. Most is expended in respiration or is excreted.

9. As a rule, about 10 percent of the energy entering a trophic level is available for consumption by the next trophic level.

10. **Energy flow diagrams** indicate the movement of energy through a community and show how **energy flows through** a community.

11. Once energy has been released in respiration, it cannot be reused.

12. The **energy balance sheet** indicates that energy leaving the community as respiration, export, and storage is equal to energy entering the community.

## REFERENCES

Krebs, C. J. 1994. *Ecology: The Experimental Analysis of Abundance and Distribution* (New York: Harper Collins College).

Odum, E. P. 1971. *Fundamentals of Ecology* (Philadelphia: W. B. Saunders).

Woodwell, G. M. 1970. "The Energy Cycle of the Biosphere." *Scientific American* 223, no. 3: 64–74.

1.  Which of the following best represents the plant growth harvested from a community and burned in a bomb calorimeter (page 261)?
    a.  GP + R
    b.  NP
    c.  NP + GP
    d.  R
    e.  GP

2.  Which of the following best describes the changes in oxygen concentration in light and dark bottles at night (page 261)?
    a.  The light bottle will not change in oxygen; the dark bottle will decrease.
    b.  The dark bottle will not change in oxygen; the light bottle will decrease.
    c.  Both bottles will decrease by the same amount.
    d.  Neither bottle will change in oxygen content.
    e.  Both bottles will increase in oxygen content by the same amount.

3.  In addition to producers, light and dark bottles would also contain microscopic consumers. How would these consumers affect the measurement of gross production and net production (page 261)?
    a.  NP will be decreased.
    b.  GP will be decreased.
    c.  Neither GP nor NP will be affected.
    d.  R will be decreased.
    e.  GP will be increased.

4.  Which of the following would most likely increase estimates of gross production in a grassland?
    a.  Harvest the grass several times during a growing season.
    b.  Conduct studies in areas where grass-consuming insects are numerous.
    c.  Cut and harvest grass on cloudy days.
    d.  Harvest plants in fields where no weeds (only grass) are growing.
    e.  Conduct studies in fields where animals graze.

5.  In the table below on efficiency and productivity, which of the values is incorrect?

    | | Kcal/M$^2$/yr |
    |---|---|
    | Solar energy | 1,435,000 |
    | Percent efficiency | 1 |
    | GP | 14,350 |
    | R | 15,175 |
    | NP | 7,684 |

    a.  Solar energy
    b.  Percent efficiency
    c.  GP
    d.  R
    e.  NP

6.  A student knew that respiration decreases when temperature decreases. He argued that the ratio of net production to gross production (NP/GP) could be increased if plants were grown at lower temperatures. Which of the following was this student most likely assuming (page 262)?
    a.  GP and NP will increase by the same amounts when temperature is lowered.
    b.  NP/GP will increase when plants are grown at lower temperatures.
    c.  NP/GP will decrease when plants are grown at higher temperatures.
    d.  Photosynthesis does not decrease when the temperature is lowered.
    e.  Respiration decreases when the temperature is lowered.

7.  A farmer proposed raising chickens to feed to minks. She would use the mink pelts for mink coats and grind up the remaining mink carcasses for chicken feed. The farmer would have to buy only enough grain to equal the weight of the mink pelts and feed this grain with the mink carcass material to the chickens to maintain a full supply of food for the minks. Which of the following best explains what is wrong with the farmer's plan (page 263)?
    a.  Neither minks nor chickens are capable of generating gross production.
    b.  Respiration by both minks and chickens will decrease their body weights.
    c.  Adding grain adds energy to the system from an outside source.
    d.  The mink/chicken system should be workable without adding grain.
    e.  In this system, the chickens will eat grain but minks will eat only meat.

8.  The data in figure 21.6 are observations on energy flow in two communities. Which of the following is the best interpretation of these data (page 263)?
    a.  In both aquatic and terrestrial communities a large part of the energy leaves the system as respiration.
    b.  In the aquatic ecosystem, 6 kcal/m$^2$/day are consumed by herbivores.
    c.  Leaves are more easily consumed by herbivores than are phytoplankton.
    d.  In both communities, GP and NP will decrease during winter.
    e.  Aquatic herbivores are more efficient at capturing and consuming prey than are terrestrial herbivores.

9. Using the general rule of 10 percent energy transfer, how much energy would be consumed by tertiary consumers if gross production of producers was 1,000 kcal/m$^2$ (page 264)?
   a. 1 kcal/m$^2$
   b. 10 kcal/m$^2$
   c. 100 kcal/m$^2$
   d. 500 kcal/m$^2$
   e. 1,000 kcal/m$^2$

10. Which of the following observations best supports the conclusion that some material from producers is utilized by the detritus food web?
   a. Secondary consumers are not present in the community.
   b. Herbivores do not consume all the producers.
   c. The efficiency of producers is less than 1 percent.
   d. The community is terrestrial rather than aquatic.
   e. The largest herbivores in the community are rabbits.

11. In figure 21.7, why is energy flow measured as grams of dry matter rather than kilocalories (page 264)?
   a. Material from all living organisms can be expressed as grams of dry matter when water is removed.
   b. Grams of dry matter and kilocalories are both units of measure.
   c. Energy from the sun originally entering the community is measured in kilocalories.
   d. The figure shows that the grams of dry matter of producers is divided between respiration and storage.
   e. Grams of dry matter are equivalent to kilocalories when material is burned in the bomb calorimeter.

12. If large herbivores were present in this community (fig. 21.7), which component of energy flow would most likely decrease (page 265)?
   a. Grazing respiration
   b. Storage
   c. Litter
   d. Gross production of producers
   e. Respiration of producers

13. Which of the following occurrences in the forest community (fig. 21.7) would be comparable to the export of particulate matter out of the aquatic community (page 266)?
   a. Increasing the rate of photosynthesis
   b. Movement of rabbits into the community
   c. Increasing the population of earthworms in the soil
   d. Cutting the largest trees for lumber
   e. Planting grasses that could grow among trees in the forest

14. Based on the data in the figure at the bottom of this page, which of the following could you most likely conclude?
   a. There are no herbivores in this community.
   b. Point 3 represents net production.
   c. There are no tertiary consumers in this community.
   d. Point 5 represents decay of herbivores.
   e. The solar capture efficiency of this community is 10 percent.

15. If a swarm of grasshoppers descended on the grassland community shown in the figure for question 14, which of the following would decrease?
   a. 1
   b. 2
   c. 3
   d. 4
   e. 5

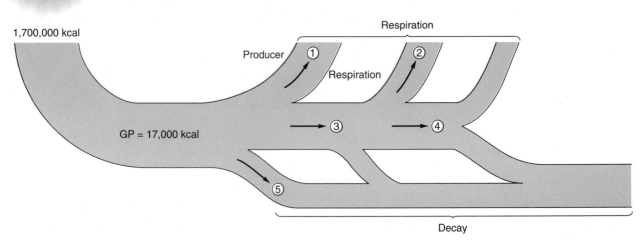

### CHAPTER 22

# MATERIAL CYCLING AND THE ENVIRONMENT

hen the material that makes up living organisms returns to the physical environment, where does it go? Can it be reused by living organisms?

A simple answer to these questions can be illustrated by the example of carbon dioxide. When carbon dioxide ($CO_2$) is released by respiration, it enters the atmosphere. This same carbon dioxide can be used in photosynthesis and again become part of the molecules of a living organism. The carbon and oxygen atoms of the carbon dioxide molecule **cycle** between the living and nonliving parts of the ecosystem and are continually "reused" (fig. 22.1).

Do carbon and oxygen cycle through the living and nonliving environments in as simple a fashion as this example suggests? Do carbon and oxygen remain in the atmosphere as carbon dioxide, or can they be converted and incorporated into other molecules? How long does carbon dioxide remain in the atmosphere?

We will examine these and other questions in this chapter, looking not only at carbon and oxygen cycling but also at nitrogen cycling. Though carbon and oxygen are chemically bonded together and cycle together in carbon dioxide, we will look at their cycles separately.

## THE CARBON CYCLE

The concentration of carbon dioxide in air is about 0.04 percent or 320 parts per million (grams of $CO_2$ per million grams of air). To understand the **carbon cycle,** we need to know how living things on earth affect the amount of carbon in the air. One important question is, how much can photosynthesis affect the atmospheric concentration of carbon dioxide?

FIGURE 22.1    Material cycling. Carbon and oxygen in $CO_2$ can cycle between the atmosphere and living organisms. $CO_2$ is removed from the atmosphere via photosynthesis and returned to the atmosphere via respiration.

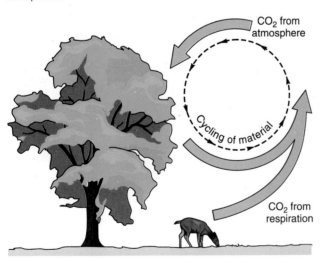

| $CO_2$ from atmosphere |
| Cycling of material |
| $CO_2$ from respiration |

**TABLE 22.1    Day and Night Atmospheric $CO_2$ Variation**

| Height Above Ground Level (m) | $CO_2$ Concentration (ppm) | |
| --- | --- | --- |
| | Day | Night |
| 0.0 | 330 | 360 |
| 0.5 | 320 | 350 |
| 7.0 | 306 | 340 |
| 30.0 | 305 | 330 |

We can investigate this question by measuring the carbon dioxide concentration in air at various times during the day and night in an area where there is heavy plant growth. What predictions do you make about these measurements?

Since photosynthesis occurs during daylight hours, we can predict that $CO_2$ will decrease during this period. At night, $CO_2$ should increase. Recall the light and dark bottle experiments in chapter 21 and the $CO_2$ concentration changes in figure 21.4.

Table 22.1 shows the carbon dioxide measurements taken at various heights above the ground in a forest. As we predicted, $CO_2$ decreases during the day and increases at night. Moreover, significant changes in carbon dioxide concentration are observed at heights well above tree-top level.

In table 22.1, why should $CO_2$ concentrations be highest near ground level?

**TABLE 22.2    Seasonal Atmospheric $CO_2$ Variation**

| Altitude (km) | $CO_2$ Concentration (ppm) | |
| --- | --- | --- |
| | September | April |
| 0 | 308 | 329 |
| 2 | 311 | 326 |
| 4 | 314 | 324 |
| 6 | 315 | 324 |
| 8 | 316 | 323 |
| 10 | 316 | 323 |
| 12 | 319 | 320 |
| 14 | 319 | 320 |

We can also investigate changes in carbon dioxide concentration during different parts of the year. Table 22.2 shows measurements taken during April and September for the region north of 30° N latitude. As you can see, there are considerable differences between these two seasons. Seasonal variation is observed as high as 10 kilometers above the earth's surface.

*Evolution and Ecology*

FIGURE 22.2 Atmospheric carbon dioxide concentrations. Data show a yearly fluctuation in carbon dioxide and an increase in the average concentration over a ten-year period. *Source: Data from C. D. Keeling, "Is Carbon Dioxide From Fossil Fuel Changing Man's Environment?" in* Proc. Am. Phil. Soc., *114:10–17, 1970.*

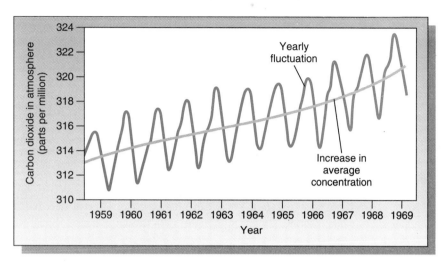

Why should carbon dioxide concentration be low in September and high in April?

A student predicted that atmospheric $CO_2$ would be near 320 parts per million in midwinter. What argument would support this prediction?

The data in these two tables indicate that photosynthesis does exert a major influence on atmospheric $CO_2$. When biologists at the mountaintop Mauna Loa Observatory in Hawaii took measurements over several years, they obtained the data graphed in figure 22.2.

Carbon dioxide clearly varies in a regular pattern during the year, indicating changes in the amount of photosynthesis occurring. Moreover, during the ten-year period when the measurements were taken, the average amount of $CO_2$ in the atmosphere steadily increased. Scientists have proposed that this increase is caused by the burning of fossil fuels such as coal and oil and could lead to an increase in the earth's temperature (Thinking in Depth 22.1).

## EARTH'S COMPARTMENTS

How much carbon is there on the earth? Biologists have made estimates of the total amount of carbon in sources such as land plants, the atmosphere, and ocean water. These separate sources are called **compartments** (fig. 22.3).

How could scientists estimate the amount of carbon in the atmosphere?

FIGURE 22.3 Carbon cycling. Land, ocean, and sediments are shown as the three major compartments for carbon. Arrows indicate rates of exchange in billions of metric tons per year. Quantities in each compartment indicate the amount of carbon in billions of metric tons. *Source: Data from* Scientific American, *p. 130, September, 1970.*

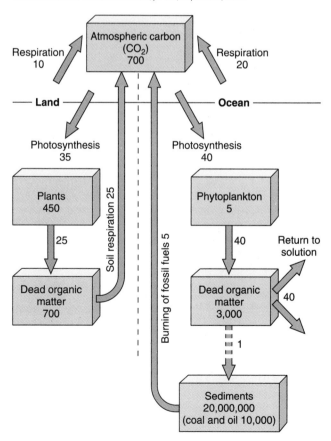

*Material Cycling and the Environment*

**271**

## GLOBAL WARMING

The data in figure 22.2 suggest that the earth's atmospheric carbon dioxide is slowly increasing. This is possibly due to the burning of fossil fuels such as coal and oil. We often hear that increasing carbon dioxide will lead to higher temperatures around the globe. How does $CO_2$ raise temperatures? What might be the consequences of this increase in $CO_2$?

The earth's atmosphere is more transparent to visible light rays than to heat rays. This difference in transparency is primarily due to $CO_2$. When visible and ultraviolet light reach the earth's surface and are absorbed, the surface is warmed. As the temperature of the surface increases, heat rays, or infrared radiation, are emitted into the atmosphere (fig. 1).

Once in the atmosphere, however, these heat rays are reflected back to earth. Heat is "trapped" by the atmosphere. Carbon dioxide is the primary atmospheric component that causes the reflection of heat rays, so the higher the $CO_2$ content, the more heat is reflected and the higher the temperature of earth rises.

Some scientists have argued that increases in atmospheric $CO_2$ will be minimized because more $CO_2$ will be removed from the atmosphere. What factors would increase $CO_2$ removal?

Scientists predict that if current trends in carbon dioxide production continue, the earth's temperature will rise significantly. This temperature increase could melt the polar icecaps and cause a consequent rise in ocean levels.

### FIGURE 1

Global warming. (*a*) Visible light penetrates the earth's atmosphere relatively easily. (*b*) Light falling on the earth results in warming and production of heat rays. (*c*) Carbon dioxide is *not* transparent to heat rays and reflects some heat rays back to earth.

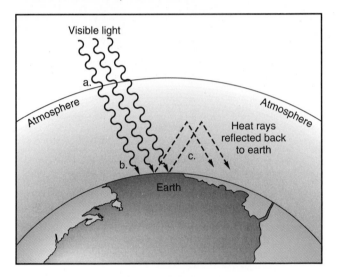

Figure 22.3 shows the rates of carbon exchange between various compartments. At best, the carbon amounts and rates of exchange are estimates with a large margin of error. Calculating compartment sizes and exchange rates is difficult and usually involves making assumptions and extending actual measurements. When we consider the size of the earth's entire atmosphere or its geological formations (fig. 22.4), we can only begin to imagine the enormous amounts of material contained in these compartments.

We do know that exchange between the atmosphere and living plants can be relatively fast. The average time for respired $CO_2$ to enter the atmosphere and again be captured by green plants is estimated as 300 years. Exchange in some other compartments, however, can be much, much slower. For example, coal and oil are thought to be the remains of plants and animals that died millions of years ago. The carbon in coal and oil has thus been "out

According to figure 22.3, how much carbon (in carbon dioxide) is generated by the respiration of decay organisms?

of circulation" for a very long time. When we burn these fuels we release this ancient carbon at an increasingly rapid rate with potentially serious consequences to the earth's climate (Thinking in Depth 22.1).

## THE OXYGEN CYCLE

If carbon dioxide is cycled through the living and nonliving environments, then we might expect that oxygen has a cycle similar to the carbon cycle. Is this the case?

FIGURE 22.4
Compartments. The white cliffs of Dover in England reveal part of a geological deposit composed almost entirely of the skeletons of plankton that accumulated on the ocean floor millions of years ago. The Ca $CO_3$ in the skeletons represents a huge reservoir of carbon, oxygen, and calcium.

Current measurements indicate that about 21 percent of the earth's atmosphere is composed of oxygen. We have evidence, however, that the earth has not always had such a high oxygen concentration. Only about 2 billion years ago sediments appeared with extensive amounts of minerals, primarily iron oxide, that contain oxygen. At periods earlier than 2 billion years ago, there was apparently little atmospheric oxygen available to combine with iron or other minerals.

Where did the earth's atmospheric oxygen come from? Was it compartmentalized in some other form? A highly likely possibility is that cells capable of photosynthesis evolved and began generating large amounts of oxygen. Such cells first appear in the fossil record about 2 billion years ago, just at the time when atmospheric oxygen began to appear.

Does the simultaneous appearance of photosynthetic cells in the fossil record and oxygen in the atmosphere *prove* that these cells generated the oxygen?

Suppose we propose the following hypothesis:

*Hypothesis*   The original atmosphere of the earth contained little, if any, free oxygen. Atmospheric oxygen was and is generated by photosynthetic organisms.

In other words, the hypothesis claims that all oxygen in the atmosphere was created by photosynthesis. However, plants also respire and use up oxygen (see chapter 6). If plant respiration is using oxygen as this oxygen is being produced by photosynthesis, then how can excess oxygen accumulate in the atmosphere? If respiration and photosynthesis occur at the same time, then the net result should be only $CO_2$ and $H_2O$ as shown in the following equations.

Photosynthesis

$$6\ CO_2 + 6\ H_2O \rightarrow C_6H_{12}O_6 + 6\ O_2$$

Respiration

$$C_6H_{12}O_6 + 6\ O_2 \rightarrow 6\ CO_2 + 6\ H_2O$$

Can excess oxygen be produced? Only if all the organic material produced in photosynthesis is not respired. Actually, this appears to be the case, since huge amounts of organic material have not been respired but have been compartmentalized in the earth as coal, oil, and gas (fig. 22.5). When we compare the amount of photosynthesis required to produce these organic materials to the amount of atmospheric oxygen and oxygen combined with other elements, such as iron, the amounts are very nearly equal! This equivalence is strong support for the hypothesis.

Geologists have proposed that much of the earth's early atmosphere was formed by gases vented by volcanoes, but volcanic gases do not contain oxygen. How does this observation affect our hypothesis regarding how oxygen was formed in the earth's atmosphere?

Figure 22.6 is a summary of the oxygen cycle, showing the compartments where oxygen is found and the pathways of exchange between them. As was the case in the carbon cycle, some exchange of oxygen, such as that between photosynthetic organisms and the atmosphere, occurs fairly rapidly. Other avenues of exchange, such as oxidative weathering, occur much more slowly.

A student argued that if all the herbivores were removed from an ecosystem, the cycling of oxygen might not be affected. Is the student's argument reasonable?

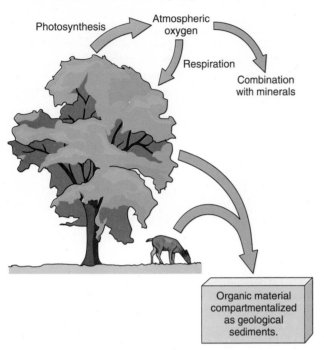

FIGURE 22.5 Possible accumulation of atmospheric oxygen. As photosynthesis produces oxygen, some of the oxygen is used in respiration and some combines with minerals such as iron. It is proposed that some oxygen is left in the atmosphere, because much organic material is not respired but is stored in geological sediments.

# THE NITROGEN CYCLE

So far we have considered two elements, carbon and oxygen, that play an essential role in the biological world. Another element, nitrogen, also exists as a primary component in the atmosphere and has a major effect on living organisms. Indeed, it is estimated that the availability of usable nitrogen is the greatest single limitation on plant production.

Nitrogen makes up 79 percent of the earth's atmosphere. It is an essential part of many biological molecules, primarily proteins (see chapter 3). The form of nitrogen found in the atmosphere is unavailable for use except to a few soil microorganisms that can convert nitrogen gas ($N_2$) to ammonia ($NH_3$). Once it is converted to ammonia, nitrogen can be absorbed by plants and used in the synthesis of molecules such as amino acids.

Amazingly, even though we are immersed in an atmospheric ocean of nitrogen and fill our lungs with this gas with every breath, only a few microorganisms can make nitrogen available for use by living organisms. An enzyme called nitrogenase that is present in these microorganisms can catalyze a chemical reaction in which nitrogen gas is converted to ammonia. This reaction is called **nitrogen fixation.**

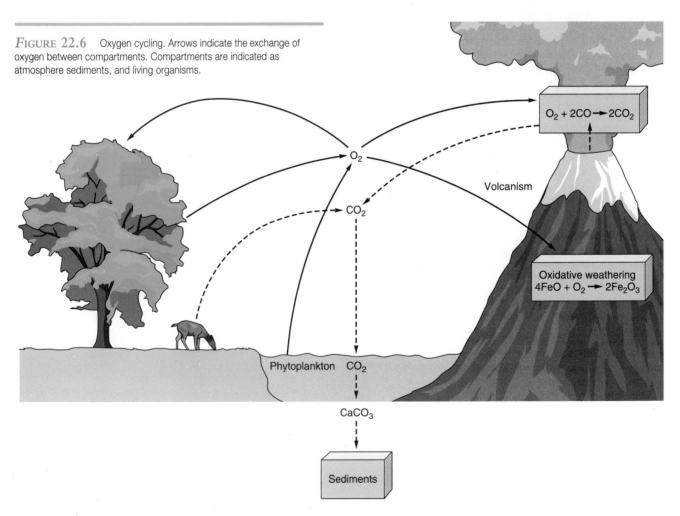

FIGURE 22.6    Oxygen cycling. Arrows indicate the exchange of oxygen between compartments. Compartments are indicated as atmosphere sediments, and living organisms.

*Evolution and Ecology*

FIGURE 22.7   Nitrogen fixation. (a) Soybean, a legume, is capable of "fixing" atmospheric nitrogen in a form available to plants. (b) Nodules on soybean roots contain symbiotic bacteria. These bacteria are the organisms that allow soybeans and other legumes to fix nitrogen.

a.

b.

Many of these microorganisms live in a mutually beneficial association with plants called **symbiosis.** The best known example is the bacterium *Rhizobium,* which invades the roots of legumes (fig. 22.7). Why these microorganisms associate only with legumes is not known; such an association with other kinds of plants is not observed. The amount of nitrogen fixed by the bacteria, which are symbiotic to the legumes, is impressive. When a crop of legumes is plowed into the soil, it adds as much as 350 kilograms of usable nitrogen per hectare.

Nitrogen can also be fixed by industrial chemical processes to produce commercial fertilizers. This process, however, requires temperatures of several hundred degrees and thousands of pounds of pressure. Nitrogenase, on the other hand, achieves this same result at ordinary temperatures and pressure. Additional sources of fixed nitrogen include "juvenile" nitrogen as ammonia from volcanic eruptions and atmospheric fixation caused by lightning. This fixed nitrogen can be used by plants in the synthesis of amino acids and other nitrogen-containing compounds.

Nitrogen is returned to the atmosphere by bacteria that convert nitrogen compounds to nitrogen gas. This is called **denitrification.** The entire nitrogen cycle for land organisms is diagrammed in figure 22.8.

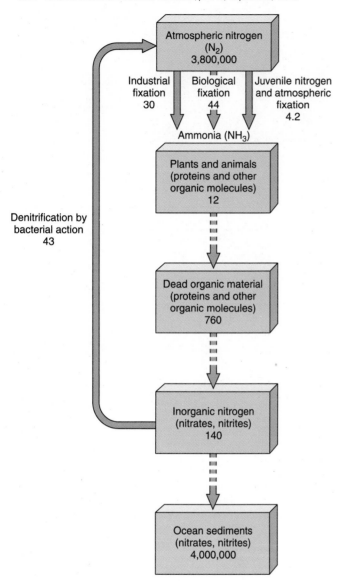

**FIGURE 22.8** Nitrogen cycling. Compartments are shown as land, ocean, atmosphere, and sediment. Compartment sizes are shown as billions of metric tons of nitrogen. Arrows indicate movement in millions of metric tons. *Source: Data from* Scientific American, *p. 140, September, 1970.*

# OTHER CYCLES

In addition to carbon, oxygen, and nitrogen, materials such as phosphorus and sulfur cycle through the living and non-living environment. These cycles follow different paths than the ones we have discussed, but they are equally complex and reveal an intimate interdependence between living organisms and the nonliving world.

If the availability of nitrogen were limited, the production of which biological molecules would be inhibited?
A student argued that continually applying fertilizers will cause nitrogen to accumulate in the soil. What was this student most likely assuming?

# SUMMARY

*Materials such as carbon, oxygen, and nitrogen cycle through the living and nonliving environments in complex ways. These cycles may occur over hundreds or millions of years, and are vital to life on earth. As you review this chapter, keep in mind the following key points.*

1.  Material such as carbon, oxygen, and nitrogen **cycles** in the living and nonliving environment and can be repeatedly "reused" by living organisms.
2.  Carbon dioxide is removed from the atmosphere by photosynthesis and synthesized into biological molecules. Respiration returns $CO_2$ to the atmosphere.
3.  Carbon dioxide exchange between geological deposits of carbon compounds and the atmosphere is much slower than the relatively rapid exchange between living organisms and the atmosphere.

4.  The earth's atmosphere now contains about 21 percent oxygen, but the very earliest atmosphere contained little, if any, oxygen. Atmospheric oxygen is thought to have been generated by photosynthesis over the past 2 billion years.
5.  Evidence supporting the photosynthetic origin of free oxygen includes the approximate balance between photosynthetically generated carbon compounds and oxygen.

6. Like carbon, oxygen cycles relatively quickly between the atmosphere and living organisms but much more slowly between the atmosphere and geological deposits.

7. Although nitrogen makes up 79 percent of the atmosphere, it must be **fixed,** or converted, to ammonia before it can be used by living organisms.

8. Most nitrogen fixation occurs by the **symbiotic** actions of bacteria and legumes.

9. Nitrogen gas is regenerated from inorganic nitrogen compounds by the action of **denitrifying** bacteria.

## References

Bolin, B. 1970. "The Carbon Cycle." *Scientific American* 223, no. 3: 124–32.

Cloud, P., and A. Gibor. 1970. "The Oxygen Cycle." *Scientific American* 223, no. 3: 110–23.

Delwiche, C. C. 1970. "The Nitrogen Cycle." *Scientific American* 223, no. 3: 136–46.

National Research Council. 1991. *Global Environmental Change: The Human Dimensions* (Washington, D.C.: National Academy Press).

## Exercises and Questions

1. In table 22.1, why should $CO_2$ concentrations be high near ground level (page 270)?
   a. Little light reaches ground level, so photosynthesis is low but respiration is high because of decay.
   b. As carbon cycles, it enters living organisms closer to ground level.
   c. The table shows that $CO_2$ is high near the ground but decreases with height.
   d. $CO_2$ is easier to measure at higher elevations than close to ground level.
   e. If the $CO_2$ close to ground level were used in photosynthesis, then $CO_2$ levels would decrease.

2. In table 22.2, why should carbon dioxide concentration be low in September and high in April (page 271)?
   a. The atmosphere is a large reservoir for $CO_2$ and small variations due to photosynthesis cannot be measured.
   b. In September, plants have stopped growing so the photosynthetic rate decreases.
   c. Photosynthesis and respiration occur at the same time in plants and cancel each other.
   d. By September, $CO_2$ has been depleted from the atmosphere by summer photosynthesis.
   e. Evergreens such as pine trees carry out photosynthesis year-round.

3. A student predicted that atmospheric $CO_2$ would be near 320 parts per million in midwinter. Which of the following arguments best support this prediction (page 271)?
   a. In midwinter, changes in $CO_2$ between day and night should not be apparent.
   b. $CO_2$ levels near 320 ppm represent the average concentration usually found for $CO_2$.
   c. Neither photosynthesis nor respiration are high in winter and, therefore, $CO_2$ levels should be near average.
   d. During the winter, more $CO_2$ is produced by burning fossil fuels.
   e. A concentration greater than 320 ppm is observed in summer.

4. Some scientists have argued that increases in atmospheric $CO_2$ will be minimized because more $CO_2$ will be removed from the atmosphere. Which of the following best supports this argument (page 272)?
   a. Energy is required to convert $CO_2$ to organic molecules.
   b. Photosynthesis increases when $CO_2$ concentration increases.
   c. When respiration increases, additional $CO_2$ is produced.
   d. When $CO_2$ is utilized in photosynthesis, oxygen is produced.
   e. If atmospheric $CO_2$ decreases, less heat is reflected back to earth.

5. Which of the following would scientists most likely use to measure the total amount of carbon in the atmosphere (page 272)?
   a. Measure the amount of $CO_2$ produced by respiration in a square meter of land and multiply by area.
   b. Measure the thickness of the atmosphere and multiply by the average density of air.
   c. Measure the increase in $CO_2$ in the atmosphere and subtract this from the average value.
   d. Measure the maximum and minimum concentrations of $CO_2$ during the year and calculate the average.
   e. Measure $CO_2$ in a cubic meter of air and multiply by the volume of the atmosphere.

6. According to figure 22.3, how much carbon (in billion metric tons of carbon dioxide) is generated by the respiration of decay organisms (page 272)?
    a. 60
    b. 25
    c. 10
    d. 35
    e. 97

7. Radioactive carbon ($C^{14}$) is constantly being produced in the atmosphere by gamma rays. At the same time, $C^{14}$ is decaying, thus atmospheric $C^{14}$ remains approximately constant. When living organisms take up $C^{14}$ in photosynthesis or by consuming other organisms, this radioactive carbon is incorporated into molecular components of their bodies. When organisms die, no more $C^{14}$ is taken up and radioactive carbon in the dead material slowly decreases. Given this information, which of the following would have the least amount of $C^{14}$?
    a. Secondary consumers
    b. Deposits of coal and oil
    c. $CO_2$ respired by decay organisms
    d. $CO_2$ respired by secondary consumers
    e. Herbivores

8. Does the simultaneous appearance of photosynthetic cells in the fossil record and oxygen in the atmosphere *prove* that these cells generated the oxygen (page 273)?
    a. No. Other organisms were also part of the fossil record and could exchange gases with the atmosphere.
    b. Yes. Photosynthetic cells *could* produce oxygen that would accumulate over a long period of time.
    c. No. The observations show that the two occurred at the same time, not that one caused the other.
    d. Yes. If photosynthetic cells did not exist, then oxygen would not be formed.
    e. No. Photosynthetic organisms also respire and would use up oxygen that is produced.

9. Geologists have proposed that much of the earth's early atmosphere was formed by gasses vented by volcanoes, but volcanic gases do not contain oxygen. How does this observation affect the hypothesis regarding the formation of oxygen (page 273)?
    a. It contradicts the hypothesis since oxygen should have been present with other gases.
    b. It supports the hypothesis since data indicate a second equally valid interpretation.
    c. It neither supports nor contradicts the hypothesis since oxygen was not present in gases from volcanoes.
    d. It contradicts the hypothesis because no data are given on the presence of photosynthetic cells.
    e. It supports the hypothesis since it eliminates a possible source of oxygen other than photosynthesis.

10. A student argued that if all the herbivores were removed from an ecosystem, the cycling of oxygen might not be affected. Is the student's argument reasonable (page 273)?
    a. No. Oxygen would not be removed from the atmosphere without herbivores.
    b. Yes. Cycling would occur through plants.
    c. No. Oxygen could be produced but it would not be removed from the atmosphere.
    d. Yes. But cycling of oxygen would occur much more slowly.
    e. No. Ultimately oxygen would be trapped by minerals (such as iron oxide) if oxygen were limited.

11. If the availability of nitrogen were limited, the production of which of the following biological molecules would be inhibited (page 276)?
    a. sugars
    b. fats
    c. starches
    d. nucleic acids
    e. carbohydrates

12. A student argued that continually applying fertilizers will cause nitrogen to accumulate in the soil. Which of the following was this student most likely assuming (page 276)?
    a. Application of fertilizers should be decreased to prevent nitrogen accumulation.
    b. If nitrogen is continually removed from the atmosphere, then the nitrogen content will decrease.
    c. The fixation of nitrogen by *Rhizobium* in association with legumes will decrease.
    d. Denitrifying bacteria will not increase activity as nitrogen concentration increases.
    e. Continued application of fertilizers will result in increasing accumulation of nitrogen.

# DIVERSITY OF LIFE

hen we see an unfamiliar flower, or any other kind of organism, we usually ask, "What kind of flower is that?" Our initial curiosity about the flower is a desire to know its *name*. Naming living organisms and grouping them into categories are natural and logical processes that enable us to organize our knowledge and understand the world around us. Of course, simply naming a plant or animal does not explain how the organism lives or functions, but it does allow us to identify and communicate information about the kinds of organisms we observe.

How do biologists assign names to living things? Do they group organisms according to categories? What are the categories? To answer these questions, we must take a look at **taxonomy,** the branch of biology that is concerned with naming and classifying organisms.

# CLASSIFICATION SYSTEMS

Many principles of modern taxonomy were first established by Carolus Linnaeus, a Swedish botanist who published his system in the early 1700s. Linnaeus recognized that the most efficient and useful classification system could be constructed as a **hierarchy,** or a ranked series. Figure 23.1 shows a hierarchy of footwear. The first divisions, such as shoes with laces and shoes without laces, are broad. As the footwear are divided into smaller and smaller groups, the distinctions become more detailed. This hierarchical structure makes it easy to organize the categories and identify individual types.

Suppose you could ask twenty questions to identify an unknown shoe. The questions have to be answerable by

**FIGURE 23.1** Classification hierarchy. A classification scheme is shown for different kinds of footwear. The initial divisions are rather broad (laces or no laces) but gradually become more detailed until a specific type of footwear is identified.

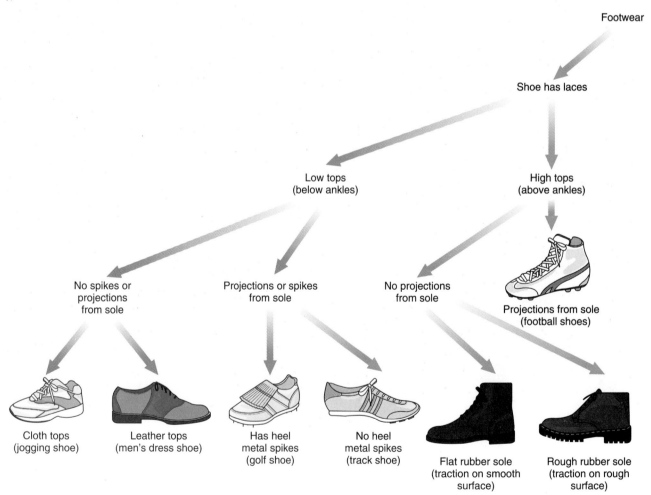

"yes" or "no." You could use your twenty questions by randomly guessing shoe types ("Is it a jogging shoe?" "Is it a house slipper?"), or you could ask questions that gradually narrow the possibilities ("Does it have laces?"). If you guess randomly, you may have to ask fourteen questions before you identify the shoe. If you ask gradually more detailed questions, however, you would need at most four questions.

Using the hierarchy in figure 23.1, we can construct a classification for each type of footwear. For example, we would classify hiking boots as follows:

1. Has laces
2. Has high tops over ankles
3. Has no projections from sole
4. Sole is indented for traction

These four characteristics uniquely identify the hiking boot among all other kinds of footwear.

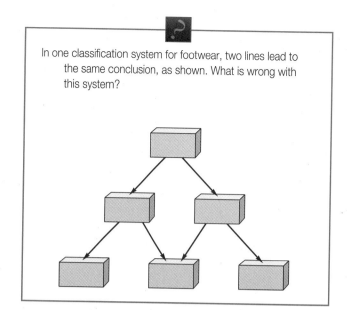

In one classification system for footwear, two lines lead to the same conclusion, as shown. What is wrong with this system?

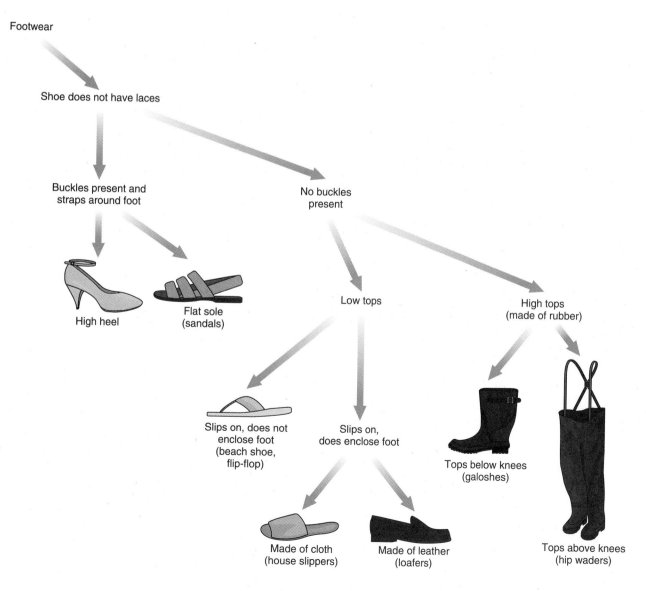

## TABLE 23.1 Characteristics Useful in Classification

| Kinds of Data | Why Useful | Limitations on Usefulness | Taxonomic Levels at Which Most Useful |
|---|---|---|---|
| Gross morphology | Basic reflection of total adaptation; many features available for study; only data available from fossils | Important biochemical and behavioral differences may not be apparent at the gross morphological level | All levels |
| Embryology | Most striking modifications affect later developmental and adult stages more than they affect early developmental stages | Adaptive requirements of larval life may lead to the evolution of pervasive larval adaptations | Relationships among phyla, also within phyla |
| Physiology and biochemistry | Are direct phenotypic expressions of underlying genetic differences | Similar physiological and biochemical mechanisms may result from different genetic bases | All levels |
| Behavior | Represents a basic method of adapting to the environment | Physiological and genetic bases of behavior poorly known; difficult to determine which characters are evolutionarily conservative | Relationships within a phylum and its subgroups |
| Breeding experiments | Direct test of degree of genetic difference | Can only be used with genetically similar forms; negative results can be produced for a variety of reasons | Genera and species |
| DNA base sequence similarities | Direct measurement of amount of genetic similarity and difference | Techniques just being developed | All levels |

*Original table in: G. Oriens, The Study of Life, 2nd Ed. Allyn & Bacon Publishers, Page 124, Table 9.4*

# CLASSIFYING LIVING ORGANISMS

Linnaeus's classification system for biological organisms is structured in the same hierarchical fashion as the footwear example. He proposed seven levels or categories of classification that are still used by biologists today.

1. kingdom
2. phylum (in plants, division)
3. class
4. order
5. family
6. genus
7. species

When biologists classify a newly discovered organism, they make a decision at each of these levels regarding where the organism should be placed. These decisions are much like the questions we asked in the hiking boot example. Of course, living things are far more complex than footwear. What characteristics of living things are most useful to taxonomists?

When biologists first began classifying organisms, they used characteristics that were readily apparent to the eye, such as physical differences. Take, for example, the characteristic of size. If we used size as a point of classification, we could put elephants and trees in the same category. This does not seem to be a very logical relationship.

## TABLE 23.2 Classification of the Gray Wolf

| Taxonomic Category | Designation | Description |
|---|---|---|
| Kingdom | Animalia | No cell walls, require energy-rich molecules from environment |
| Phylum | Chordata | Hollow nervous system, gill slits during development |
| Class | Mammalia | Have hair, feed young on milk, maintain high body temperatures |
| Order | Carnivora | Have meat-eating teeth, long canine teeth |
| Family | Canidae | Dogs and doglike animals; three lower molars; four toes on forefoot |
| Genus | *Canis* | Upper incisors lobed; post orbital processes thickened dorsally |
| Species | *lupus* | Large, carries tail horizontally when it runs |

*Original data from Oriens (See table 23.1)*

As the theory of evolution developed, taxonomists paid greater attention to evolutionary relationships when they categorized organisms. They used such characteristics as similarities in DNA, proteins, and bone structure, which we covered in chapter 19.

The most useful characteristics for classifying organisms are listed in table 23.1. The way in which taxonomists use these characteristics is illustrated in table 23.2, which is a

## DISCOVERING A NEW SPECIES

In the early 1990s, an infestation of whiteflies (fig. 1) struck California and Arizona, causing an estimated half-billion dollars in damage. The whitefly has been a common crop pest in the American south and southwest for many years, but this new infestation was more voracious and attacked crop plants that had been previously unaffected. This led biologists to suspect that the infestation was a new whitefly species.

The "new" and the "old" whiteflies were identical in appearance, so biologists used two tests to determine whether the new bugs were indeed a new species. In the first test, males and females from the old and new populations were paired. These whiteflies did not mate, indicating that they were of two different species.

In the second test, biologists compared DNA in the two groups. They found 80 percent similarity in DNA between individuals of the same group and only 10 percent similarity between individuals in the two different groups.

Using the results of their tests, scientists argued that the new whitefly was indeed a new species. Their conclusion was important, because up to that point agriculturalists had been unable to control the pest with insecticides. Knowing that the fly was a new species, biologists could concentrate their efforts on finding a natural enemy, which would be a more effective means of controlling the damaging flies. And since this whitefly was a new species, it might not have the same predators as the old whitefly species. The search for a natural enemy would be an expensive and time-consuming project. Knowing that they were dealing with two different species was a great help to the biologists.

Suppose the two types of whitefly were different in appearance but were able to mate and produce offspring. Would this affect the biologists' conclusion about whether the two flies were different species?

**FIGURE 1** The whitefly is a tiny insect that can cause significant damage to a variety of vegetable crops in the United States.

---

classification of the gray wolf. Notice that almost all the characteristics listed in table 23.1 are used to classify this single animal.

The names used in each of the taxonomic categories are always Latin (e.g., Animalia). Linnaeus initiated this practice, arguing that since Latin is a "dead" language, the definitions of terms will not likely change. In addition, taxonomists usually italicize genus and species names that give the scientific name of the organism (*Canis lupus*); higher order names are capitalized and in roman type (Canidae).

The smallest and most detailed of Linnaeus's categories is the **species,** a category we considered in some detail in

chapter 18. We defined a species as a group of organisms that is incapable of exchanging genes with other groups under natural conditions.

You might expect that all living organisms have been classified and described, but this is not the case. Even among large and easily visible organisms, such as birds and mammals, new species are discovered every year. On the average, biologists discover three to five new bird species, ten new mammal species, and one new genus each year. Among less easily visible organisms, such as insects and fungi, the yearly rate of discovery is much greater. Many times these new species have practical and economic importance for humans (Thinking in Depth 23.1).

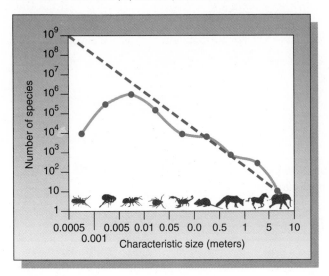

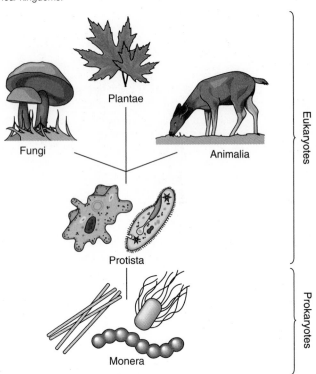

How many species exist in the world? This question is difficult to answer because new organisms are continually being discovered. Biologists estimate that there are currently 1.5 million to 1.8 million *known* species on earth. How do they arrive at an estimate for all existing species? One method is illustrated in figure 23.2.

The data in this figure suggest a relationship between species number and the size of organisms; that is, the largest animals have the fewest species. This relationship does not hold for organisms smaller than 1 centimeter. However, for animals larger than 1 centimeter, if the numbers in the graph are carried out to the vertical axis, we obtain an estimate of about 10 million species. This figure includes all those species yet to be discovered.

According to figure 23.2, most new animal species that are discovered will be of what size?

What does figure 23.2 assume about the numbers of species of animals 0.1 meter long and longer?

Using other methods, biologists have put the number of species on earth even higher. Considering the variation caused by different ways of calculating, it is not surprising that a great deal of uncertainty exists about the total number of existing species.

# THE TAXONOMIC SYSTEM

Some description of the different taxonomic categories is necessary to understand the classification of living organisms. Obviously, we cannot examine every genus or species; however, we can study the larger categories and examine representative species to get an idea of their distinguishing characteristics and the enormous diversity in the living world.

The most general category Linnaeus proposed was the **kingdom.** Early biologists simply divided organisms into plants or animals, but since that time biologists have found divisions into further kingdoms useful. Most scientists today use the five-kingdom classification illustrated in figure 23.3. The kingdoms are divided into many **phyla,** some of which we will review in the next sections.

Biologists distinguish between prokaryotic and eukaryotic organisms, a distinction that has proven particularly useful in the five-kingdom system. In chapter 2, we saw that eukaryotic cells have a well-defined nucleus and complex, membrane-bounded cellular organelles. Prokaryotic cells, on the other hand, do not exhibit this level of structural complexity (see fig. 2.10). The prokaryotic structure distinguishes **Monera** from the other four kingdoms that all exhibit eukaryotic cell structure. **Protista, Animalia, Plantae,** and **Fungi** are differentiated by cell structure, the ways they obtain nourishment, and whether they are multicellular or unicellular.

Biologists generally agree that the differences between prokaryotes and eukaryotes are greater than the differences between plants and animals. Given this information, would you conclude that the differences between fish and humans are greater or less than the differences between prokaryotes and eukaryotes?

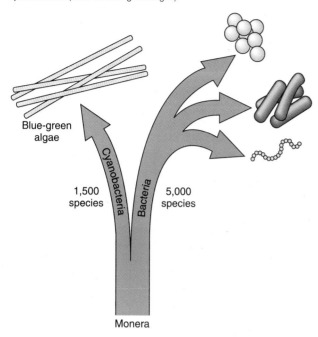

## Monera

All members of the Monera kingdom are prokaryotes, which comprise all the bacteria and cyanobacteria (photosynthetic bacteria). Figure 23.4 indicates the different shapes biologists use to distinguish different forms. Chemical properties and food sources are also used to classify the approximately 6,500 known species in this kingdom.

These organisms exert an enormous influence, both beneficial and harmful, on humans. As disease-causing agents, bacteria are destructive. As nitrogen-fixing bacteria (chapter 22), inhabitants of our digestive tract (chapter 9), and decomposers (chapter 22), they are essential to our well-being.

At one time, the cyanobacteria were called blue-green algae. All of the approximately 1,500 species are capable of photosynthesis.

## Protista

Protista are the simplest eukaryotes and include a variety of organisms. As you can see in figure 23.5, the largest phyla are the algae and protozoans. Organisms in all the Protista phyla

FIGURE 23.5   Protista. The largest phyla are represented by the algae and the protozoans.

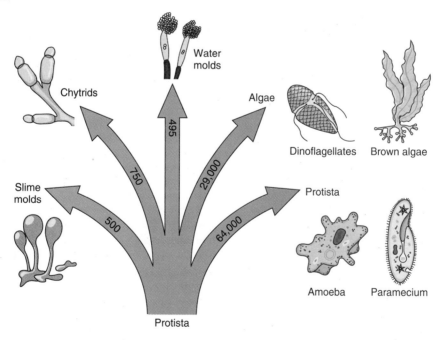

are usually single-celled, but some, such as algae, can form extremely large organisms. For example, a brown alga like marine kelp can form multicellular structures up to 100 meters long. These large algae, however, do not have differentiated tissues or organs characteristic of more advanced plants.

One of the ways biologists classify organisms in the Protista kingdom is by the way they move. Protozoans move by a variety of mechanisms. Amoeba move by cytoplasmic projections, paramecium by waving hairlike projections, and others by rapid movement of a single whiplike appendage called a flagellum. Few algae are capable of movement, but the single-celled Euglena and dinoflagellates achieve movement using flagella. Smaller numbers of species are included in the slime and water mold phyla. Slime molds are found in single or multicellular forms. Among the water molds is the organism that causes late blight infestation in potatoes.

A student claimed that the numbers of species in the Monera and Protista kingdoms indicated in figures 23.4 and 23.5 are incorrect. Should the student's claim be taken seriously?

## *Fungi*

The Fungi kingdom includes three phyla (fig. 23.6) that differ in their method of reproduction. With the exception of yeast, all grow as long, threadlike structures called hyphae. The zygomycetes form sexual zygospores when hyphal strands fuse to create the diploid spore. The common bread mold is a zygomycete. The ascomycetes include yeast and species that bear spores in saclike structures called asci. The genus Penicillum belongs in this phyla and includes the organism producing the antibiotic penicillin and the organism that causes athlete's foot. The basidiomycetes include the familiar mushrooms, toadstools, and puffballs. The spores of these fungi are produced on clublike structures called basidia.

## *Plantae*

The Plantae kingdom comprises all the plants. These are multicellular, generally terrestrial organisms that obtain energy by photosynthesis. Recall in chapter 13 that the reproductive strategies of plants can be quite varied and are characterized by the alternation of generations. The classification of plants is shown in figure 23.7.

Plants can be divided into two main categories: the **bryophytes** or nonvascular plants, and the **tracheophytes** or vascular plants. Bryophytes (mosses and liverworts) have

**FIGURE 23.6**  Fungi. The three phyla are differentiated by the method of reproduction. Zygospores are formed in zygomycetes, spores are formed in saclike structures in ascomycetes, and in basidiomycetes spores are formed on clublike structures (basidia).

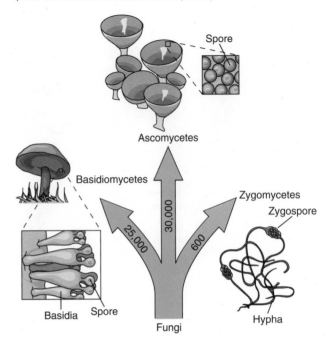

no specialized conducting tissue of xylem and phloem (chapter 10). Tracheophytes can be further divided into seed-bearing plants and non-seed-bearing plants.

The largest division among the non-seed-bearing plants is the ferns. Relatively few species are classed as horsetails, club mosses, and wisk ferns. All ferns reproduce by releasing sperm directly into water around the gametophyte, as we described in chapter 13.

A more efficient mode of gamete transfer is found in the seed plants where sperm cells are transferred within the protective structure of pollen grains. Seed plants can be further divided into **gymnosperms** and **angiosperms.**

Gymnosperms are sometimes called the naked seed plants; they include the familiar conifers such as pine, spruce, and cedar. Angiosperms are the flowering plants such as roses, apple trees, and violets. In a numerical sense, the plant world is dominated by the angiosperms, which have 235,000 different species. Although evolutionary studies indicate that angiosperms have evolved more recently than any other types of plants, this division has by far the largest number of species.

In terms of evolutionary theory, why should there be such a large number of angiosperm species?

FIGURE 23.7 Plantae. Approximately 24,000 species are classed
as nonvascular plants. Among the seedless plants, the ferns are the
largest group, while the flowering plants are the largest single group
in the seed plants with 235,000 species.

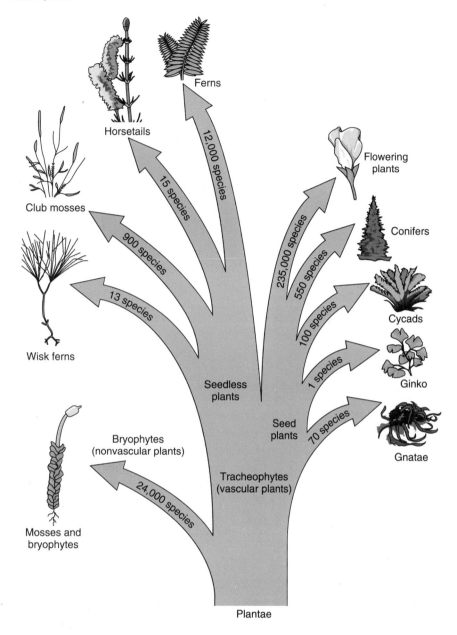

Ferns

Horsetails

12,000 species

Club mosses

15 species

900 species

13 species

Wisk ferns

Flowering plants

235,000 species

Conifers

550 species

Cycads

100 species

Ginko

1 species

Gnatae

70 species

Seedless plants

Seed plants

Bryophytes (nonvascular plants)

Tracheophytes (vascular plants)

24,000 species

Mosses and bryophytes

Plantae

FIGURE 23.8 Symmetry. (a) Asymmetrical organisms cannot be divided by a plane separating the organisms into nearly identical parts. (b) Bilaterally symmetric organisms can be divided into two nearly identical parts. (c) Animals that are radially symmetric can be divided into several identical parts.

Amoeba

a.

c.

b.

## Animalia

While plants obtain energy by photosynthesis, animals obtain energy from complex organic molecules. Obtaining these molecules requires animals to expend energy by actively seeking out food. To be successful at obtaining food, animals need a nervous system that gathers information from the environment and coordinates body movement. Because evolutionary selection has strongly favored the development of nervous systems in animals, biologists use the differences in nervous systems to classify animals.

Two other characteristics used to classify animals are **symmetry** and **body cavity type.** Animals can be bilateral, radial, or asymmetrical, as illustrated in figure 23.8. The body cavity, called the coelom, is characteristic of body form.

There are thirty-two animal phyla. We have chosen to show only the most familiar groups in figure 23.9. The simplest of the animal phyla are sponges, which are asymmetrical and live their lives attached to a solid surface. Even though these animals do not move about seeking food, they expend energy by moving flagella to obtain nourishment. This movement forces water through pores in their body wall, filtering out microorganisms and particulate material as food sources.

Another simple phyla, the cnidarians, are radially symmetrical, more complex than sponges, and possess muscles and nervous systems. All the other major groups are bilaterally symmetrical, a body plan that lends itself well to mobility. The sense organs of these animals are normally located at one end where the organism first contacts the environment. This improves the reception of information from the animal's surroundings. The differences in these animals' nervous systems place them in different groups within the animal kingdom.

FIGURE 23.9 Animalia. Nine of the 32 animal phyla are shown. The largest group by far is the arthropods that include insects, spiders, and crustaceans.

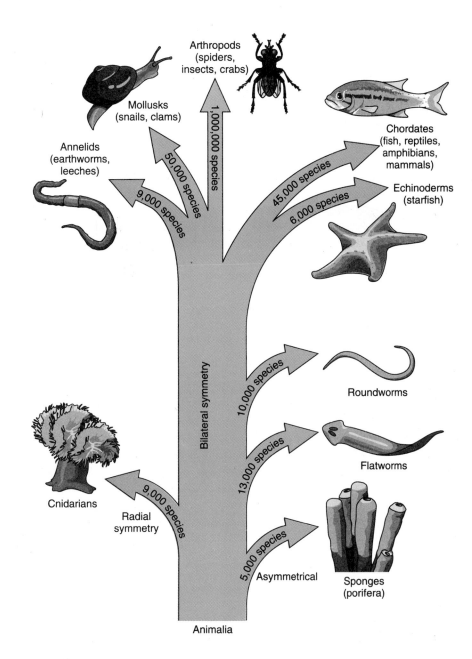

## SUMMARY

*Biologists name and categorize living organisms in order to organize our knowledge and increase our understanding of the world around us. The classifications described in this chapter only begin to introduce the incredible variety in the living world. The basic physiological processes we have discussed throughout this text provide the basis for classifying all organisms on earth. As you review this chapter on the diversity of life, keep in mind the following key points.*

1. **Taxonomy** is the branch of biology concerned with identifying, naming, and classifying organisms.
2. Modern principles of taxonomy were established by Carolus Linnaeus in the early 1700s.
3. The most useful classification system is constructed as a **hierarchy** in which categories become progressively more detailed. Categories are identified with scientific names given in Latin.
4. The seven basic categories used in classification are kingdom, phylum, class, order, family, genus, and species.
5. Many different factors, including structural, biochemical, and behavioral characteristics, are used to classify organisms.
6. New species are constantly being discovered, especially among smaller organisms. Biologists estimate that there may be as many as 10 million species, though only about 1.8 million species are known. If evolution is true, then new species are constantly being formed while others are becoming extinct.

7. Most taxonomists use the five-kingdom system of classification.
   a. **Monera** includes all prokaryotic cells, divided into bacteria and cyanobacteria.
   b. **Protista** are the simplest eukaryotes, which include algae and protozoans.
   c. **Fungi** feed on organic material from other organisms. Except for yeast, they grow as long, threadlike hyphae.
   d. **Plantae** includes eukaryotic, primarily terrestrial organisms that obtain energy by photosynthesis—i.e., plants. They are divided into **bryophytes** and **tracheophytes.** Seed plants are divided into **gymnosperms** and **angiosperms.**
   e. The **Animalia** kingdom includes all animals, which obtain energy from complex organic molecules by expending energy. **Symmetry** and **body cavity** are characteristics used to classify animals.

## REFERENCES

Barinaga, M. 1993. "Is the Devastating Whitefly Invader Really a New Species?" *Science* 259: 30.

May, R. M. 1992. "How Many Species Inhabit the Earth?" *Scientific American* 267, no. 4: 42–48.

Perring, T. M., et al. 1993. "Identification of a Whitefly Species by Generic and Behavioral Studies." *Science* 259: 74–77.

## EXERCISES AND QUESTIONS

1. In one classification system for footware, two lines lead to the same conclusion, as shown. Which of the following is most likely wrong with this system (page 281)?

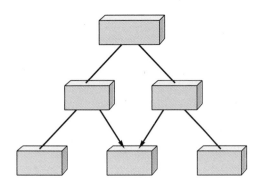

   a. More types of footwear should have been included in the system.
   b. The person constructing the system assumed that all types of footwear were being identified.
   c. Characteristics used did not sufficiently distinguish between different types of footwear.
   d. The system must have included objects other than footwear to be identified.
   e. The three types of footwear shown in this system must all have flat soles.

2. Suppose the two types of whitefly in Thinking in Depth 23.1 were different in appearance but were able to mate and produce offspring. Would this affect the biologists' conclusion about whether the two flies were different species (page 283)?
   a. No. A difference in appearance indicates a different species.
   b. Yes. If the two types appear different, then these differences could be due to environmental effects.
   c. No. Different species must inhabit different environments.
   d. Yes. The two types would not be reproductively isolated.
   e. No. Similarities in DNA had not been tested.

3. According to figure 23.2, most new animal species that are discovered will be of what size (page 284)?
   a. greater than 10 meters
   b. less than 0.005 meters
   c. greater than 0.05 meters
   d. between 1.0 and 0.1 meters
   e. greater than 1.0 meter

4. Which of the following does figure 23.2 assume about the numbers of species of animals 0.1 meter long and longer (page 284)?
   a. Foxes and mice are smaller than horses and elephants.
   b. About 10 million total species exist on the earth.
   c. Animals 0.001 meter in length are less abundant than animals between 0.005 and 0.01 meters in length.
   d. Some larger animals are becoming extinct at an increasing rate.
   e. Almost all these species have been discovered and described.

5. Biologists generally agree that the differences between prokaryotes and eukaryotes are greater than the differences between plants and animals. Given this information, would you conclude that the differences between fish and humans are greater or less than the differences between prokaryotes and eukaryotes (page 284)?
   a. Less, because fish and humans are in a common category of chordates.
   b. Greater, because fish and humans live in very different environments.
   c. Less, because eukaryotes and prokaryotes can live in very different environments.
   d. Greater, because fish evolved much earlier than humans.
   e. Less, because prokaryotic and eukaryotic cells can make up the bodies of fish and humans.

6. A student claimed that the numbers of species in the Monera and Protista kingdoms indicated in figures 23.4 and 23.5 are incorrect. Which of the following would most likely support the student's claim (page 286)?
   a. Sometimes species are reclassified into different categories.
   b. New species are constantly being discovered.
   c. The protista are extremely diverse, with many different body forms.
   d. The five-kingdom system includes all known species.
   e. Prokaryotes and eukaryotes differ greatly in the number of species in these categories.

7. In terms of evolutionary theory, why should there be such a large number of angiosperm species (page 286)?
   a. There may be more species of angiosperms, but gymnosperms have evolved into larger plants.
   b. Fertilization in angiosperms occurs by the transfer of pollen.
   c. Selection in angiosperm species has resulted in many species becoming extinct.
   d. Angiosperms have successively adapted to a large variety of environments.
   e. For speciation to occur in angiosperms, each species must be reproductively isolated.

## WRITING ACROSS THE DISCIPLINES

*The exercises presented in this section provide an opportunity for you to explore relationships between biology and other academic disciplines such as law, literature, and psychology. There are no "right" answers to the questions posed in each exercise. Your response may be a short viewpoint or an in-depth research paper, depending on your teacher's instructions. References are provided with each exercise, should you want or need to refer to them as part of your assignment.*

## Public Policy
### TRANSGENIC PREDATORS

Biologists have been field-testing genetically engineered (transgenic) crop plants for several years. There is relatively little concern that these plants will "escape" into the natural environment and have unanticipated effects because transgenic plants typically survive only in managed agricultural systems.

The possibility of developing transgenic predators,* however, has raised more serious concerns. Unlike plants, predators can move; they might possibly escape and invade natural populations. Biologists have considered the possibility of genetically altering arthropods. For example, certain mites that prey on other harmful arthropods could be implanted with a gene for pesticide resistance.

The gene would allow the predators to survive pesticide applications and more effectively control harmful pests. Such beneficial effects, however, may be outweighed by unknown and possibly damaging effects to other beneficial arthropods. We cannot possibly know all the implications of creating a transgenic predator, including potentially damaging effects.

How should we make decisions about the potential benefits and risks associated with transgenic predators?

If there is *any* risk, should transgenic predators be developed?

*B. Goodman, 1993, "Debating the Use of Transgenic Predators," *Science* 262: 1507.

# Sociology
## Population Growth

The growth rate (fertility) of the world's population has declined consistently since 1950. Birth rates and population growth rates are falling in the developing countries, as well as in Europe, North America, and Australasia. These changes have probably been the result of global family planning efforts to control population growth and major advances in contraceptive technology such as intrauterine devices.

There is one striking exception, however, to this worldwide trend. In the African countries south of the Sahara, population growth has not decreased by any measurable amount. What accounts for this difference between the sub-Saharan countries and the rest of the world? In their article in *Scientific American,* John and Pat Caldwell offer an explanation.

> The socioreligious system that accounts for sub-Saharan Africa's persistently high fertility differs greatly from those that pervade Europe, Asia and the Americas. It is not more traditional, primitive or backward; it is simply very different, and the differences have profound implications.*

The belief system the authors refer to places primary emphasis on having descendants who will ensure survival of the family lineage. Large families are considered good, while barrenness is considered immoral and even evil.

African governments have been extremely hesitant to impose family planning policies in the face of these socioreligious beliefs, but there are signs of change. In 1984, at the world population conference in Mexico City, most African nations supported family planning, in marked contrast to ten years earlier when these same nations vehemently opposed such policies.

What might account for the African nations' change in outlook between 1974 and 1984?

What effect might better health care and higher infant survival rates have on the policies of these nations?

*John C. Caldwell and Pat Caldwell, 1990, "High Fertility in Sub-Saharan Africa," *Scientific American* 262, no. 5: 118–25.

# Public Policy
## Lyme Disease

Twenty years ago, Lyme disease was unknown in the United States. Today it is spreading rapidly, and thousands of cases are reported each year. Lyme disease is caused by a bacterium that is spread to humans by tick bites. The bacterium is found in wild animals, such as deer. Ticks feeding on the blood of these animals are infected with the bacterium and pass the disease to humans.

A recent report* attributes the increasing incidence of Lyme disease to factors that people generally regard as improvements in their quality of life.

Ironically, the emergence of Lyme disease as a health problem is attributable in part to the "greening" of the United States: forests are regaining those lands formerly devoted to agriculture. Citizens value ever more highly the propinquity of wildlife to their residences. Deer, once close to elimination in many parts of the United States, are now as commonly noted as squirrels in some suburban communities.

Should local governments enact laws that limit wildlife populations and reduce green space in order to control the spread of Lyme disease?

*A. G. Barbour and D. Fish, 1993, "The Biological and Social Phenomenon of Lyme Disease," *Science* 260: 1610–16.

# Literature
## What's in a Name?

In his play *Romeo and Juliet,* William Shakespeare wrote:

> What's in a name? That which we call a rose
> By any other name would smell as sweet.

What would a taxonomist think of Shakespeare's statement? Could the taxonomist and Shakespeare both be right?

# CREDITS

## PHOTOGRAPHS

*Part Openers*
**1:** © Matt Meadows/Peter Arnold, Inc.; **2:** © David Newman/Visuals Unlimited; **3:** © Mark Boulton/Photo Researchers, Inc.; **4:** © Tom McHugh/Photo Researchers, Inc.

*Chapter 1*
**Opener:** © Matt Meadows/Peter Arnold, Inc.; **1.2 A,B:** © Eric Greene

*Chapter 2*
**Opener:** © K. G. Murti/Visuals Unlimited; **2.1A:** © Bernd Wittich/Visuals Unlimited; **2.1B:** © John Durham/SPL/Photo Researchers, Inc.; **2.3A:** © Alfred Owczarzak/Biological Photo Service; **2.3B:** © Dwight Kuhn; **2.4:** © Leonard Lessin/Peter Arnold, Inc.; **2.6A:** © K. G. Murti/Visuals Unlimited; **2.6B:** © Richard H. Gross; **Page 18 (bottom middle):** © Fred E. Hossler/Visuals Unlimited; **Page 18 (bottom left, bottom right):** © D. W. Fawcett/Visuals Unlimited; **Page 18 (top left):** Science VU/Visuals Unlimited; **Page 18 (top middle):** K. G. Murti/Visuals Unlimited; **Page 18 (top right):** © David M. Phillips/Visuals Unlimited; **2.9A:** © Micrograph by W. P. Wergin & E. H. Necomb, University of Wisconsin/Biological Photo Service; **2.9C:** © David M. Phillips/Visuals Unlimited; **2.10:** © Dr. T. J. Beveridge, University of Guelph/Biological Photo Service

*Chapter 3*
**Opener:** © Leonard Lessin/Peter Arnold, Inc.

*Chapter 4*
**Opener:** © Irving Geis/Peter Arnold, Inc.

*Chapter 5*
**Opener:** © David M. Phillips/Visuals Unlimited; **5.10:** © M. Schliwa/Visuals Unlimited

*Chapter 6*
**Opener:** © Alice J. Garik/Peter Arnold, Inc.; **6.2:** © Don Fawcett/Visuals Unlimited

*Chapter 7*
**Opener:** © Ed Reschke/Peter Arnold, Inc.; **7.5B:** © Micrograph by W. P. Wergin & E. H. Newcomb, University of Wisconsin/Biological Photo Service

*Chapter 8*
**Opener:** © John D. Cunningham/Visuals Unlimited

*Chapter 9*
**Opener:** © S. L. Palay, D. Fawcett/Visuals Unlimited; **9.2A:** © Fred E. Hossler/Visuals Unlimited; **9.6B:** © Jack Dermid/Photo Researchers, Inc.

*Chapter 10*
**Opener:** © Fred E. Hossler/Visuals Unlimited; **10.11A:** © John N. A. Lott, McMaster University/Biological Photo Service

*Chapter 11*
**Opener:** © Manfred Kage/Peter Arnold, Inc.; **11.5:** © M. Schilwa/Visuals Unlimited; **11.6:** © W. Rosenberg, Iona College/Biological Photo Service; **11.8 (all), 11.9 (both):** Courtesy of Dr. H. E. Huxley; **11.10 A-1:** © David M. Phillips/The Population Council/Science Source/Photo Researchers, Inc.; **11.10 B-1:** © John D. Cunningham/Visuals Unlimited; **11.10 C-1:** © Ed Reschke/Peter Arnold, Inc.

*Chapter 12*
**Opener:** © John D. Cunningham/Visuals Unlimited; **12.1:** © Cabisco/Visuals Unlimited; **12.2:** © Ed Reschke/Peter Arnold, Inc.; **12.3 A-1:** © Cabisco/Visuals Unlimited; **12.3 B-1:** © J. Robert Waaland, University of Washington/Biological Photo Service; **12.3 C-1:** © Cabisco/Visuals Unlimited; **12.3 D-1, E-1:** © Jack M. Bostrack/Visuals Unlimited; **12.3 F-1:** © J. Robert Waaland, University of Washington/Biological Photo Service; **12.4A:** © G. F. Bahr, M.D., Armed Forced Institute of Pathology/Biological Photo Service; **Box 12.1B:** © L. Lisco, D. Fawcett/Visuals Unlimited; **12.5:** © Biophoto Associates/Science Source/Photo Researchers, Inc.; **12.6:** Courtesy of Dr. Valerie Lindgren, University of Chicago; **12.7A, 12.8A:** © Lee D. Simon/Photo Researchers, Inc.

*Chapter 13*
**Opener:** © David M. Phillips/Visuals Unlimited; **13.1B:** © C. Shih & R. Kessel/Visuals Unlimited

*Chapter 14*
**Opener:** © Scott Camazine/Photo Researchers, Inc.; **14.2:** From *The Double Helix* by J. D. Watson, Atheneum Press, New York, 1968, pg. 72, Cold Spring Harbor Laboratory; **14.11 A,B:** Courtesy of Dr. David M. Prescott; **14.14:** Dr. O. L. Miller

*Chapter 15*
**Opener:** © Dwight Kuhn; **15.5A:** Field Museum, #118, Chicago; **15.5B:** Courtesy of Dr. Ralph L. Brinster; **15.8A:** © David M. Phillips/Photo Researchers, Inc.; **15.8B:** © Omikron/Photo Researchers, Inc.

*Chapter 16*
**Opener:** © Kevin Schafer/Peter Arnold, Inc.

*Chapter 17*
**Opener:** © Dwight Kuhn; **17.4 A,B:** © Michael Tweedie/Photo Researchers, Inc.; **17.5A:** © Bruce Coleman; **17.5B:** © Kim Taylor/Bruce Coleman

*Chapter 18*
**Opener:** © Dwight Kuhn

*Chapter 19*
**Opener:** © Ed Reschke/Peter Arnold, Inc.; **19.7:** Senckenbergmuseum Frankfurt A. M. Abt. Messel, Foto: E. Haupt

*Chapter 20*
**Opener:** © Len Rue, Jr./Visuals Unlimited

*Chapter 21*
**Opener:** © Max & Bea Hunn/Visuals Unlimited

*Chapter 22*
**Opener:** © Kent Wood/Photo Researchers, Inc.; **22.4:** © George Whiteley/Photo Researchers, Inc.; **22.7A:** © D. Newman/Visuals Unlimited; **22.7B:** © David M. Dennis/Tom Stack & Associates

*Chapter 23*
**Opener:** © Gregory G. Dimijian/Photo Researchers, Inc.; **Box 23.1:** © Dwight Kuhn

# LINE ART

### Chapter 2

**2.2:** From Stephen A. Miller and John P. Harley, *Zoology,* 2d edition. Copyright © 1994 Wm. C. Brown Communications, Inc., Dubuque, Iowa. All Rights Reserved. Reprinted by permission; **2.5:** From Stephen A. Miller and John P. Harley, *Zoology,* 2d edition. Copyright © 1994 Wm. C. Brown Communications, Inc., Dubuque, Iowa. All Rights Reserved. Reprinted by permission; **2.8 (art):** From John W. Kimball, *Biology,* 6th edition. Copyright © 1994 Wm. C. Brown Communications, Inc., Dubuque, Iowa. All Rights Reserved. Reprinted by permission; **Box 2.1a:** From John W. Kimball, *Biology,* 6th edition. Copyright © 1994 Wm. C. Brown Communications, Inc., Dubuque, Iowa. All Rights Reserved. Reprinted by permission.

### Chapter 4

**4.1:** From Eldon D. Enger, et al., *Concepts in Biology,* 5th edition. Copyright © 1988 Wm. C. Brown Communications, Inc., Dubuque, Iowa. All Rights Reserved. Reprinted by permission.

### Chapter 5

**5.7:** From John W. Kimball, *Biology,* 6th edition. Copyright © 1994 Wm. C. Brown Communications, Inc., Dubuque, Iowa. All Rights Reserved. Reprinted by permission.

### Chapter 8

**8.8:** From John W. Kimball, *Biology,* 6th edition. Copyright © 1994 Wm. C. Brown Communications, Inc., Dubuque, Iowa. All Rights Reserved. Reprinted by permission.

### Chapter 9

**9.1:** From Eldon D. Enger, *Concepts in Biology,* 5th edition. Copyright © 1988 Wm. C. Brown Communications, Inc., Dubuque, Iowa. All Rights Reserved. Reprinted by permission; **9.8:** From John W. Kimball, *Biology,* 6th edition. Copyright © 1994 Wm. C. Brown Communications, Inc., Dubuque, Iowa. All Rights Reserved. Reprinted by permission; **9.9 (top art):** From Eldon D. Enger, et al., *Concepts in Biology,* 7th edition. Copyright © 1994 Wm. C. Brown Communications, Inc., Dubuque, Iowa. All Rights Reserved. Reprinted by permission.

### Chapter 10

**10.3:** From John W. Kimball, *Biology,* 6th edition. Copyright © 1994 Wm. C. Brown Communications, Inc., Dubuque, Iowa. All Rights Reserved. Reprinted by permission; **10.6:** From E. Peter Volpe, *Biology and Human Concerns,* 4th edition. Copyright © 1993 Wm. C. Brown Communications, Inc., Dubuque, Iowa. All Rights Reserved. Reprinted by permission.

### Chapter 11

**11.1:** From John W. Kimball, *Biology,* 6th edition. Copyright © 1994 Wm. C. Brown Communications, Inc., Dubuque, Iowa. All Rights Reserved. Reprinted by permission; **11.2:** From Schottelius and Schottelius, *Textbook of Physiology,* 17th edition, C. V. Mosby Company, St Louis, MO. Reprinted by permission of Dorothy D. Schottelius; **11.4:** From Schottelius and Schottelius, *Textbook of Physiology,* 17th edition, C. V. Mosby Company, St Louis, MO. Reprinted by permission of Dorothy D. Schottelius; **11.11B:** From John W. Kimball, *Biology,* 6th edition. Copyright © 1994 Wm. C. Brown Communications, Iowa. All Rights Reserved. Reprinted by permission; **11.12:** From E. Peter Volpe,

*Biology and Human Concerns,* 4th edition. Copyright © 1993 Wm. C. Brown Communications, Inc., Dubuque, Iowa. All Rights Reserved. Reprinted by permission; **11.16b:** From John W. Kimball, *Biology,* 6th edition. Copyright © 1994 Wm. C. Brown Communications, Inc., Dubuque, Iowa. All Rights Reserved. Reprinted by permission.

### Chapter 14

**14.7:** From Helena Curtis, *Biology,* 4th edition. Copyright © 1983 Worth Publishers, New York, NY. Reprinted by permission; **14.8:** From Helena Curtis, *Biology,* 4th edition. Copyright © 1983 Worth Publishers, New York, NY. Reprinted by permission.

### Chapter 18

**18.3 (map):** Reprinted with the permission of Macmillan College Publishing Company from *Process and Pattern in Evolution* by Terrell H. Hamilton. Copyright © 1967 by Terrell H. Hamilton.

### Chapter 19

**Box 19.1:** From: *Evolution* by Dobzhansky. Copyright © 1977 by W. H. Freeman and Company. Reprinted with permission; **19.6:** From Starr and Taggart, *Biology,* 5th edition. Wadsworth Inc., Belmont, CA. Reprinted by permission; **19.8:** *Biology* 3d edition by Hardin and Bajema. Copyright © 1978 by W. H. Freeman and Company. Reprinted with permission.

### Chapter 20

**20.1:** From Ralph D. Bird, "Biotic Communities of the Aspen Packland of Central Canada" in *Ecology II,* (2):410, figure 14. Reprinted by permission of the Center for Environmental Studies, Ecological Society of America, Arizona State University; **20.11:** From Eldon D. Enger, et al., *Concepts in Biology,* 7th edition. Copyright © 1994 Wm. C. Brown Communications, Inc., Dubuque, Iowa. All Rights Reserved. Reprinted by permission.

## ILLUSTRATORS

**Diphrent Strokes:** 1.1, BOX 1.1, 1.3, 1.4, 1.5, BOX 2.1B, 2.7, 2.9B, 3.1, 3.1BX, 3.2, BOX 3.2, 3.3, 3.4, 3.5, 3.6, 3.7, 3.8, 3.9, 3.10, 3.11, 4.1, BOX 4.1, 4.2, BOX 4.2, 4.3, 4.4, 4.5, 4.6, 4.7, 4.8, 4.9, 4.10, 4.11, 5.1, BOX 5.1, 5.2, 5.3, 5.4, 5.5, 5.6, 5.8, 5.9, 5.11, 5.12, 5.13, 6.1, BOX 6.1, BOX 6.2, 6.3, 6.4, 6.5, 6.6, 6.7, 7.1, BOX 7.1, 7.2, BOX 7.2, 7.3, 7.4, 7.5A, 7.6, 7.7, 7.8, 7.9, 7.10, 7.11, 8.1, BOX 8.1, 8.2, 8.3, 8.4, 8.5, 8.6, 8.7, 8.8, 8.9, 8.10, 9.1, BOX 9.1, 9.2B, 9.3, 9.4, 9.5, 9.6A, 9.7, 9.8, 9.9, 10.1, BOX 10.1, 10.2, 10.3, 10.4, 10.5, 10.6, 10.7, 10.8, 10.9, 10.10, 10.11B–C, 10.12, 10.13, 11.1, BOX 11.1, 11.2, 11.3, 11.4, 11.7, 11.8, 11.10, 11.11A, 11.12, 11.13, 11.14, 11.15, 11.16A–B, BOX 12.1A, 12.3, 12.4B, 12.7B, 12.8B, 12.9, 12.10, 12.11, BOX 13.1, 13.1A, 13.2, BOX 13.2A–B, 13.3, 13.4, 13.5, 13.6, 13.7, 13.8, 13.9A–B, 13.10, 13.11, 13.12, 13.13, 13.14, 14.1, BOX 14.1, 14.3, 14.4, 14.5, 14.6, 14.7, 14.8, 14.9, 14.10, 14.12, 14.13, 14.15, 15.1, BOX 15.1, 15.2, 15.3, 15.4, 15.6, 15.7, 15.9, 15.10, 15.11, 15.12, 15.13, 15.14, 16.1, BOX 16.1, 16.2, 16.3, 16.4, 16.5, 16.6, 16.7, 17.1, 17.2, 17.3, 17.6, 17.8A, 18.1, 18.2, 18.3, 18.4, 18.5, 18.6, 18.7, 18.8, 19.1, BOX 19.1, 19.2, 19.3, 19.4, 19.5, 19.6, 19.8, 20.1, BOX 20.1, 20.2, 20.3, 20.4, 20.5, 20.6, 20.7, 20.8, 20.9, 20.10, 20.11, 21.1, 21.2, 21.3, 21.4, 21.5, 21.6, 21.7, 21.8, 22.1, BOX 22.1, 22.2, 22.3, 22.5, 22.6, 22.8, 23.1, 23.2, 23.3, 23.4, 23.5, 23.6, 23.7, 23.8, 23.9

**Precision Graphics:** BOX 2.1A, 5.7

**Rolin Graphics:** 2.2, 2.5, 2.8, 11.11B

# INDEX

Competitive exclusion, 252
Complementary, bases, DNA, 170
Concentration gradient, 50
Consumers, 246
Controlled variables, 3
Corpus luteum, 159
Cortical layer, 159
Covalent bonds, 27
Crick, Francis, 169–70, 171
Crop, 100
Crossing-over, 190–91
Cytokinesis, 144
Cytosol, 17

## D

Darwin, Charles, 232–33, 236
Data, 4
Decomposers, 248
Dendrites, 131, 134
Denitrification, 275
Deoxyribonucleic acid (DNA), 148–50
  control of cell function, 172–73
  DNA code and theory of
    evolution, 234
  and enzyme synthesis, 173–77
  hybrid DNA and theory of evolution,
    235–36
  replication, 170–72
  structure of, 168–70
  transcription, 175
Depolarization, 132
Detritus food web, 263, 263–66
  in aquatic community, 265–66
  in forest community, 264–65
Diaphragm, 88
Diastolic pressure, 117
Diffusion, 50
  and boundary layers, 91
  facilitated diffusion, 52
Diffusion distance, and gas exchange, 88
Digestion
  and excretion, 104–108
  meaning of, 100
  process in humans, 100–103
Digestive system, structures of, 100
Diploid number, 151
DNA. See Deoxyribonucleic acid (DNA)
DNA fingerprinting, 172
Dominant traits, 182
Down syndrome, 148
Dynamic equilibrium, of chemical
  reactions, 38

## E

Ecology, meaning of, 246
Electron microscope, 16
Electrons, 26
Electron transport, 68–69
Electrophoresis, 216
Elements, 26
  structural formula, 27
Embedded material, observation with
  microscope, 15
Embryo, development of, 160
Endocytosis, 103
Endoplasmic reticulum, 17
Energy, 32–33
  chemical bond energy, 32

consumption by herbivores, 262–64
  heat energy, 32–33
  meaning of, 32
  measurement of, 260–61
Energy flow
  aquatic community, 265–66
  energy balance sheet, 266
  energy flow diagram, 260
  forest community, 264–65
Enzymes, 38–40
  as catalysts, 38–39, 60
  and chemical reactions, 38–39
  digestive, 102, 104
  in metabolic pathways, 44–45
  pH, effects on, 41–44
  processing rates, 44
  structure of, 40
Esophagus, 101
Estrogen, 159
Eukaryotic cell, 17
Evolution
  common descent, 232
  evidence for Darwin's theory, 234–38
  inheritance of acquired characteristics,
    232–33
  meaning of, 232
  natural selection, 232–34
  objections to theory, 238–40
Excretion, 104–108
  insects, 105
  kidney, 106–108
  waste products, 104–105
Experimental design, 3
  variables in, 3
Experimental variables, 3
External fertilization, 158

## F

Facilitated diffusion, 52
Feedback inhibition, 45
Female reproductive cycle, and
  hormones, 159
Fermentation, 69–71
  alcoholic, 69–70
  lactate, 69–70
Ferns, reproduction of, 161
Fertilization, 157, 158–60
  external and internal, 158
  self-fertilization, 158
Fertilization membrane, 159
Filtrate, 106
Fish, heart of, 117
Fitness, of genotype, 203
Five-kingdom system, 284–89
  Animalia kingdom, 288–89
  Fungi kingdom, 286
  Monera kingdom, 285
  Plantae kingdom, 286–87
  Protista kingdom, 285–86
Flowering plants, reproduction in, 162
Fluorescence, 81
Food chain, 246
Food web, 246
  detritus food web, 263
Forest community, energy flow, 264–65
Fossil record, and theory of evolution,
  237–38, 239–40
Fungi kingdom, 286

## G

Gametes, 157, 158
Gametophyte, 160
Gas exchange, 88–95
  affecting factors, 88
  and gills, 90–92
  and lungs, 88–90
  meaning of, 88
  plants, 92–95
  and tracheae, 92
Gause, G. F., 251–52
Gene flow, 202
Gene frequency, 198–99
  calculation of, 199–201
Genetic drift, 202
Genetic equilibrium, 201
Genetics
  alleles, 184, 185
  blending hypothesis, 182
  crossing-over, 190–91
  dihybrid cross, 188, 189
  homozygous and heterozygous, 184
  linked genes, 188–90
  Mendelian genetics, 182–84
  mutations, 187–88
  particulate hypothesis, 182
  pedigree chart, 186, 187
  phenotype and genotype, 184
  sex-linked traits, 191–92
  speciation, 222–27
  two traits, inheritance of, 188
  See also Chromosomes; Population
    genetics
Genotypes, 184
  fitness of, 203
  and selection, 203
Geographic isolation, 224
Gills, 90–92
  gas exchange, 90–92
Global warming, 272
Glomerulus, 106
Glycolysis, 69–71
Golgi complex, 17
Graded response, 126
Grana, 78
Guard cells, 93
Gymnosperms, 286

## H

Habitat isolation, 227
Haploid number, 151
Hardy, Godfrey, 199
Hardy-Weinberg law, 199, 201–203
Harvey, William, 114
Heart, 114, 116
  amphibians, 117
  chambers of, 116
  fish, 117
Heat energy, 32–33, 71
Hemodialysis, 108
Hemoglobin, 118–20
Herbivores, 246
Hermaphroditic organisms, 158
Hershey, Alfred, 148–50
Heterotrophs, 76
Heterozygosity, 184, 216
Heterozygote superiority, 216
Hierarchy, in classification systems,
  280–81

Hiesey, William, 222
Homozygous, 184
Hooke, Robert, 15
Hormones, and female reproductive
  cycle, 159
Hybrids, 227
  DNA and theory of evolution,
    235–36
Hydrogen bonds, 28
Hydrophobic amino acids, 40
Hydrophylic amino acids, 40
Hypothesis
  additional testing of, 6–7
  meaning of, 2
  proof for, 5
  statement of, 2–3

## I

Impermeability, 51
Independent assortment, traits, 184
Independent techniques, 20
Inheritance. See Chromosomes; Genetics
Inheritance of acquired characteristics,
  232–33
Insects
  circulatory system, 114
  excretion, 105
  respiration, 92
Internal fertilization, 158
Interphase, 144
Interpretation, in scientific investigation,
  4–5
In vitro fertilization, 207
Ionic bonds, 27
Ions, 27
Isolation, and speciation, 223–25,
  226–27

## K

Karyotypes, 145–47
  preparation of, 146
Keck, David, 222
Kettlewell, H. B. D., 213–14
Kidney, 106–108
  hemodialysis, 108
  structures of, 106
  urine production, 106–108
Krebs, Hans Adolf, 68
Krebs cycle, 68–69
Kymograph, 126

## L

Lactase, 102
Lactate fermentation, 69–70
Lactose intolerance, 102
Large intestine, 103
Lederberg, Joshua, 212–13
Left atrium, 116
Left ventricle, 116
Legal factors, scientific investigation, 59
Light
  and absorption spectrum, 79, 80
  fluorescent light, 81
  and photosynthesis, 78–83
  wavelength, 78–79
Light microscope, 14–15
Linked genes, 188–90
Linnaeus, Carolus, 282–84
Lipases, 102

Lipids, 29, 31
solubility of, 51
Liver, 101
Lungs, 88–90
and gas exchange, 88–95
structures of, 88–89
water loss from, 90
Lysosomes, 17

## M

Malpighian tubules, 105
Material cycling
carbon cycle, 270–72
nitrogen cycle, 274–75
oxygen cycle, 272–73
Matter, 26
Meiosis, 151
stages in, 151, 152
Mendel, Gregor, 182–84
Mendelian genetics, 182–84
dominant and recessive traits, 182, 185–86
independent assortment, 184
plant experiments, 182–84
Meselson, Matthew, 171
Messenger RNA (mRNA), 175–77
Metabolic pathway, 38
enzymes in, 44–45
tracing of, 45–46
Metaphase, 144
Microelectrodes, 132
Microscope, 14–16
electron microscope, 16
light microscope, 14–15
preparation of material for observation, 15
Microvilli, 102
Migration, 202
Mitochondria, 17
and energy production, 65–66, 69–70
Mitosis, 144–45
stages of, 144–45
Molecules, 26–30
biological molecules, 28–30
chemical bonds, 27–28
Monera kingdom, 285
Monomers, 28
Morgan, Thomas Hunt, 191–92
Mosses, reproduction of, 161
Muscle
cardiac muscle, 130–31
skeletal muscle, 130
smooth muscle, 131
Muscle contraction, 126–28
and adenosine triphosphate (ADT), 130
all-or-none response, 127–28
measurement of, 126
process of, 126
single cell contraction, 126–28
sliding-filament hypothesis, 128
Muscle tissue, components of muscle cells, 128–30
Mutant, 187
Mutation, 173, 187–88, 202–203
back mutation, 202
Mutualism, 253

## N

Natural selection, 232–34
Nephrons, 106
Nerve cells, 131–35
action potential, 132–34
detection of stimuli, 135
and muscle fiber stimulation, 132
parts of, 131–32
synapses, 134–35
Nervous system, nerve cells, 131–35
Net electrical charge, 26
Net production, 260
Net result, 260
Neuro-muscular junction, 132, 135
Neutrons, 26
Nitrogen cycle, 274–75
Nitrogen fixation, 275
Nucleic acids, 29
deoxyribonucleic acid (DNA), 148–50
ribonucleic acid (RNA), 174–77
Nucleotides, 29
Nucleus, cell, 17

## O

Occult, belief in, 60
Open circulatory system, 114
Orbitals, 26
Organelles, types of, 17
Organ transplantation, 139
Osmosis, 56
Ostia, 114
Oxidation, 67
Oxygen
and hemoglobin, 118–19
and respiration, 64
Oxygen cycle, 272–73
Oxygen transport, blood, 119

## P

Pancreas, 101
Particulate hypothesis, 182
Pedigree chart, 186, 187
Permeability
affecting factors, 51
of cell membrane, 51
and gas exchange, 88
pH, 41–44
acidic and basic solutions, 41–42
pH scale, 42
Phagocytes, 103
Phenotype, 184
Phenylketonuria (PKU), 45
Phloem cells, 120
Phloem transport, 121
Photosynthesis
and adenosine triphosphate (ADP), 81–83
and carbon dioxide, 83
and chlorophyll, 78–81
early experiments, 76–78
formula for, 76
and light, 78–83
net result of, 260
and plant cells, 78
Phyla, 284
Phytol, 78
Plantae kingdom, 286–87
Plant cells, 17
structures of, 17

Plants
absorption, 103–104
gas exchange, 92–95
photosynthesis, 76–83
sexual reproduction, 160–62
taxonomic categories of, 286
transpiration, 120–21
transport in, 120–21
vascular plants, 120
water loss in, 95
Plasma membrane, 17
Polarity, 30
water, 30, 32
Polymers, 28
Polypeptides, 102
Polyploidy, 227–28
Population
growth curves, 252
*See also* Communities
Population genetics
gene flow, 202
gene frequency, 198–99
gene frequency calculation, 199–201
gene pool, 198–99
genetic drift, 202
genetic equilibrium, 201
Hardy-Weinberg law, 199, 201–203
migration, 202
mutation, 202–203
random mating, 201–202
selection, 203, 212–17
Porphyrin, 78
Predator-prey oscillations, 250
Predator-prey relationship, 250
Prediction, in scientific investigation, 3
Pressure, and water movement, 55–56
Pressure-flow, 121
Presynaptic vesicles, 134
Primary consumers, 246
Probes, 172
Producers, 246
Production
net production, 260
production efficiency of communities, 261–62
Products, of chemical reactions, 38
Progesterone, 159
Prokaryotic cells, 18–19
Prophase, 144
Proteins, 28
Prothallus, 161
Protista kingdom, 285–86
Protons, 26
Pulmonary artery, 116
Pulmonary vein, 116
Purines, 168
Pyramid of biomass, 248, 249
Pyramid of numbers, 246
Pyrimidines, 168

## R

Races, 222
Radioactive tracers, 53
Random mating, 201–202
Reactants, 38
Reasoning, line of reasoning, 4–5
Recessive traits, 182
Reduction, 67
Replication, DNA, 170–72

Reproduction
asexual reproduction, 156
cell differentiation, 163–64
sexual reproduction, 151, 157–64
Reproductive isolation, 224
Respiration
cellular respiration, 64–71
gas exchange, 88–95
insects, 92
measurement of, 65
oxygen uptake, 64
rate of, 64
water-breathing animals, 90–92
Resting potential, 132, 134
Reversibility, of chemical reactions, 38
Ribonucleic acid (RNA), 174–77
messenger RNA (mRNA), 175–77
ribosomal RNA (rRNA), 175–76
and transcription, 175
transfer RNA (tRNA), 175–77
translation, 175–76
Ribosomal RNA (rRNA), 175–76
Right atrium, 116
Right ventricle, 116
Root hairs, plants, 103
Root pressure, 120
Roots, plants, 103

## S

Salivary glands, 100
Scientific investigation, 2–7
additional hypothesis testing, 6–7
assumptions in, 7
experimental design, 3
general question in, 2
goal of, 2
hypothesis statement in, 2–3
interpretation of results, 4–5
legal factors, 59
prediction in, 3
proving hypothesis, 5
results of investigation, 4
specific question in, 2
Secondary consumers, 246
Sectioned material, observation with microscope, 15
Selection, 212–17
and adaptation, 222
balanced polymorphism in, 214–15
and genotypes, 203
Lederberg's experiment, 212–13
at molecular level, 215–16
in peppered moth, 213–14
selecting factor, 212
selective advantage, 214
and variability, 216–17
Sex cells, meiosis, 151, 152
Sex chromosomes, 145
Sex-linked traits, 191–92
Sexual reproduction, 157–64
development of embryo, 160
fertilization, 157, 158–60
meaning of, 151
plants, 160–62
Sickle-cell anemia, 187, 188, 216
Sinks, 121
Skeletal muscle, 130
Sliding-filament hypothesis, 128
Small intestine, 101

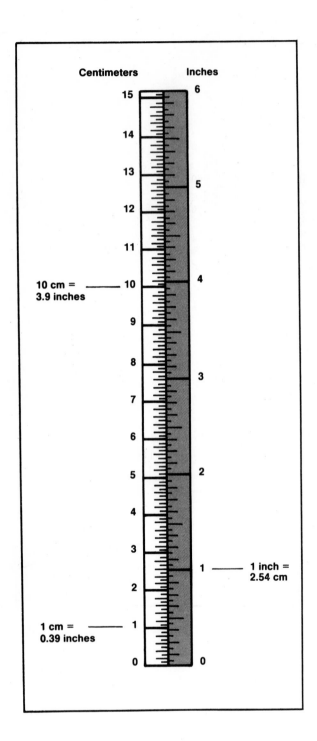

## The Metric System

| Standard Metric Units | | Abbreviations |
|---|---|---|
| Standard unit of mass | gram | g |
| Standard unit of length | meter | m |
| Standard unit of volume | liter | l |

| Common Prefixes | | Examples |
|---|---|---|
| kilo | 1,000 | a kilogram is 1,000 grams |
| centi | 0.01 | a centimeter is 0.01 meter |
| milli | 0.001 | a milliliter is 0.001 liter |
| micro ($\mu$) | one-millionth | a micrometer is 0.000001 (one-millionth) of a meter |
| nano (n) | one-billionth | a nanogram is $10^{-9}$ (one-billionth) of a gram |
| pico (p) | one-trillionth | a picogram is $10^{-12}$ (one-trillionth) of a gram |

## Units of Length

| Unit | Abbreviation | Equivalent |
|---|---|---|
| meter | m | approximately 39 in |
| centimeter | cm | $10^{-2}$ m |
| millimeter | mm | $10^{-3}$ m |
| micrometer | $\mu$m | $10^{-6}$ m |
| nanometer | nm | $10^{-9}$ m |
| angstrom | Å | $10^{-10}$ m |

### Length Conversions

| | | | | | |
|---|---|---|---|---|---|
| 1 in | = | 2.5 cm | 1 mm | = | 0.039 in |
| 1 ft | = | 30 cm | 1 cm | = | 0.39 in |
| 1 yd | = | 0.9 m | 1 m | = | 39 in |
| 1 mi | = | 1.6 km | 1 m | = | 1.094 yd |
| | | | 1 km | = | 0.6 mi |

| To Convert | Multiply by | To Obtain |
|---|---|---|
| inches | 2.54 | centimeters |
| feet | 30 | centimeters |
| centimeters | 0.39 | inches |
| millimeters | 0.039 | inches |